T0074867

LONDON MATHEMATICAL SOCIETY LECTURE NOTE SERIES

Managing Editor: Professor Endre Süli, Mathematical Institute, University of Oxford,
Woodstock Road, Oxford OX2 6GG, United Kingdom

The titles below are available from booksellers, or from Cambridge University Press at
www.cambridge.org/mathematics

Computational Cryptography

Algorithmic Aspects of Cryptology

Edited by

JOPPE W. BOS
NXP Semiconductors, Belgium

MARTIJN STAM
Simula UiB, Norway

CAMBRIDGE
UNIVERSITY PRESS

University Printing House, Cambridge CB2 8BS, United Kingdom

One Liberty Plaza, 20th Floor, New York, NY 10006, USA

477 Williamstown Road, Port Melbourne, VIC 3207, Australia

314-321, 3rd Floor, Plot 3, Splendor Forum, Jasola District Centre, New Delhi - 110025, India

103 Penang Road, #05-06/07, Visioncrest Commercial, Singapore 238467

Cambridge University Press is part of the University of Cambridge.

It furthers the University's mission by disseminating knowledge in the pursuit of education, learning and research at the highest international levels of excellence.

www.cambridge.org
Information on this title: www.cambridge.org/9781108795937
DOI: 10.1017/9781108854207

First published 2021

A catalogue record for this publication is available from the British Library

Library of Congress Cataloging in Publication data
Names: Bos, Joppe W., editor. | Stam, Martijn, editor.
Title: Computational cryptography : algorithmic aspects of cryptology /
edited by Joppe W. Bos, NXP Semiconductors, Belgium, Martijn Stam,
Simula UiB, Norway.
Description: Cambridge, United Kingdom ; New York, NY, USA : Cambridge
University Press, 2021. | Series: London mathematical society lecture
note series | Includes bibliographical references and index.
Identifiers: LCCN 2021012303 (print) | LCCN 2021012304 (ebook) |
ISBN 9781108795937 (paperback) | ISBN 9781108854207 (epub)
Subjects: LCSH: Cryptography. | BISAC: MATHEMATICS / Number Theory |
MATHEMATICS / Number Theory
Classification: LCC QA268 .C693 2021 (print) | LCC QA268 (ebook) |
DDC 005.8/24–dc23
LC record available at https://lccn.loc.gov/2021012303
LC ebook record available at https://lccn.loc.gov/2021012304

ISBN 978-1-108-79593-7 Paperback

Contents

Contributors

Martin R. Albrecht   *Information Security Group, Royal Holloway, University of London, United Kingdom*

Joppe W. Bos   *NXP Semiconductors, Leuven, Belgium*

Léo Ducas   *Centrum Wiskunde & Informatica (CWI), Amsterdam, The Netherlands*

Robert Granger   *University of Surrey, Guildford, United Kingdom*

Nadia Heninger   *University of California, San Diego, USA*

Antoine Joux   *CISPA Helmholtz Center for Information Security, Saarbrücken, Germany*

Thorsten Kleinjung   *EPFL, Lausanne, Switzerland*

Alexander May   *RUB, Bochum, Germany*

Dan Page   *University of Bristol, Bristol, United Kingdom*

Victor Shoup   *New York University, New York, USA*

Nigel P. Smart   *imec-COSIC, KU Leuven, Leuven, Belgium*

Martijn Stam   *Simula UiB, Bergen, Norway*

Marc Stevens   *Centrum Wiskunde & Informatica (CWI), Amsterdam, The Netherlands*

Emmanuel Thomé   *Université de Lorraine, CNRS, INRIA, Nancy, France*

Samuel S. Wagstaff, Jr   *Purdue University, West Lafayette, USA*

Preface

This book is a tribute to the scientific research career of Professor Arjen K. Lenstra, on the occasion of his 65th birthday. Its main focus is on computational cryptography. This area, which he has helped to shape during the past four decades, is dedicated to the development of effective methods in algorithmic number theory that improve implementation of cryptosystems or that further their cryptanalysis. Here, cryptanalysis of cryptosystems entails both the assessment of their overall security and the evaluation of the hardness of any underlying computational assumptions. In the latter case, the area intersects non-trivially with high-performance scientific computing. The technical chapters in this book are inspired by his achievements in computational cryptography.

Arjen is best known for his seminal work on the algorithmic aspects of various factorisation problems. In the early 1980s, he started with efficient factorisation of polynomials with rational coefficients. This work led to the celebrated Lenstra–Lenstra–Lovász lattice reduction algorithm. Furthermore, he devised factorisation techniques for polynomials defined over other algebraic structures, such as finite fields or number fields. Towards the end of the decade, his focus shifted to integer factorisation methods, particularly development of the number field sieve, and its impact on the selection of strong cryptographic keys for widely deployed cryptographic standards. His honours include the RSA Award for Excellence in Mathematics in 2008 and his lifetime appointment as Fellow of the International Association for Cryptologic Research (IACR) in 2009.

In addition to his rich research career, Arjen is a great educator and he has provided lasting inspiration to many of his students. We both were lucky enough to have him as our PhD supervisor and we will come back to our respective experiences momentarily. This book is intended for students in security and cryptography as well as for security engineers and architects in industry who want to develop a deeper understanding about the algorithms used in computational cryptography.

When I (Martijn) started my PhD studies at the TU Eindhoven, Arjen was not yet appointed as a part-time professor there, but as soon as he did, there was an immediate click. Although he wasn't physically in Eindhoven that often, his availability and generosity with his time always struck me. We often met at conferences, where he would invariably join the front row, providing me with a running commentary, but also where he would introduce me to his wider academic network. One peculiarity when working with Arjen is his absolute aversion of footnotes, which he enthusiastically weeded out of early drafts of papers and also discouraged by expecting some friendly 'compensation' for each footnote remaining in my final PhD thesis. When he was appointed as a full professor at EPFL, a few years after my graduation, he asked me whether I wanted to join as a post-doctoral researcher. During those years he encouraged me to explore my own research agenda and helped me to mature as an independent academic.

After I (Joppe) obtained my master's degree in Amsterdam, the required funding for a PhD position related to integer factorisation failed to materialise. When Arjen learned about this, he arranged for me to come over and eventually start my PhD study in Lausanne. There we had the coolest equipment to brag about at birthday parties: a cluster of PlayStation 3 game consoles. More seriously, with his broad academic network he ensured I could collaborate with the brightest minds in public-key cryptology. This led to a summer internship under the supervision of Peter Montgomery at Microsoft Research, where I eventually became a post-doctoral researcher which paved the way for me to join the Competence Center Crypto & Security at NXP Semiconductors. I learned a lot from Arjen's direct but honest way of conducting research and to always ask critical questions when people skim over the sometimes complicated but necessary details.

It was a genuine pleasure for both of us (being PhD siblings) to honour Arjen's scientific career. We would like to thank all the contributors who are leading researchers in the various fields for their participation and hope you (the reader) will enjoy reading this book and be as enthusiastic about the fascinating and interesting field of computational cryptography as we are.

1

Introduction

Joppe W. Bos and Martijn Stam

This introductory chapter provides a sketch of Arjen Klaas Lenstra's scientific career, followed by a preview of the technical chapters. We are indebted to Ronald Cramer and Herman te Riele for information about Arjen's longstanding connections to CWI, Monique Amhof for providing additional information about Arjen's time at EPFL and Peter van Emde Boas for his kind permission to use a photograph from his private collection.

1.1 Biographical Sketch

Arjen Klaas Lenstra was born on 2 March 1956 in Groningen, the Netherlands. He received a BA and an MA degree in Mathematics and Physics at the University of Amsterdam in 1975 and 1980, respectively. The picture in Figure 1.1 shows Arjen defending his master's thesis. Afterwards, he obtained a research position at CWI – the Dutch national research institute for mathematics and computer science – in Amsterdam and started conducting his PhD research. Actually, when Arjen started in 1980, the institute was still called Mathematisch Centrum, but in 1983 it would be renamed CWI, an abbreviation for 'Centrum voor Wiskunde en Informatica', using the Dutch words 'wiskunde' for mathematics and 'informatica' for computer science. This renaming reflected the emergence of computer science as a discipline in its own right, separate from mathematics. In retrospect, the new name CWI turned out to be indicative for Arjen's research career as well. Throughout, his work has provided exemplary proof of the impressive powers summoned by harnessing both mathematics and computer science. Since his PhD days, he has always maintained close links with CWI, for instance, as an officially appointed advisor from 2004 until 2017, but also by co-supervision of various PhD students from CWI over the years.

Figure 1.1   Arjen defending his master's thesis on 10 December 1980. The committee, seated behind the desk, was formed by Th. J. Dekker, Peter van Emde Boas and Hendrik W. Lenstra, Jr. Photo from the private collection of van Emde Boas.

Arjen's own PhD research at CWI was conducted under the external supervision of Peter van Emde Boas and, in 1984, he received his PhD degree at the University of Amsterdam for his work on 'polynomial time algorithms for the factorization of polynomials' [312, 362, 363, 364, 380], which was also the title of his PhD thesis [365]. One key ingredient of this research became one of his most widely known results: the Lenstra–Lenstra–Lovász (LLL) lattice basis reduction algorithm, a polynomial time lattice basis reduction algorithm [380]. The LLL algorithm found countless applications, as described in more detail in the LLL book [457] that was published to commemorate its 25th birthday. Accurately predicting the effectiveness and efficiency of lattice reduction algorithms continues to be a key tool in the cryptanalytic toolbox. Indeed, with the ongoing post-quantum cryptographic standardisation effort, LLL and similar algorithms play an essential role in determining the practical parameters for lattice-based cryptography. This is detailed further in Chapter 2.

From 1984 to 1989, Arjen was a visiting professor at the computer science department of the University of Chicago. During this time he retained a position as visiting researcher at CWI and he conducted multiple summer research visits to Digital Equipment Corporation (DEC) in Palo Alto, CA. These latter visits resulted in the famous distributed effort using factoring by electronic mail together with Mark S. Manasse [371]. Such computational cryptanalysis

Figure 1.2 Reward for the factorisation of the *Scientific American* RSA challenge.

is essential to understand the practical security strength of the Rivest–Shamir–Adleman (RSA) cryptosystem [501], which is related to the presumed hardness of integer factorisation. It marked the start of Arjen's research in all computational aspects of integer factorisation, leading to his involvement in virtually all integer factorisation records. His records include the factorisation of the ninth Fermat number [382] using the first implementation of the number field sieve [381], solving the famous *Scientific American* RSA challenge [24] by using the quadratic sieve algorithm [478], the kilo-bit [22] special number field sieve factorisation [474], new elliptic-curve integer factorisation [388] records by using game consoles [90] and the factorisation of the RSA-768 challenge [327] (but see also [90, 117, 118, 143, 156, 165, 329, 371, 372, 384]). His work on the *Scientific American* RSA challenge was even featured on the front page of the *New York Times* on 12 October 1988 and, after some delay, he and his collaborators received their due reward for solving this challenge, as evidenced by Figure 1.2. His brother Hendrik would eventually refer to him as 'world champion in factoring'.

Arjen's involvement led to many practical optimisations of these algorithms [167, 224, 372, 392, 428] and most notably to the development of the number field sieve [381]. A historical overview of integer factorisation techniques, including the asymptotically best methods to which Arjen contributed, is provided in Chapter 3.

In order to run integer factorisation algorithms on a large variety of computer architectures, a portable and fast arbitrary-length integer arithmetic software library was needed. Arjen developed the software library FreeLIP (1988–1992), later maintained by Paul Leyland. FreeLIP was used in the early integer factorisation records and formed the early backbone of Shoup's Number Theory Library (NTL). More details about the relationship between FreeLIP, NTL, and other high-performance, portable software libraries for doing number theory are outlined and explained in Chapter 9.

Recall that Arjen's PhD research on polynomial factoring included the LLL algorithm for lattice reduction. In 1996, Coppersmith [136, 137] showed how the LLL algorithm can be used to factor poorly generated RSA keys in polynomial time. Lattice-based integer factorisation methods are the main topic of Chapter 4. In this chapter, the details behind Coppersmith's method and how to use this for assessing the security of the RSA cryptosystem are explained.

Showing the practical impact of various asymptotic methods and determining, for given circumstances, which is superior is one of the main ingredients when recommending cryptographic key sizes in standards. Arjen pioneered the concrete extrapolation of known factoring and DLP methods to determine meaningful bit-level security estimates for common cryptosystems, including their long-term security. His contributions in this field are widely known and valued by academia, industry and government agencies alike [328, 368, 376, 379] and served as the foundation to determine the exact key sizes for different security levels for virtually all public-key cryptographic standards. More details about the history of selecting appropriate cryptographic keys and current recommendations are described in Chapter 11.

From 1989 to 1996, Arjen held various positions at Bell Communications Research, Morristown, NJ, in the Mathematics and Cryptology Research Group. From 1996 to 2002 he was Vice President of Emerging Technologies in the Corporate Technology Office of Citibank and from 2002 to 2004 Vice President, Information Security Services, still at Citibank. He joined Lucent Technologies' Bell Labs in 2004 where he stayed until the end of 2005.

In addition to his corporate positions at Citibank and later Bell Labs, from 2000 to the end of 2005 Arjen held a position as part-time professor at the Department of Mathematics and Computer Science at the Technische Universiteit Eindhoven in the Netherlands. When the appointment was still being negotiated, the Dean had to keep the Faculteitsraad ('Departmental Council') informed, although without mentioning any concrete names. At the time, the Dean happened to be his brother Jan Karel, who appeared to take some pleasure in referring to an excellent candidate for a part-time professorship as a reknowned Dutch cryptographer who – tongue in cheek – had turned banker.

At the beginning of this period, Arjen invented XTR, together with Eric Verheul [375, 377, 378]. In Eindhoven, Arjen remotely supervised a number of PhD students, including this book's second editor, whose topic was related to XTR [559, 560]. An overview of XTR and subsequent developments is provided in Chapter 10.

At the start of 2006, Arjen was appointed a professor at the École

Figure 1.3   Photo of Arjen in front of the PS3 cluster taken in 2014. Copyright @Tamedia Publications romandes/Sabine Papilloud.

Polytechnique Fédérale de Lausanne (EPFL) in Lausanne, Switzerland. At EPFL, it is customary for professors to name their group a 'laboratory' with associated acronym. Arjen proudly settled on LACAL, as it worked both in French and in English. Of course, for any abbreviation to work in both those languages, it helps if this is a palindrome and LACAL fitted the bill perfectly. In English it stands for LAboratory for Cryptologic ALgorithms, whereas in French it reads LAboratoire de Cryptologie ALgorithmique.

At LACAL, Arjen continued his focus on the design and analysis of algorithms used in cryptographic protocols or security assessments. Given Arjen's interest in computational cryptography, in hindsight it was no surprise that he started to investigate cheap and powerful alternatives to conventional computer clusters, culminating in the purchase of 215 first generation PlayStation 3 (PS3) game consoles in 2007. The purchase of this exotic computer cluster coincided with the start of the PhD of this book's first editor in Arjen's group at EPFL. Arjen and his cluster were featured in a local newspaper in 2014; a photo as used by the newspaper of Arjen posing in front of the cluster is shown in Figure 1.3.

The main motivation for using the PS3 game console for computational cryptography was the Cell processor: a powerful general-purpose processor, which on the first generation PS3s could be accessed using Sony's hypervisor. What made a PS3's Cell processor especially interesting were its eight Synergistic Processing Units (SPUs), although when running Linux only six of these SPUs can be used (as one is disabled and one is reserved by the hypervisor).

Each SPU runs independently from the others at 3.2GHz, using its own 256 kilobyte of fast local memory for instructions and data. It is cache-less and has 128 registers of 128 bits each, allowing Single Instruction, Multiple Data (SIMD) operations on 16 8-bit, 8 16-bit, or 4 32-bit integers. This opens up a myriad of algorithmic possibilities to enhance cryptographic as well as crypt-analytic applications. Indeed, a wide variety of projects were executed on the PS3 cluster.

A particular notorious one was a continuation of Arjen's previous work with Xiaoyun Wang and Benne de Weger related to the possibility of constructing meaningful hash collisions for public keys [369, 385]. This research effort, led by Marc Stevens, resulted in short chosen-prefix collisions for MD5 and, more impactful, rogue CA certificates [570, 573, 574]. The details are described in Chapter 7.

The PS3 cluster was also used by the first author for a PhD project to compute a 112-bit elliptic curve discrete logarithm, an effort roughly equivalent to 14 full 56-bit DES key searches and, at the time, a new ECDLP record [89, 91]. It is worth highlighting that members of Arjen's lab LACAL played an important role in the fall of discrete logarithms in finite fields of fixed characteristic over the past decade. Arjen's involvement in computing discrete logarithms [91, 330, 366] as well as recent progress is detailed in Chapter 5.

Other computational research sparked by the cluster include fast modular multiplication [85, 86, 90]. Chapter 8 provides a detailed description on fast modular reduction.

Working with the cluster was a fun and unique experience: after all, not every PhD student gets the opportunity to work with such challenging computer equipment. The bespoke installation of the cluster also led to some hilarious moments. At some point, after a global power glitch at EPFL, the PhD students (first author included) had to climb the racks in order to manually switch on every PS3 again, all much to Arjen's entertainment.

In an original twist on 'dual-use cryptography', Arjen had part of the cluster reconfigured as PlayLab: a room of 25 PS3s equipped with game controllers and headsets (see Figure 1.4). This room was used to host groups of children between 7 and 18 years old. There were various kinds of demos and presentations, ranging from explaining the Cell processor as an eight-headed dragon to a live demo on MD5 password cracking. The presentations in the PlayLab always ended with the opportunity for the children to play some games. Arjen himself is not in any sense a gamer, a fact that is emphasised in an interview in 2011 [607]:

Figure 1.4   PlayLab: a room of 25 PS3s equipped with game controllers and headsets for visits of groups of children aged 7 to 18.

My first exposure to computer games was in 1974, in the basement of the mathematics department of the university of Amsterdam, where about half a dozen terminals were remotely connected to Amsterdam's only computer (a CDC with 12-bit bytes, and computationally speaking probably less powerful than a current microwave oven). It may have been the dungeon-like environment or the unhealthy look of those who were playing, but it did not attract me and that never changed. But, 25 of our playstations are equipped with monitors and students can come and play, one of the many games that we have.

Besides these computational results, Arjen has a keen interest in the security assessment of widely deployed security systems. One year he enthusiastically returned from the RSA conference with a sizeable stash of security dongles that, upon pressing a button, would reveal a supposedly random six-digit security code. He had been using this exact same device for some application already and, after an eery feeling of not-quite-randomness, he had started keeping track of the numbers, which indeed revealed a pattern. Now, the question was whether the problem was device specific or not, so upon spotting a large bowl of such devices at the RSA conference, he quietly set upon collecting a few more samples to confirm his suspicions. After a responsible disclosure to those hosting the stand with the bowl, the bowl promptly disappeared for the rest of the conference. Yet, enough devices had been secured to task a master's student with pressing the button sufficiently often to reverse-engineer and successfully cryptanalyse the device.

More seriously, the interest in deployed security system extended to implementation mistakes or misuse [374, 386]. An age-old adage in cryptanalysis is 'to look for plaintext'. In the context of RSA public keys, specifically the public moduli, this can be translated to look for primes (indeed!) or common factors. It turns out RSA moduli are often not as random as one might wish. Progress in this fascinating area is described in Chapter 6.

Arjen's expertise, broad interest and entertaining way of asking questions, make him a regular member of PhD committees. Memorable are the moments where he asks seemingly naive questions which, in the heat of a defense, can catch a candidate off guard. In one instance, a candidate working on lattice reduction to find the shortest vector was asked why one would not look for the longest vector in lattice instead. In another instance, a candidate working on weaknesses of RSA when either the public exponent e or the private exponent d is small, was asked what would happen if e and d are small simultaneously.

Among EPFL students, Arjen is well known for his clear, fun and interesting lectures. To illustrate examples, he often uses characters from popular television series, for example from the American television sitcom *The Big Bang Theory*. The students' appreciation is highlighted by him twice receiving the 'polysphère de faculté IC' (the teaching award handed out by the students from the computer science department), first in 2008 and then again in 2018. Even more impressively, in 2011 he received the overall, EPLF-wide best teaching award (the 'polysphère d'or').

The Big Bang Theory also featured at LACAL's movie lunches, where the team watched an episode of the series, or 30 to 45 minutes of a movie, every day while eating: one summer we consumed a substantial part of the best Spaghetti Westerns. Perhaps it is a tradition inspired by an earlier one, where Arjen would watch the latest movies with Professor Johannes Buchmann while visiting the flagship conference 'Crypto' in Santa Barbara, USA. One such movie was *Pulp Fiction*, an all-time favorite. Indeed, Arjen's preference for movies from the director Quentin Tarantino or the Coen brothers is no secret. His ideal table partner would be Jules Winnfield (a character in *Pulp Fiction* played by Samuel L. Jackson) 'to make philosophy lessons less unbearable' [177]. His adoration for Tarantino even made it into a scientific publication: see e.g. 'that's a bingo' [327, Section 2.5] from the movie *Inglourious Basterds*.

1.2 Outline

Computational cryptography has two sides: a potentially destructive side and a constructive one. The destructive side challenges the hardness assumptions underlying modern cryptographic systems by the development of suitable number-theoretic algorithms; the constructive side offers techniques and recommendations that help implement and deploy cryptosystems. Although the two sides are intricately intertwined, this book is divided into two parts accordingly. The first part is dedicated to cryptanalysis, whereas the second part focusses on the implementation aspects of computational cryptography.

In Chapter 2, 'Lattice Attacks on NTRU and LWE: A History of Refinements', Martin R. Albrecht and Léo Ducas provide an overview of the advances and techniques used in the field of lattice reduction algorithms. Four decades after its invention, the LLL algorithm still plays a significant role in cryptography, not least because it has become one of the main tools to assess the security of a new wave of lattice-based cryptosystems intended for the new post-quantum cryptographic standard. The runtime of the LLL algorithm was always well understood, but the quality of its output, i.e., how short its output vectors were, could be hard to predict, even heuristically. Yet, an important aspect in the evaluation of the new lattice schemes are accurate predictions of the hardness of the underlying lattice problems, which crucially relies on estimating the 'shortness' of the vectors that can be efficiently found using lattice reduction and enumeration. Albrecht and Ducas have been on the forefront of improving such estimators and build upon their expertise in their chapter.

In Chapter 3, 'History of Integer Factorisation', Samuel S. Wagstaff, Jr, gives a thorough overview of the hardness of one of the cornerstones of modern public-key cryptography. The history starts with the early effort by Eratosthenes and his sieve, eventually leading to the modern number field sieve, currently the asymptotically fastest general-purpose integer factorisation method known. Also included are 'special' integer factorisation methods like the elliptic curve method, where the run-time depends mainly on the size of the unknown prime divisor. Modern factorisation efforts often include a gradual escalation of different methods, so it is essential to be familiar with a wide range of methods and the essence of all relevant algorithms is explained clearly. Wagstaff's chapter is based on his far more extensive book on the same topic [611].

In Chapter 4, 'Lattice-Based Integer Factorisation: An Introduction to Coppersmith's Method', Alexander May investigates the use of LLL to factor integers as pioneered by Coppersmith. Conceptually, Coppersmith's method can be deceptively simple: given additional information about an integer to factor (e.g., the knowledge that an RSA key pair (N, e) has a small corresponding private exponent d), derive a system of equations with a small root that reveals the factorisation and use LLL to find the small root. As a result, it becomes possible to explore exponentially sized search spaces, while preserving polynomial time using the famous LLL lattice reduction algorithm. Yet, exploiting Coppersmith's method in a cryptographic context optimally often involves a number of clever choices related to which system of equations to consider. At first, a tantalisingly annoying problem where the choice may appear obvious only in retrospect. May uses his extensive experience in improving the state of the art to explain the reasoning behind various applications in his chapter.

In Chapter 5, 'Computing Discrete Logarithms', Robert Granger and Antoine Joux discuss the question 'how hard is it to compute discrete logarithms in various groups?'. The key ideas and constructions behind the most efficient algorithms for solving the discrete logarithm problem are detailed, with a focus on the recent advances related to finite fields of extension degree >1. A highlight is the rapid development, in the period 2012–2014, of quasi-polynomial time algorithms to solve the DLP in finite fields of fixed chararterstic. Both Granger and Joux contributed significantly to this development, albeit on competing teams. For this book, they join forces and explain how different ideas eventually led to the fall of the fixed characteristic finite field discrete logarithm problem.

In Chapter 6, 'RSA, DH and DSA in the Wild', Nadia Heninger outlines the various cryptographic pitfalls one can – but really should not – make in practice. Often it is possible to bypass the 'hard' mathematical problem a cryptosystem is based upon, and instead take advantage of implementation, deployment or protocol mistakes to extract the private key. Often, the techniques used are excellent examples of the interplay of mathematics and computer science, requiring a combination of ingenuity to find the core idea and perseverance to exploit the weakness in practice. Heninger gives a wide-ranging overview of the multitude of cryptographic implementation vulnerabilities that have been found in the past decades and their impact in practice, including a fair number where she was personally involved in identifying the vulnerability. In her chapter, she wonders whether after several decades of implementation chaos and catastrophic vulnerabilities, we are doomed, but concludes that there is hope yet by bringing into practice the lessons learned.

In Chapter 7, 'A Survey of Chosen-Prefix Collision Attacks', Marc Stevens surveys the technical advances, impact and usage of collision attacks for the most widely used cryptographic hash functions. Cryptographic hash functions are the Swiss army knives within cryptography and are used in many applications including digital signature schemes, message authentication codes, password hashing, cryptocurrencies and content-addressable storage. Stevens was one of the driving forces in turning initial weaknesses in the compression function into practical attacks against various widely deployed protocols relying on hash function collision resistance for their security. In his chapter, he explains how each scenario involves the development of slightly different ideas to exploit weaknesses in especially MD5.

Chapters 2 to 7 constitute the 'Cryptanalysis' part of the book. They directly affect the vast majority of currently deployed public-key cryptography, as well as an important family of future, post-quantum schemes. In the second part of this book, 'Implementations', the emphasis shifts to techniques and recommendations to deploy public-key cryptosystems.

In Chapter 8, 'Efficient Modular Arithmetic', Joppe W. Bos, Thorsten Kleinjung and Dan Page discuss how to realise efficient modular arithmetic in practice on a range of platforms. They cover generic modular multiplication routines such as the popular Montgomery multiplication as well as special routines for primes of a specific shape. Moreover, especially fast methods are outlined that might produce errors with a very small probability. Such faster 'sloppy reduction' techniques are especially beneficial in cryptanalytic settings. Bos, Kleinjung and Page have all been involved in various cryptographic and cryptanalytic record-setting software implementations. In their chapter, they explain how to select the most efficient modular arithmetic routines for the cryptologic implementation task at hand.

In Chapter 9, 'Arithmetic Software Libraries', Victor Shoup provides a peek under the hood of NTL, a software library for doing number theory, as well as its relation to a few other related software libraries. NTL supports data structures and arithmetic operations manipulating signed, arbitrary-length integers, as well as extensions dealing with vectors, matrices, finite fields, and polynomials over the integers. These mathematical objects are essential building blocks for any efficient cryptologic implementation. As Shoup explains, he started development of NTL around 1990 in order to implement novel number-theoretic algorithms and determine at what point theoretical, asymptotic gains turned practical. NTL has become an essential tool for number-theoretic cryptanalysis to answer how well algorithms perform in practice and contains several algorithms for lattice basis reduction, including LLL. As explained in Shoup's chapter, NTL might have been initially based on FreeLIP, and it continues to evolve to make best use of GMP and modern hardware to provide cutting-edge performance.

In Chapter 10, 'XTR and Tori', Martijn Stam discusses how to work efficiently and compactly in certain multiplicative subgroups of finite fields. The emphasis is on XTR, which works in the cyclotomic subgroup of $\mathbb{F}_{p^6}$. Stam

worked extensively on speeding up XTR and related systems based on algebraic tori and uses his first-hand knowledge to provide a historical perspective
of XTR in his chapter.

In Chapter 11, 'History of Cryptographic Key Sizes', Nigel P. Smart and
Emmanuel Thomé provide an overview of improvements over the years in estimating the security level of the main cryptographic hardness assumptions.
These improvements heavily rely on the algorithmic innovations discussed in
Chapters 2 to 5, but combining all those ideas into concrete key-size recommendations is not immediate. Smart and Thomé have been at the forefront of
cryptanalytic research and recommending cryptographic key sizes and conclude with current recommendations in their chapter.

PART I

CRYPTANALYSIS

2

Lattice Attacks on NTRU and LWE:
A History of Refinements

Martin R. Albrecht and Léo Ducas

2.1 Introduction

Since its invention in 1982, the Lenstra–Lenstra–Lovász (LLL) lattice reduction algorithm [380] has found countless applications. In cryptanalysis, the two most prominent applications of LLL and its generalisations,
e.g., Slide [205], Block-Korkine–Zolotarev (BKZ) [512, 520] and Self-Dual
BKZ (SD-BKZ) [425], are factoring RSA keys with extra information on the
secret key via Coppersmith's method [136, 451] (see the chapter by Alexander
May) and the cryptanalysis of lattice-based schemes.

After almost 40 years of cryptanalytic applications, predicting and optimising lattice reduction algorithms remains an active area of research. While we
do have theorems bounding the worst-case performance of these algorithms,
those bounds are asymptotic and not necessarily tight when applied to practical or even cryptographic instances. Reasoning about the behaviour of those
algorithms relies on heuristics and approximations, some of which are known
to fail for relevant corner cases.

Recently, decades after Arjen Lenstra and his co-authors gave birth to this
fascinating and lively research area, this state of affairs became a more pressing issue. Motivated by post-quantum security, standardisation bodies, governments and industries started to move towards deploying lattice-based cryptographic algorithms. This spurred the refinement of those heuristics and approximations, leading to a better understanding of the behaviour of these algorithms
over the past few years.

Lattice reduction algorithms, such as LLL and BKZ, proceed with repeated local improvements to the lattice basis, and each such local improvement means solving the short(est) vector problem in a lattice of a smaller dimension. Therefore, two questions arise: how costly is it to find those local

improvements and what is the global behaviour when those improvements are applied.

While these two questions may not be perfectly independent, we will, in this chapter, survey the second one, namely, the global behaviour of such algorithms, given oracle access for finding local improvements. Our focus on the global behaviour is motivated by our intent to draw more of the community's attention to this aspect. We will take a particular interest in the behaviour of such algorithms on a specific class of lattices, underlying the most popular lattice problems to build cryptographic primitives, namely the Learning with Errors (LWE) problem and the NTRU problem. We will emphasise the approximations that have been made, their progressive refinements and highlight open problems to be addressed.

2.1.1 LWE and NTRU

The LWE problem and the NTRU problem have proven to be versatile building blocks for cryptographic applications [104, 218, 274, 493]. For both of these problems, there exist ring and matrix variants. More precisely, the original definition of NTRU is the ring variant [274] and the matrix variant is rarely considered whereas for LWE the original definition is the matrix variant [494] with a ring variant being defined later [401, 561]. In this chapter, we generally treat the matrix variants since our focus is on lattice reduction for general lattices.

Definition 2.1 (LWE [494]). Let n, q be positive integers, χ be a probability distribution on $\mathbb{Z}$ and $\mathbf{s}$ be a uniformly random vector in $\mathbb{Z}_q^n$. We denote by $L_{\mathbf{s},\chi}$ the probability distribution on $\mathbb{Z}_q^n \times \mathbb{Z}_q$ obtained by choosing $\mathbf{a} \in \mathbb{Z}_q^n$ uniformly at random, choosing $e \in \mathbb{Z}$ according to χ and considering it in $\mathbb{Z}_q$, and returning $(\mathbf{a}, c) = (\mathbf{a}, \langle \mathbf{a}, \mathbf{s} \rangle + e) \in \mathbb{Z}_q^n \times \mathbb{Z}_q$.

Decision-LWE is the problem of deciding whether pairs $(\mathbf{a}, c) \in \mathbb{Z}_q^n \times \mathbb{Z}_q$ are sampled according to $L_{\mathbf{s},\chi}$ or the uniform distribution on $\mathbb{Z}_q^n \times \mathbb{Z}_q$.

Search-LWE is the problem of recovering $\mathbf{s}$ from pairs $(\mathbf{a}, c) = (\mathbf{a}, \langle \mathbf{a}, \mathbf{s} \rangle + e) \in \mathbb{Z}_q^n \times \mathbb{Z}_q$ sampled according to $L_{\mathbf{s},\chi}$.

We note that the above definition puts no restriction on the number of samples, i.e., LWE is assumed to be secure for any polynomial number of samples. Further, since for many choices of n, q, χ solving Decision-LWE allows solving Search-LWE [105, 494] and vice versa, it is meaningful just to speak of the LWE problem (for those choices of parameters). By rewriting the system in systematic form [23], it can be shown that the LWE problem, where each component of the secret $\mathbf{s}$ is sampled from the error distribution χ, is as secure as the problem for uniformly random secrets. LWE with such a secret, following the error distribution, is known as normal form LWE. We will consider normal form LWE in this chapter. Furthermore, in this note, the exact specification of the distribution χ will not matter, and we may simply specify an LWE instance by giving the standard deviation σ of χ. We will, furthermore, implicitly assume that χ is centred, i.e., has expectation 0. We may also write LWE in matrix form as $\mathbf{A} \cdot \mathbf{s} + \mathbf{e} \equiv \mathbf{c} \mod q$. The NTRU problem [274] is defined as follows.

Definition 2.2 (NTRU [274]). Let n, q be positive integers, $f, g \in \mathbb{Z}_q[x]$ be polynomials of degree n sampled from some distribution χ, subject to f being invertible modulo a polynomial ϕ of degree n, and let $h = g/f \mod (\phi, q)$. The NTRU problem is the problem of finding f, g given h (or any equivalent solution $(x^i \cdot f, x^i \cdot g)$ for some $i \in \mathbb{Z}$).

Concretely, the reader may think of $\phi = x^n + 1$ when n is a power of two and χ to be some distribution producing polynomials with small coefficients. The matrix variant considers $\mathbf{F}, \mathbf{G} \in \mathbb{Z}_q^{n \times n}$ such that $\mathbf{H} = \mathbf{G} \cdot \mathbf{F}^{-1} \mod q$.

2.2 Notation and Preliminaries

All vectors are denoted by bold lower case letters and are to be read as column vectors. Matrices are denoted by bold capital letters. We write a matrix $\mathbf{B}$ as $\mathbf{B} = (\mathbf{b}_0, \ldots, \mathbf{b}_{d-1})$ where $\mathbf{b}_i$ is the ith column vector of $\mathbf{B}$. If $\mathbf{B} \in \mathbb{R}^{m \times d}$ has full-column rank d, the lattice Λ generated by the basis $\mathbf{B}$ is denoted by $\Lambda(\mathbf{B}) = \{\mathbf{B} \cdot \mathbf{x} \mid \mathbf{x} \in \mathbb{Z}^d\}$. A lattice is q-ary if it contains $q\mathbb{Z}^d$ as a sublattice, e.g., $\{\mathbf{x} \in \mathbb{Z}_q^d \mid \mathbf{x} \cdot \mathbf{A} \equiv \mathbf{0}\}$ for some $\mathbf{A} \in \mathbb{Z}^{d \times d'}$. We denote by $(\mathbf{b}_0^\star, \ldots, \mathbf{b}_{d-1}^\star)$ the Gram–Schmidt (GS) orthogonalisation of the matrix $(\mathbf{b}_0, \ldots, \mathbf{b}_{d-1})$. For $i \in \{0, \ldots, d-1\}$, we denote the orthogonal projection to the span of $(\mathbf{b}_0, \ldots, \mathbf{b}_{i-1})$ by π_i; π_0 denotes 'no projection', i.e., the identity. We write $\pi_{\mathbf{v}}$ for the projection orthogonal to the space spanned by $\mathbf{v}$. For $0 \le i < j \le d$, we denote by $\mathbf{B}_{[i:j]}$ the local projected block $(\pi_i(\mathbf{b}_i), \ldots, \pi_i(\mathbf{b}_{j-1}))$, and when the basis is clear from context, by $\Lambda_{[i:j]}$ the lattice generated by $\mathbf{B}_{[i:j]}$. We write $\lg(\cdot)$ for the logarithm to base two.

The Euclidean norm of a vector $\mathbf{v}$ is denoted by $\|\mathbf{v}\|$. The volume (or determinant) of a lattice $\Lambda(\mathbf{B})$ is $\mathrm{vol}(\Lambda(\mathbf{B})) = \prod_i \|\mathbf{b}_i^\star\|$. It is an invariant of the lattice. The first minimum of a lattice Λ is the norm of a shortest non-zero vector, denoted by $\lambda_1(\Lambda)$. We use the abbreviations $\mathrm{vol}(\mathbf{B}) = \mathrm{vol}(\Lambda(\mathbf{B}))$ and $\lambda_1(\mathbf{B}) = \lambda_1(\Lambda(\mathbf{B}))$.

The Hermite constant γ_β is the square of the maximum norm of any shortest vector in all lattices of unit volume in dimension β:

$$\gamma_\beta = \sup\left\{\lambda_1^2(\Lambda) \mid \Lambda \in \mathbb{R}^\beta, \mathrm{vol}(\Lambda) = 1\right\}.$$

Minkowski's theorem allows us to derive an upper bound $\gamma_\beta = O(\beta)$, and this bound is reached up to a constant factor: $\gamma_\beta = \Theta(\beta)$.

2.3 Lattice Reduction: Theory

All lattices of dimension $d \geq 2$ admit infinitely many bases, and two bases $\mathbf{B}, \mathbf{B}'$ generate (or represent) the same lattice if and only if $\mathbf{B} = \mathbf{B}' \cdot \mathbf{U}$ for some unimodular matrix $\mathbf{U} \in \mathrm{GL}_d(\mathbb{Z})$. In other words, the set of (full-rank) lattices can be viewed as the quotient $\mathrm{GL}_d(\mathbb{R})/\mathrm{GL}_d(\mathbb{Z})$. Lattice reduction is the task of finding a good representative of a lattice, i.e., a basis $\mathbf{B} \in \mathrm{GL}_d(\mathbb{R})$ representing $\Lambda \in \mathrm{GL}_d(\mathbb{R})/\mathrm{GL}_d(\mathbb{Z})$.

While there exists a variety of formal definitions for what is a good representative, the general goal is to make the Gram–Schmidt basis $\mathbf{B}^\star$ as small as possible. Using the simple size-reduction algorithm (see [454, Algorithm 3]), it is possible to also enforce the shortness of the basis $\mathbf{B}$ itself.

It should be noted that because we have an invariant $\prod_i \|\mathbf{b}_i^\star\| = \mathrm{vol}(\Lambda)$, we cannot make all GS vectors small at the same time, but the goal becomes to balance their lengths. More pictorially, we consider the log profile of a basis as the graph of $(\ell_i = \lg \|\mathbf{b}_i^\star\|)_{i=0\ldots d-1}$ as a function of i. By the volume invariant, the area under this graph is fixed, and the goal of reduction is to make this graph flatter.

A very strong[1] notion of reduction is the Hermite–Korkine–Zolotarev (HKZ) reduction, which requires each basis vector $\mathbf{b}_i$ to be a shortest non-zero vector of the remaining projected lattice $\Lambda_{[i:d]}$. The Block-Korkine–Zolotarev (BKZ) reduction relaxes HKZ, only requiring $\mathbf{b}_i$ to be close-to-shortest in a local 'block'. More formally, we have the following.

Definition 2.3 (HKZ and BKZ [454]). The basis $\mathbf{B} = (\mathbf{b}_0, \ldots, \mathbf{b}_{d-1})$ of a lattice Λ is said to be HKZ reduced if $\|\mathbf{b}_i^\star\| = \lambda_1(\Lambda(\mathbf{B}_{[i:d]}))$ for all $i < d$. It is said BKZ reduced with block size β and $\epsilon \geq 0$ if $\|\mathbf{b}_i^\star\| \leq (1 + \epsilon) \cdot \lambda_1(\Lambda(\mathbf{B}_{[i:\min(i+\beta,d)]}))$ for all $i < d$.

In practice, the BKZ algorithm [512, 520] and its terminated variant [257]

---

[1]   HKZ should nevertheless not be considered to be the strongest notion of reduction. Indeed HKZ is a greedy definition, speaking of the shortness of each vector individually. One could go further and require, for example, $\Lambda_{[0:d/2]}$ to be a densest sublattice of Λ [491].

Algorithm 2.1 High-level description of the BKZ algorithm.

Input: LLL-reduced lattice basis **B** and block size β

1: **repeat**
2:      **for** $i \leftarrow 0$ **to** $d-2$ **do**
3:          LLL on $\mathbf{B}_{[i:\min(i+\beta,d)]}$
4:          $\mathbf{v} \leftarrow$ find a short vector in $\Lambda\left(\mathbf{B}_{[i:\min(i+\beta,d)]}\right)$
5:          insert $\mathbf{v}$ into **B** at index i and handle linear dependencies with LLL
6: **until** until no more change

---

are commonly employed to perform lattice reduction. BKZ is also the algorithm we will focus on in this chapter.

The BKZ algorithm will proceed by enforcing the condition $\|\mathbf{b}_i^\star\| \leq (1 + \epsilon) \cdot \lambda_1(\Lambda(\mathbf{B}_{[i:\min(i+\beta,d)]}))$ cyclically for $i = 0, \ldots, d-2, 0, \ldots, d-2, 0 \ldots$, see Algorithm 2.1. However, each modification of $\mathbf{b}_i^\star$ may invalidate the same condition for $j \neq i$. The value of ϵ, which allows to account for numerical instability, is typically chosen very close to 0 (say 0.01); we may sometimes omit it and just speak of a BKZ-β reduced basis. Overall, we obtain the following guarantees for the BKZ algorithm.

Theorem 2.4 (BKZ). *If a basis* **B** *is BKZ-β reduced with parameter $\epsilon > 0$ it satisfies*

- $\|\mathbf{b}_0\| \leq \sqrt{(1 + \epsilon) \cdot \gamma_\beta}^{\frac{d-1}{\beta-1}+1} \cdot \operatorname{vol}(\Lambda(\mathbf{B}))^{1/d}$ *(Hermite factor) and*
- $\|\mathbf{b}_0\| \leq \left((1 + \epsilon) \cdot \gamma_\beta\right)^{\frac{d-1}{\beta-1}} \cdot \lambda_1(\Lambda(\mathbf{B}))$ *(approximation factor).*

Remark. The approximation factor is established in [517], the Hermite factor bound is claimed in [206]. In [257] a bound of $2 \cdot \sqrt{\gamma_\beta}^{\frac{d-1}{\beta-1}+3}$ is established for the terminating variant. In [258] this bound is improved to $K \cdot \sqrt{\beta}^{\frac{d-1}{\beta-1}+0.307}$ for some universal constant K.

Asymptotically, the lattice reduction algorithm with best, known worst-case guarantees is Slide reduction [205]. We refer to its introduction by Gama and Nguyen [205] for a formal definition, which requires the notion of duality, and only state some of its guarantees concerning Gram–Schmidt length here.

Theorem 2.5 (Slide reduction [205]). *If a basis* **B** *is Slide reduced for parameters $\beta \mid d$ and $\epsilon > 0$ it satisfies*

- $\|\mathbf{b}_0\| \leq \sqrt{(1 + \epsilon) \cdot \gamma_\beta}^{\frac{d-1}{\beta-1}} \cdot \operatorname{vol}(\Lambda(\mathbf{B}))^{1/d}$ *(Hermite factor) and*
- $\|\mathbf{b}_0\| \leq \left((1 + \epsilon) \cdot \gamma_\beta\right)^{\frac{d-\beta}{\beta-1}} \cdot \lambda_1(\Lambda(\mathbf{B}))$ *(approximation factor).*

In practice, BKZ is not implemented as in Algorithm 2.1. Most notably, stronger preprocessing than LLL is applied. A collection of improvements to the algorithm (when enumeration is used to instantiate the SVP oracle) are collectively known as BKZ 2.0 [122] and implemented, e.g., in FPLLL [587] and thus Sage [562]. Slide reduction is also implemented in FPLLL.

2.4 Practical Behaviour on Random Lattices

2.4.1 Shape Approximation

The Gaussian heuristic predicts that the number $|\Lambda \cap \mathcal{B}|$ of lattice points inside a measurable body $\mathcal{B} \subset \mathbb{R}^n$ is approximately equal to $\mathrm{vol}(\mathcal{B})/\mathrm{vol}(\Lambda)$. Applied to Euclidean d-balls, it leads to the following prediction of the length of a shortest non-zero vector in a lattice.

Definition 2.6 (Gaussian heuristic)**.** We denote by $\mathrm{gh}(\Lambda)$ the expected first minimum of a lattice Λ according to the Gaussian heuristic. For a full-rank lattice $\Lambda \subset \mathbb{R}^d$, it is given by

$$\mathrm{gh}(\Lambda) = \left(\frac{\mathrm{vol}(\Lambda)}{\mathrm{vol}(\mathcal{B})}\right)^{1/d} = \frac{\Gamma\left(1 + \frac{d}{2}\right)^{1/d}}{\sqrt{\pi}} \cdot \mathrm{vol}(\Lambda)^{1/d} \approx \sqrt{\frac{d}{2\pi e}} \cdot \mathrm{vol}(\Lambda)^{1/d} \,,$$

where $\mathcal{B}$ denotes the d-dimensional Euclidean ball. We also denote by $\mathrm{gh}(d)$ the quantity $\mathrm{gh}(\Lambda)$ of any d-dimensional lattice Λ of volume 1: $\mathrm{gh}(d) \approx \sqrt{d/2\pi e}$. For convenience we also denote $\mathrm{lgh}(x)$ for $\lg(\mathrm{gh}(x))$.

Combining the Gaussian heuristic with the definition of a BKZ reduced basis, after BKZ-β reduction we expect

$$\ell_i = \lg\left(\lambda_1(\Lambda(\mathbf{B}_{[i:\min(i+\beta,d)]}))\right) \approx \mathrm{lgh}(\min(\beta, d-i)) + \frac{\lg\left(\mathrm{vol}(\Lambda(\mathbf{B}_{[i:\min(i+\beta,d)]}))\right)}{\min(\beta, d-i)}$$

$$= \mathrm{lgh}(\min(\beta, d-i)) + \frac{\sum_{j=i}^{\min(i+\beta,d)-1} \ell_j}{\min(\beta, d-i)}.$$

If $d \gg \beta$ this linear recurrence implies a geometric series for the $\|\mathbf{b}_i^\star\|$. Considering one block of dimension β and unit volume, we expect $\ell_i = (\beta - i - 1) \cdot \lg(\alpha_\beta)$ for $i = 0, \dots, \beta - 1$ and some α_β. We obtain

$$\ell_0 = (\beta - 1) \cdot \lg(\alpha_\beta) \approx \mathrm{lgh}(\beta) + \frac{1}{\beta} \sum_{j=0}^{\beta-1} j \cdot \lg(\alpha_\beta)$$

$$= \mathrm{lgh}(\beta) + (\beta - 1)/2 \cdot \lg(\alpha_\beta).$$

Solving for α_β assuming equality we obtain $\alpha_\beta = \text{gh}(\beta)^{2/(\beta-1)}$.

Applying the same argument to a basis in dimension $d \gg \beta$ with $\ell_i = (d - i - 1) \cdot \lg(\alpha_\beta)$ for $i = 0, \ldots, d - 1$, we get $\|\mathbf{b}_0\|/\text{vol}(\Lambda)^{1/d} = \alpha_\beta^{d-1}/\alpha_\beta^{(d-1)/2} = \alpha_\beta^{(d-1)/2} = \text{gh}(\beta)^{(d-1)/(\beta-1)}$. This is known as the geometric series assumption (GSA).

Definition 2.7 (GSA [518]). Let $\mathbf{B}$ be a BKZ-β reduced basis of a lattice of volume V. The geometric series assumption states that

$$\lg \|\mathbf{b}_i^\star\| = \ell_i = \frac{d - 1 - 2i}{2} \cdot \lg(\alpha_\beta) + \frac{1}{d} \lg V,$$

where $\alpha_\beta = \text{gh}(\beta)^{2/(\beta-1)}$.

The above assumption is reasonably accurate in the case $\beta \ll d$ (and $\beta \gg 50$), but it ignores what happens in the last $d - \beta$ coordinates. Indeed, the last block is HKZ reduced, and should therefore follow the typical profile of an HKZ reduced basis.

Under the Gaussian heuristic, we can predict the shape $\ell_0 \ldots \ell_{d-1}$ of an HKZ reduced basis, i.e., the sequence of expected norms for the vectors $\mathbf{b}_i^\star$. This, as before, implicitly assumes that all the projected lattices Λ_i also behave as random lattices. The sequence is inductively defined as follows.

Definition 2.8. The (unscaled) HKZ shape of dimension d is defined by the following sequence for $i = 0, \ldots, d - 1$:

$$h_i = \lg \text{gh}(d - i) - \frac{1}{d - i} \sum_{j < i} h_j.$$

This leads to the following refinement of the GSA.

Definition 2.9 (Tail-adapted geometric series assumption (TGSA)). Let $\mathbf{B}$ be a BKZ-β reduced basis of a lattice of volume V. The TGSA states that

$$\ell_i = \frac{d - 1 - 2i}{2} \cdot \lg \alpha_\beta + s \qquad \text{if } 0 \le i \le d - \beta,$$

$$\ell_i = h_{i-(d-\beta)} + \ell_{d-\beta} - h_0 \qquad \text{if } d - \beta \le i < d,$$

where $s \in \mathbb{R}$ is the scaling term such that $\sum \ell_i = \lg V$.

We plot an example for a basis after BKZ reduction under the GSA and the TGSA in Figure 2.1 to illustrate their respective shapes. In Figure 2.1 we chose $d = 2\beta$ to highlight the difference between the two models. As can be seen from that figure, the first few indices of the HKZ shape drop slower than predicted by the GSA and the last indices drop faster.

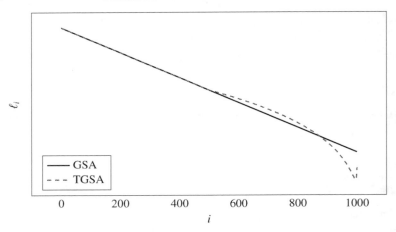

Figure 2.1 GSA and TGSA for $d = 1000$ and $\beta = 500$.

Appealing to the Gaussian heuristic, we may also replace $\sqrt{\gamma_\beta}$, i.e., worst-case bounds, with $\mathrm{gh}(\beta)$, i.e., average-case expectations, in Theorems 2.4 and 2.5. This suggests the following heuristics.

Definition 2.10 (Estimates for block reductions). If a basis **B** is BKZ-β reduced for $50 \ll \beta \ll d$ we expect

$$\|\mathbf{b}_0\| \lessapprox \min \begin{cases} \sqrt{\alpha_\beta}^{d-1} \cdot \mathrm{vol}(\Lambda(\mathbf{B}))^{1/d} & \text{(Hermite factor)} \\ \alpha_\beta^{d-1} \cdot \lambda_1(\Lambda(\mathbf{B})) & \text{(approximation factor)}. \end{cases}$$

If a basis **B** is Slide reduced with parameter β we expect

$$\|\mathbf{b}_0\| \lessapprox \min \begin{cases} \sqrt{\alpha_\beta}^{d-1} \cdot \mathrm{vol}(\Lambda(\mathbf{B}))^{1/d} & \text{(Hermite factor)} \\ \alpha_\beta^{d-\beta} \cdot \lambda_1(\Lambda(\mathbf{B})) & \text{(approximation factor)}. \end{cases}$$

The cases over which the minimum is taken define two regimes: the 'Hermite regime' and the 'approximation regime'.

If the lattice is random, then $\lambda_1 \approx \mathrm{gh}(\Lambda)$ and we expect to be in the Hermite regime; the approximation regime is only triggered by the presence of an unusually short vector. In the Hermite regime, we can replace $\lessapprox$ by $\approx$ and we will discuss what happens in the approximation regime further in Section 2.5.4.

We note that the literature usually writes the above approximate equations in terms of the so-called root-Hermite factor $\delta_\beta := \left(\|\mathbf{b}_0\|/\mathrm{vol}(\Lambda)^{1/d}\right)^{1/d}$. We can therefore establish that $\delta_\beta = \sqrt{\alpha_\beta}^{1-1/d} \approx \sqrt{\alpha_\beta}$. We note that making this approximation or not leads to the '-1 discrepancy' blamed on [19] in a footnote

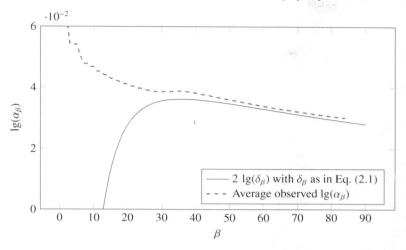

Figure 2.2   Experimentally observed slopes of 16 lattices compared with δ_β as predicted in Eq. (2.1). The input lattices are q-ary lattices in dimension $d = 170$ with $q = 2^{20} - 3$; the experimental $\lg(\alpha_\beta)$ are established using a least-square fit of the log Gram–Schmidt vectors.

of [15]: the analysis of [19] simply did not apply this approximation step. In [121] an expression for δ_β is given as

$$\lim_{\beta \to \infty} \delta_\beta = \left(\frac{\beta}{2\pi e} \cdot (\pi \beta)^{\frac{1}{\beta}} \right)^{\frac{1}{2(\beta-1)}} \tag{2.1}$$

assuming $d \gg \beta$. Experimentally, Eq. (2.1) also holds with good accuracy for $\beta > 50$ and typical d used in cryptography (say, $d \geq c \cdot \beta$ for some $c > 1$). We compare experimentally observed $\lg(\alpha_\beta)$ with the right-hand side of Eq. (2.1) in Figure 2.2.

2.4.2 Simulators

While the (T)GSA provides a first rough approximation of the shape of a basis, it is known to be violated in small dimensions [122]. Indeed, it also does not hold exactly for larger block sizes when d is a small multiple of β, the case most relevant to cryptography. Furthermore, it only models the shape after the algorithm has terminated, leaving open the question of how the quality of the basis improves throughout the algorithm. To address these points, Chen and Nguyen [122] introduced a simulator for the BKZ algorithm which is often referred to as the 'CN11 simulator'. It takes as input a list of ℓ_i representing the

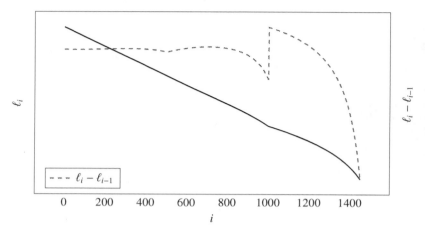

Figure 2.3 CN11 simulator output for $\beta = 500$ on random q-ary lattice in dimension $d = 1500$.

shape of the input basis and a block size β. It then considers blocks $\ell_i, \ldots, \ell_{i+\beta-1}$ of dimension β, establishes the expected norm of the shortest vector in this block using the Gaussian heuristic and updates ℓ_i. To address that the Gaussian heuristic does not hold for $\beta < 50$, the simulator makes use of a precomputed list of the average norms of a shortest vector of random lattices in small dimensions. The simulator keeps on going until no more changes are made or a provided limit on the number of iterations or 'tours' is reached.

The simulator is implemented, for example, in FPyLLL [588] and thus in Sage. In Figure 2.3 we plot the output of the simulator for a basis in dimension 1500 with block size 500 (solid line). We also plot the derivative (dotted line) to illustrate that the GSA also does not hold for $i < d - \beta$. In fact, we observe a ripple effect, with the tail shape exhibiting a damped echo towards the left of the basis. The TGSA is in some sense only a first-order approximation, only predicting the first ripple.

A further simulation refinement was proposed in [27]. Building upon [631], the authors confirmed that the CN11 simulator can be pessimistic about the norm of the first vector output by BKZ. This is because it assumes that the shortest vector in a lattice always has the norm that is predicted by the Gaussian heuristic. By, instead, modelling the norm of the shortest vector as a random variable, the authors were able to model the 'head concavity' behaviour of BKZ as illustrated in Figure 2.4 after many tours and in small block sizes. They also proposed a variant of the BKZ algorithm (pressed-BKZ) that is tailored to exploit this phenomenon. For example, they manage to reach a basis

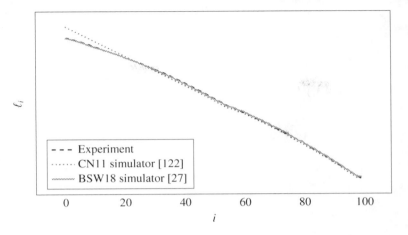

Figure 2.4 Head concavity: dimension $d = 2000$ and block size $\beta = 45$ after 2000 tours, reproduced from [27].

reduction equivalent to BKZ-90 while only using block size 60. The authors note, though, that the head concavity phenomenon does not significantly affect cryptographic block sizes. Indeed, exploiting luck on this random variable seems to be interesting for small block sizes only.

2.4.3 q-ary Lattices and the Z-Shape

Recall that both NTRU and LWE give rise to q-ary lattices. These lattices always contain the vector $(q, 0, \ldots, 0)$ and all its permutations. These so-called 'q-vectors' can be considered short, depending on the parameters of the instance being considered, and might be shorter than what we would expect to obtain following predictions such as the GSA or the TGSA. Furthermore, some of those q-vectors naturally appear in the typical basis construction of q-ary lattices. Even when this is not the case, they can be made explicit by computing the Hermite Normal Form.

To predict lattice reduction on such bases, we may observe that one of the guarantees of the LLL algorithm is that the first vector $\mathbf{b}_0$ never gets longer. For certain parameters this can contradict the GSA. In fact, if $\mathbf{b}_i^*$ does not change for all $i < j$, then $\mathbf{b}_j^*$ cannot become longer either, which means that after the reduction algorithm has completed we may still have many such q-vectors at the beginning of our basis, unaffected by the reduction. It is therefore tempting to predict a piecewise linear profile, with two pieces. It should start with a flat line at $\lg q$, followed by a sloped portion following the predicted GSA slope.

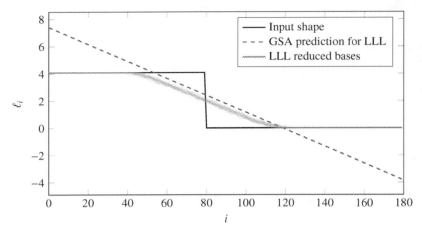

Figure 2.5 GSA and q-ary lattice contradiction. Norms of Gram–Schmidt vectors of 180-dimensional random q-ary lattices with $q = 17$ and volume q^{80}. The grey, blurry lines plot ℓ_i for LLL reduced bases of 16 independent lattices.

In fact, the shape has three pieces, and this is easy to argue for LLL, since LLL is a self-dual algorithm.[2] This means in particular that the last Gram–Schmidt vector cannot get shorter, and following the same argument, we can conclude that the basis must end with a flat piece of 1-vectors. All in all, the basis should follow a Z-shape, and this is indeed experimentally the case [280, 625], as depicted in Figure 2.5, where we picked a small q to highlight the effect. We shall call such a prediction [169, 625] the ZGSA.

It is tempting to extend such a ZGSA model to other algorithms beyond LLL and this has been used for example in [169]. We might also attempt to refine it to a ZTGSA model, where we put an HKZ tail just before the flat section of Gram–Schmidt vectors of norm 1. However, this is a questionable way of reasoning, because BKZ, unlike LLL, is not self-dual. However, it is worth noting that it seems possible to force BKZ to behave in such a way, simply by restricting BKZ to work on the indices up $i < j$, where j is carefully calibrated so that $\|\mathbf{b}_j^\star\| \approx 1$. This is not self-dual, but up to the tail of BKZ, it would produce a Z-shape as well.

Yet, we could also let BKZ work freely on the whole basis, and wonder what would happen. In other words, we may ask whether it is preferable to apply such a restriction to BKZ or not. A natural approach to answering this

---

[2] This is not entirely true, as the size-reduction condition is not self-dual, but the constraints on the Gram–Schmidt vectors themselves are, which is enough for our purpose.

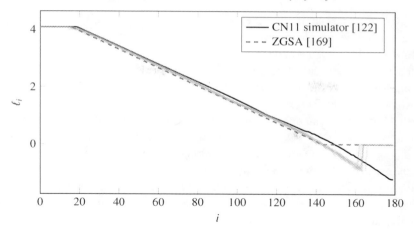

Figure 2.6 BKZ behaviour on q-ary lattice bases with small q. Norms of Gram–Schmidt vectors (grey, blurry lines) after BKZ-65 reduction of 16 180-dimensional q-ary lattices with $q = 17$ and volume q^{80} compared with models from the literature.

question would be to simply use the CN11 simulator, however, it appears that the Z-shape is very poorly simulated. Indeed, while the simulator can easily maintain q-vectors when they are shorter than the one locally predicted by the Gaussian heuristic, the phenomenon on the right end of the Z seems more complicated: some 1-vectors are replaced by Gram–Schmidt vectors of norm strictly less than 1, but not all, see Figure 2.6. Thus, we see the Z-shape known from the literature but with the addition of a kink in the tail block.

Simulating or predicting the behaviour of BKZ on q-ary lattices is still open, but it would allow addressing the question if it can be exploited. A partial answer seems obtainable by defining a specialised variant of the Gaussian heuristic that takes orthogonal sublattices into account. Although we are not certain that a deeper study of this phenomenon would lead to cryptanalytic advances, it is nevertheless quite frustrating to have to resort to Z(T)GSA without a perfect understanding of the behaviour of lattice reduction on this class of lattices.

2.4.4 Random Blocks?

The heuristic analysis of BKZ is based on the assumption that each sublattice considered by the algorithm 'behaves like a random lattice' (strong version), or at least that the expectation or distribution of its shortest vector is the same as for a random lattice (weak version).

More formally, we would have to define the notion of a random lattice, invoking the Haar measure. However, we can nevertheless interrogate this heuristic without going into those details here. Indeed, as we can see in Figure 2.2, the predicted slopes below dimension 30 are far from the actual behaviour. In fact, the predictions for small block sizes are nonsensical as they predict a flatter slope as β decreases below 30 and even an inversion of the slope below block size ≈ 10.

Although we can observe the prediction and the observation converging for block sizes above 50, what level of precision do we attribute to those predictions? Given the phenomena perturbing the GSA surveyed in this chapter (heads, tails, ripples), how pertinent are the data from Figure 2.2? Pushing experimental evidence a bit further would be reassuring here: although we do not expect surprises, it would be good to replace this expectation with experimental evidence.

But, more conceptually, we note that making the strong version of the heuristic assumption (each block behaves like a random lattice) is self-contradictory. Indeed, the model leads us to conclude that the shape is essentially a line, at least when $\beta \ll d$ and the considered block $\mathbf{B}_{[\kappa:\kappa+\beta]}$ is far from the head and the tail, i.e., $\kappa \gg \beta, d - \kappa \gg \beta$. But this block, like all other blocks, is fully HKZ-reduced: since $\mathbf{b}^\star_{\kappa+i}$ is a shortest vector of $\Lambda(\mathbf{B}_{[\kappa+i:\kappa+i+\beta]})$, it is also a shortest vector of $\Lambda(\mathbf{B}_{[\kappa+i:\kappa+\beta]})$. Yet, HKZ-reduced bases of random lattices have a concave shape not a straight slope.

We do not mean to discredit the current methodology to predict attacks on lattice-based schemes; current evidence does suggest predictions such as Eq. (2.5) in Section 2.5.4 are reasonably precise. In particular, the above argument does not rule out the weak version of the hypothesis: the shortest vector of those non-random blocks may still have an expected length following the Gaussian heuristic. In fact, for random lattices, it is known that the length of the shortest vector is increasingly concentrated around the Gaussian heuristic; there may be increasingly fewer lattices that fall far from it, which may explain why a bias in the distribution of the lattices themselves does not translate to a bias on the length of its shortest vector.

However, we wish to emphasise that the question of the distribution of those local blocks is at the centre of our understanding of lattice reduction algorithms but remains open. While even formulating specific yet relevant questions seems hard, this phenomenon suggests itself as a challenging but pressing area to study.

2.5 Behaviour on LWE Instances

We can reformulate the matrix form of the LWE equation $\mathbf{c} - \mathbf{A} \cdot \mathbf{s} \equiv \mathbf{e} \bmod q$ as a linear system over the integers as

$$\begin{pmatrix} q\mathbf{I} & -\mathbf{A} \\ 0 & \mathbf{I} \end{pmatrix} \cdot \begin{pmatrix} * \\ \mathbf{s} \end{pmatrix} + \begin{pmatrix} \mathbf{c} \\ \mathbf{0} \end{pmatrix} = \begin{pmatrix} \mathbf{e} \\ \mathbf{s} \end{pmatrix}$$

or homogeneously as

$$\mathbf{B} = \begin{pmatrix} q\mathbf{I} & -\mathbf{A} & \mathbf{c} \\ 0 & \mathbf{I} & 0 \\ 0 & 0 & t \end{pmatrix}, \qquad \mathbf{B} \cdot \begin{pmatrix} * \\ \mathbf{s} \\ 1 \end{pmatrix} = \begin{pmatrix} \mathbf{e} \\ \mathbf{s} \\ t \end{pmatrix}, \tag{2.2}$$

where t is some chosen constant and $*$ stands in for an arbitrary vector. In other words, there exists an element in the lattice spanned by $\mathbf{B}$ with expected norm $\sqrt{(n + m) \cdot \sigma^2 + t^2}$. Let $d = n + m + 1$. If we have $\sqrt{(n + m) \cdot \sigma^2 + t^2} <$ $\mathrm{gh}(\Lambda(\mathbf{B})) \approx \sqrt{\frac{d}{2\pi \cdot e}} \cdot q^{n/d}$ then $\mathbf{B}$ admits an unusually short vector. With a slight abuse of notation, we will refer to the (column) vector $(\mathbf{e}^T, \mathbf{s}^T, t)^T$ simply as $(\mathbf{e}, \mathbf{s}, t)$.

Remark. We note that when $t \nmid q$ then $\Lambda(\mathbf{B})$ is not a q-ary lattice as, in this case, $(0, \ldots, 0, q)^T \notin \Lambda$. The reader may think $t = 1$, which is commonly used in practice albeit being slightly worse compared to $t = \sigma$, which maximises $\lambda_2(\Lambda)/\lambda_1(\Lambda)$ and which makes the problem easier.

2.5.1 Kannan Embedding

More generally, we can consider this approach to solving LWE as solving an instance of the bounded distance decoding problem (BDD) using a solver for the unique shortest vector problem.

Definition 2.11 (α-Bounded Distance Decoding (BDD$_\alpha$)). Given a lattice basis $\mathbf{B}$, a vector $\mathbf{t}$, and a parameter $0 < \alpha < 1/2$ such that the Euclidean distance $\mathrm{dist}(\mathbf{t}, \mathbf{B}) < \alpha \cdot \lambda_1(\mathbf{B})$, find the lattice vector $\mathbf{v} \in \Lambda(\mathbf{B})$ that is closest to $\mathbf{t}$.

Remark. In our definition above we picked $\alpha < 1/2$, which guarantees a unique solution. The problem can be generalised to $1/2 < \alpha \leq 1$ where we expect a unique solution with high probability.

We can view LWE with a fixed number of samples as an instance of BDD (with overwhelming probability over the choice of the samples). Asymptotically, for any polynomially bounded $\gamma \geq 1$ there is a reduction from BDD$_{1/(\sqrt{2}\gamma)}$ to uSVP$_\gamma$ [26]. The unique shortest vector problem (uSVP) is defined as follows.

Definition 2.12 (γ-unique Shortest Vector Problem (uSVP$_\gamma$)). Given a lattice Λ such that $\lambda_2(\Lambda) > \gamma \cdot \lambda_1(\Lambda)$ find a non-zero vector $\mathbf{v} \in \Lambda$ of length $\lambda_1(\Lambda)$.

This reduction is essentially the embedding technique, due to Kannan [311], presented at the beginning of this section, combined with some tricks to improve the parameters of the reduction. For the remaining of this section, we will discuss how strong we require lattice reduction to be to find a unique shortest vector which can then be used to recover the secret values of an LWE instance.

2.5.2 Asymptotic Handwaving

Recall that in Definition 2.10 two regimes are defined, the Hermite regime and the approximation regime. Now, consider decision LWE. On the one hand, when $\mathbf{c}$ is just a random vector then the lattice spanned by $\mathbf{B}$ is a random q-ary lattice and we are in the Hermite regime, i.e., $\lambda_1(\Lambda(\mathbf{B})) \approx \mathrm{gh}(\Lambda(\mathbf{B}))$. On the other hand, when $\mathbf{c}$ is formed as in LWE then $\Lambda(\mathbf{B})$ contains $(\mathbf{e}, \mathbf{s}, t)$ and we expect $\lambda_1(\Lambda(\mathbf{B})) = \|(\mathbf{e}, \mathbf{s}, t)\|$. Now, if this is sufficiently smaller than $\mathrm{gh}(\Lambda(\mathbf{B}))$ then we are in the approximation regime. Thus, one way to distinguish LWE from uniform is to detect the 'phase transition' between the two regimes, the point when the approximation regime 'kicks in', i.e., when

$$\sqrt{\alpha_\beta}^{2d-2} \cdot \lambda_1(\Lambda(\mathbf{B})) < \sqrt{\alpha_\beta}^{d-1} \cdot \mathrm{vol}(\Lambda(\mathbf{B}))^{1/d} \text{ for BKZ and}$$
$$\sqrt{\alpha_\beta}^{2d-2\beta} \cdot \lambda_1(\Lambda(\mathbf{B})) < \sqrt{\alpha_\beta}^{d-1} \cdot \mathrm{vol}(\Lambda(\mathbf{B}))^{1/d} \text{ for Slide reduction.}$$

Rearranging we obtain the following success conditions

$$\lambda_1(\Lambda(\mathbf{B})) < \sqrt{\alpha_\beta}^{1-d} \cdot \mathrm{vol}(\Lambda(\mathbf{B}))^{1/d} \text{ with BKZ and} \tag{2.3}$$
$$\lambda_1(\Lambda(\mathbf{B})) < \sqrt{\alpha_\beta}^{2\beta-d-1} \cdot \mathrm{vol}(\Lambda(\mathbf{B}))^{1/d} \text{ with Slide reduction} \tag{2.4}$$

for solving decision LWE in block size β.

2.5.3 The 2008 Estimates

Gama and Nguyen [206] performed experiments in small block sizes to establish when lattice reduction finds a unique shortest vector. They considered two classes of semi-orthogonal lattices and Lagarias–Odlyzko lattices [350] which permit to estimate the gap $\lambda_2(\Lambda)/\lambda_1(\Lambda)$ between the first and second minimum of the lattice. For all three families, it was observed in [206] that LLL and BKZ seem to recover a unique shortest vector with high probability whenever $\lambda_2(\Lambda)/\lambda_1(\Lambda) \geq \tau_\beta \cdot \sqrt{\alpha_\beta}^d$, where $\tau_\beta < 1$ is an empirically determined constant that depends on the lattice family, algorithm and block size used.

In [11] an experimental analysis of solving LWE based on the same estimate was carried out for lattices of the form of Eq. (2.2). This lattice contains an unusually short vector $\mathbf{v} = (\mathbf{e}, \mathbf{s}, t)$ of squared norm $\|\mathbf{v}\|^2 = \|\mathbf{s}\|^2 + \|\mathbf{e}\|^2 + t^2$ and we expect $\lambda_1(\Lambda)^2 = \|\mathbf{v}\|^2$. Thus, when $t^2 \approx \|\mathbf{e}\|^2 + \|\mathbf{s}\|^2$ respectively $t = 1$ this implies $\lambda_1(\Lambda) \approx \sqrt{2(n+m)} \cdot \sigma$ respectively $\lambda_1(\Lambda) \approx \sqrt{n+m} \cdot \sigma$. The second minimum $\lambda_2(\Lambda)$ is assumed to correspond to the Gaussian heuristic for the lattice (a more refined argument would consider the Gaussian heuristic of $\Lambda' = \pi_{\mathbf{v}}(\Lambda)$, but these quantity are very close for relevant parameters). Experiments in [11] using LLL and BKZ with small block sizes (5 and 10) were interpreted to matched the 2008 estimate, providing constant values for τ_β for lattices of the form of Eq. (2.2), depending on the chosen algorithm, for a 10 per cent success rate. Overall, τ_β was found to lie between 0.3 and 0.4 when using BKZ.

We note that we may interpret this observation as being consistent with Inequality (2.3).

2.5.4 The 2016 Estimate

The 2008 estimates offer no insight into why the algorithm behaves the way it does but only provide numerically established constants that seem to somewhat vary with the algorithm or the block size. In [19] an alternative estimate was outlined. The estimate predicts that $(\mathbf{e}, \mathbf{s}, t)$ can be found if

$$\sqrt{\beta/d} \cdot \|(\mathbf{e}, \mathbf{s}, t)\| \approx \sqrt{\beta \cdot \sigma^2} < \sqrt{\alpha_\beta}^{2\beta-d-1} \cdot \text{Vol}(\Lambda(\mathbf{B}))^{1/d} , \qquad (2.5)$$

under the geometric series assumption (until a projection of the unusually short vector is found). The right-hand side of the inequality is the expected norm of the Gram–Schmidt vector at index $d-\beta$ (see Definition 2.7). The left-hand side is an estimate for $\|\pi_{d-\beta}((\mathbf{e}, \mathbf{s}, t))\|$. If the inequality holds then $\pi_{d-\beta}((\mathbf{e}, \mathbf{s}, t))$ is a shortest vector in $\mathbf{B}_{[d-\beta:d]}$ and will thus be found by BKZ and inserted at index $d - \beta$. This is visualised in the top part of Figure 2.7. Subsequent calls to an SVP oracle on $\mathbf{B}_{[d-2\beta+1:d-\beta+1]}$ would insert $\pi_{d-2\beta+1}((\mathbf{e}, \mathbf{s}, t))$ at index $d - 2\beta + 1$ etc.

The 2016 estimate was empirically investigated and confirmed in [15]. The authors ran experiments in block sizes up to 78 and observed that a BKZ managed to recover the target vector with good probability as predicted in [19]. An example is given in the bottom part of Figure 2.7. Furthermore, they showed (under the assumption that vectors are randomly distributed in space) that once BKZ has set $\mathbf{b}_i^\star = \pi_{d-\beta}((\mathbf{e}, \mathbf{s}, t))$, calls to LLL are expected to suffice to recover $(\mathbf{e}, \mathbf{s}, t)$ itself.

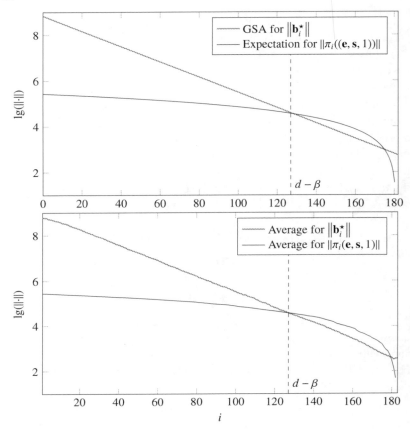

Figure 2.7 (The 2016 estimate.) Expected and observed norms for lattices of dimension $d = 183$ and volume q^{m-n} after BKZ-β reduction for LWE parameters $n = 65, m = 182, q = 521$, standard deviation $\sigma = 8/\sqrt{2\pi}$ and $\beta = 56$ (minimal (β, m) such that Inequality (2.5) holds). Average of Gram–Schmidt lengths is taken over 16 BKZ-β reduced bases of random q-ary lattices, i.e., without an unusually short vector. Reproduced from [15].

Comparing Inequality (2.5) with Inequalities (2.3) and (2.4) we note that it more closely resembles the prediction for Slide reduction rather than for BKZ, despite the rationale and experimental evidence being obtained for BKZ. This suggests that the average behaviour of BKZ and Slide reductions in the approximation factor regime is roughly the same, despite different worst-case bounds being proven. Furthermore, we note that Inequality (2.5) gains an additional factor of $\sqrt{\beta/d}$ compared with Inequality (2.4).

2.5.5 Further Refinements

On the other hand, the authors of [15] also observed that the algorithm behaves somewhat better than predicted. That is, they managed to solve the underlying instances using block sizes somewhat smaller than required to make Inequality (2.5) hold.

This is attributed to a 'double intersection' in [15]. As illustrated in Figure 2.7, the projection of the target vector and the norms of the Gram–Schmidt vectors may intersect twice: once at index $d - \beta$ and once close to index d, say at index $d - o$ for some small o. Applying the same reasoning as above, we expect $\pi_{d-o}((\mathbf{e}, \mathbf{s}, t))$ to be inserted as $\mathbf{b}^\star_{d-o}$. Thus, we expect a subsequent SVP call at index $d - \beta - o$ to recover and insert $\pi_{d-\beta-o}((\mathbf{e}, \mathbf{s}, t))$. Alternatively, an SVP call in dimension $\beta - o$ at index $d - \beta$ could now recover $\pi_{d-\beta}((\mathbf{e}, \mathbf{s}, t))$ since this vector is $\in \Lambda_{[d-\beta:d-o]}$. However, it is noted in [15] that this 'double intersection' phenomenon does not occur for typical cryptographic parameters.

Another source of imprecision when applying Inequality (2.5) is that it assumes the GSA (before an unusually short vector is found), replacing this assumption with a BKZ simulator produces refined estimates.

But there seems to be another subtle phenomenon at play. In [148] it is noted that, for very small block sizes β, the prediction of [15] is, on the contrary, too optimistic. The reason is that, while the projected vector $\pi_{d-\beta}((\mathbf{e}, \mathbf{s}, t))$ may be detected with good probability at position $d - \beta$, we require a bit more luck to lift correctly, i.e., to recover the full vectors $(\mathbf{e}, \mathbf{s}, t)$ from its projection. Instead, a probabilistic model is proposed, to account for both initial detection and lifting, and this prediction seems to fit very well with experiments; see Figure 2.8.

Balancing Costs It should be mentioned that just running BKZ is not the optimal strategy to solve uSVP instances. Indeed, having spent $O(d)$ many SVP-β calls pre-processing the whole basis, this strategy hopes for the last such SVP call to essentially produce the solution. An improved strategy instead balances the cost of the pre-processing step and the final search step. Therefore, it could, for example, be natural to do a last call to SVP-β' for β' slightly larger than β; this has for example been implemented with sieving in [17] to break Darmstadt LWE challenges [230] and was already standard in the enumeration literature [396].

The optimal strategy is, therefore, more difficult to predict, and hardness estimates often rely on scripts that numerically optimise the various parameters of the algorithm based on assumptions such as the relative costs of running SVP in slightly larger or smaller dimension, the number of calls to an

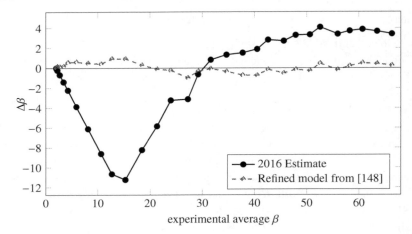

Figure 2.8   The difference $\Delta\beta$ = real − predicted, as a function of the average experimental β. The experiment consists in running a single tour of BKZ-β for $\beta = 2, 3, 4, \ldots$ until the secret short vector is found. This was averaged over 256 many LWE instances per data point, for parameters $q = 3301$, $\sigma = 20$ and $n = m \in \{30, 32, 34, \ldots, 88\}$. Reproduced from [148].

SVP oracle required to achieve a given root-Hermite factor, etc. To avoid this complication, some designers instead opt for accounting only for the cost of a single call to SVP-β when even considering several tours of BKZ-β (a simplification introduced as the 'core SVP hardness' in [19]). In this model, the issue of balancing costs between β' and β does not arise, i.e., $\beta' = \beta$ is optimal and the attack cost is bounded from below by the cost of one call to SVP-β on a BKZ-β reduced basis.

2.6 Behaviour on NTRU Instances

To solve NTRU (Definition 2.2) we may consider the lattice

$$\Lambda_{\mathbf{H}}^q = \left\{ (\mathbf{x}, \mathbf{y}) \in \mathbb{Z}^{2n} \text{ s.t. } \mathbf{H} \cdot \mathbf{x} - \mathbf{y} = \mathbf{0} \bmod q \right\}, \tag{2.6}$$

where $\mathbf{H}$ is the matrix associated with multiplication by h modulo ϕ, i.e., the columns of $\mathbf{H}$ are spanned by the coefficients of $x^i \cdot h \bmod \phi$ for $i = 0, \ldots, n - 1$. The lattice $\Lambda_{\mathbf{H}}^q$ is spanned by

$$\mathbf{B} = \begin{pmatrix} q\mathbf{I} & \mathbf{H} \\ 0 & \mathbf{I} \end{pmatrix}$$

and contains a short vector $(\mathbf{f}, \mathbf{g})$. This can be observed by multiplying the basis by $(\mathbf{f}, *)$ from the right, where $*$ represents the vector performing modular reduction modulo q and where $\mathbf{f}$ respectively $\mathbf{g}$ is the coefficient vector of f respectively g. If $\|(\mathbf{f}, \mathbf{g})\|$ is much smaller than $\mathrm{gh}(\Lambda_{\mathbf{H}}^{q}) \approx \sqrt{n/(\pi e)} \cdot \sqrt{q}$ then this lattice contains an unusually short vector. Indeed, it also contains all vectors corresponding to 'rotations' of (f, g), i.e., $(x^i \cdot f \bmod \phi, x^i \cdot g \bmod \phi)$ for $i = 0, \ldots, n - 1$ and their integral linear combinations. In other words, the NTRU lattice contains a dense sublattice.

2.6.1 NTRU as uSVP

Considering NTRU as the problem of recovering an unusually short vector in the NTRU lattice was already done in the initial NTRU paper [275]. Also, the original NTRU paper [275] discussed an observation from [139] (analysing [274]) that an attacker does not need to recover f, g exactly, but that any sufficiently small multiple of f suffices to break the scheme. For the uSVP case the hardness of the problem was related to $\mathrm{gh}(\Lambda)/\lambda_1(\Lambda)$ where $\lambda_1(\Lambda) = \|(\mathbf{f}, \mathbf{g})\|$. When considering message recovery instead of key recovery, a related quantity is considered. We may a posteriori reinterpret this as framing attacks on NTRU in the framework of the '2008 estimate' (see Section 2.5.3) but replacing $\lambda_2(\Lambda)$ by $\mathrm{gh}(\Lambda)$. This approach became a common way of reasoning about NTRU lattices; see, e.g. [168]. Yet the validity of this approach is doubtful, as in NTRU lattices we have $\lambda_2(\Lambda) = \lambda_1(\Lambda)$ in contrast to the lattices arising for LWE. In this context, we note that the early study of May and Silverman [410] massaged the lattice to decrease the NTRU lattice dimension while also eliminating all but one of the NTRU short vectors.

The 2016 estimate (see Section 2.5.4) sidesteps this discussion on whether $\lambda_2(\Lambda)$ matters, as the heuristic reasoning here does not involve this quantity. This estimate also ended up being used for estimating the hardness of breaking NTRU [633, Section 6.4.2]. More recently the framework proposed in [148] allowed us to revisit the tricks of May and Silverman [410], and it was concluded that this trick was slightly counterproductive. Indeed, the probabilistic model permits to account for the cumulated probabilities of detecting any of those short vectors in the full lattice, and this is slightly easier than finding the (up to signs) unique short vector of the massaged lattice.

Indeed, another line of works showed that the presence of many short vectors can make the problem exponentially easier, at least in some 'overstretched' regimes. These works [14, 123, 217, 324] seem to suggest that simple encryption schemes should not be affected at all, but we will argue that the exact crossover point remains to be determined.

2.6.2 Attacks on Overstretched NTRU

In this last section, we cover an attack that exploits the fact that NTRU lattices hide not one but many unexpectedly short vectors, yielding an unexpectedly dense sublattice. If the right conditions are met then it turns out that this dense sublattice is easier to uncover than the individual vectors spanning it.

This, however, is a fairly a-posteriori view of this discovery. At first, this weakness was associated not primarily with a density property, but more with an algebraic structure property: namely, the presence of subfields in NTRU. The idea of exploiting this structure had been considered as soon as 2002, by Gentry, Szydly, Jonsson, Nguyen and Stern [217, Section 6]; but it was quickly abandoned: yes, NTRU keys can be normed down to a subfield and still yield valid NTRU keys, but this trade-off of dimension versus approximation factor did not seem advantageous for the actual NTRUEncrypt parameters.

When Bai and ourselves explored this idea again [14] (independently, Cheon, Jeong and Lee [123] also explored a closely related idea), the situation was rather different: NTRU was not just a single scheme with a few parameter sets, it was a parameterised assumption with increasing popularity for building homomorphic encryption schemes. In these newly considered regimes the trade-off mentioned above seemed on the contrary quite advantageous. We, therefore, claimed asymptotic improvements over the natural lattice reduction attack, which – depending on the parameters – could decrease the costs of the attacks from exponential to sub-exponential or even polynomial.

This claimed improvement was soon challenged by Kirchner and Fouque [324]. Our mistake was not the complexity of our new algorithm but rather the fact that the complexity of straight-up lattice reduction attacks was much better than expected on such overstretched NTRU instances. They claimed that the old attack should behave as well as the new one, and – with minor performance-enhancing tricks – were able to demonstrate this in practice. In conclusion, the new algorithm we invented was completely useless, and old algorithms performed just as well, if not better, and were more generally applicable. We found solace in the belief that the results of Kirchner and Fouque may not have been discovered without our algebraic detour.

The Subfield Attack

The key idea of this attack is as follows: the relation $h = f/g \mod q$ between the public key h and the private key (f, g) can be normed down to a smaller field; furthermore, if f and g are short enough, their norms in a smaller field will also be somewhat short. Therefore, one may hope to attack the problem in a subfield and lift back the solution. We note that in the case of cyclotomic number fields, there is always at least one non-trivial subfield, namely the maximal totally real subfield $\mathbb{K}^+$, of relative rank $r = [\mathbb{K} : \mathbb{K}^+] = 2$. In the case of power of two cyclotomic number fields ($n = 2^k$), one chooses the subfield to tune r to any power of 2 less than n. On the contrary, this approach is not directly applicable to fields as chosen in [55].

In more detail, let $\mathbb{K}$ be a number field ($\mathbb{K} = \mathbb{Q}(x)/(\phi(x))$, where ϕ comes from Definition 2.2), and for simplicity let us assume that $\mathbb{K}$ is a cyclotomic number field. Let $\mathbb{L}$ be a subfield with relative rank $r = [\mathbb{K} : \mathbb{L}]$, and let N denote the relative norm $N : \mathbb{K} \to \mathbb{L}$, defined by $N(x) = \prod_a a(x)$, where a ranges over all the automorphisms of $\mathbb{K}$ that are identity over $\mathbb{L}$. Defining $f' = N(f)$, $g' = N(g)$ and $h' = N(h)$, we note that $h' = f'/g' \bmod q$ still holds over $\mathbb{L}$. Furthermore, if f, g have lengths roughly $\sqrt{n} \cdot \sigma$, we expect f', g' to have lengths roughly $(\sqrt{n} \cdot \sigma)^r$.

On the other hand, the dimension of the normed-down NTRU lattice is $2n/r$ and its volume is $q^{n/r}$. The original article [14] reasons more formally, using the approximate factor bound of lattice reduction; however, here we will give a simplified and more heuristic exposition. Roughly, using either the 2008 estimate or the 2016 estimate, we expect to solve this instance using a block size β such that

$$(\sqrt{n} \cdot \sigma)^r \cdot \delta_\beta^{2n/r} \leq \sqrt{q}.$$

For $\sigma = \mathrm{poly}(n)$, the subfield attack [14] obtains the asymptotic success condition

$$\frac{\beta}{\lg \beta} = \Theta\left(\frac{n}{r \lg q - r^2 \lg n}\right)$$

assuming $r \lg q - r^2 \lg n > 0$.

Parameterising the attack to not use a subfield ($r = 1$) should therefore require $\beta = \tilde{\Theta}(n/\log q)$, while choosing a relative rank $r = \Theta(\log q/\log n)$ leads to $\beta = \tilde{\Theta}(n/\log^2 q)$. For schemes that use large moduli such as fully homomorphic schemes [94, 399] or candidate cryptographic multi-linear maps [208], this therefore makes a significant difference; both in practice and in theory.

Full Secret Reconstruction It should be noted that finding f', g' does not lead to a full recovery of the original secret. However, we can still reconstruct a small multiple $\alpha(f, g)$ of the original secret key (f, g), by constructing $(f', g' \cdot h/h')$. This is typically enough to break encryption schemes. If we insist on recovering the original key (f, g), this intermediate information is still helpful. For example, repeating the attack with a rerandomised initial basis, we may recover the exact lattice generated by the secret key $(f, g)^T \cdot O_{\mathbb{K}}$. Recovering (f, g) is now much easier; it can be done with an algorithm for the Principal Ideal Problem, and this is classically sub-exponential time [60], and quantumly polynomial time [59].

The Dense Sublattice Attack

We will now explain why the above subfield attack was a detour to the discovery of a much more general result by Kirchner and Fouque [324]. In a sense, LLL and BKZ are rather clever algorithms and what we can try to make more visible to them via algebraic massaging of the lattice at hand was already geometrically obvious to them: there is a particularly dense sublattice to be found inside NTRU instances. This version of the attack is therefore not prevented by choosing a number field as in [55], or even by going for a matrix version of NTRU without any underlying number field.

To prove that LLL can indeed uncover this hidden dense sublattice, let us first go back to the (worst-case) argument to prove that LLL can solve a unique-SVP instance when $\lambda_2(\Lambda)/\lambda_1(\Lambda) > (4/3 + \epsilon)^{d/2}$.

It follows from the inequality $\lambda_1(\Lambda) \geq \min_i \|\mathbf{b}_i^\star\|$, which is obtained by writing a shortest vector $\mathbf{v}$ as $\mathbf{v} = \sum v_i \mathbf{b}_i^\star$ and noting that $\mathbf{v}$ must be longer than $\mathbf{b}_j^\star$ where j is the largest index such that $v_j \neq 0$. From there, we argue that

$$\|\mathbf{b}_1\| \leq (4/3 + \epsilon)^{d/2} \min_i \|\mathbf{b}_i^\star\| \leq (4/3 + \epsilon)^{d/2} \lambda_1(\Lambda) < \lambda_2(\Lambda) \,.$$

Recall that we can make an even simpler case that LLL or BKZ must distinguish this lattice from random without having to go through the full argument. Indeed, let us simply note that, for a random lattice, we expect a particular shape for the basis, say following ZGSA or ZTGSA. But for a large enough β, the prediction for the shape becomes incompatible with the constraint that $\lambda_1(\Lambda) \geq \min_i \|\mathbf{b}_i^\star\|$. In such cases, LLL and BKZ must, therefore, behave differently, and this is easily seen by just looking at the shape: the NTRU lattice has been distinguished from random.

The analysis of Kirchner and Fouque follows essentially from the same kind of argument, generalising the invariant $\lambda_1(\Lambda) \geq \min_i \|\mathbf{b}_i^\star\|$. Here, we can read '$\lambda_1(\Lambda)$' as the determinant of the densest one-dimensional sublattice; a k-dimensional variant of the inequality was given by Pataki and Tural.

Lemma 2.13 ([469, Lemma 1]). *Let Λ be a d-dimensional lattice, and $\mathbf{b}_0, \ldots,$ $\mathbf{b}_{d-1}$ be any basis of Λ, and let $k \leq d$ be a positive integer. Then, for any k-dimensional sublattice $\Lambda' \subset \Lambda$, it holds that*

$$\mathrm{vol}\,(\Lambda') \geq \min_J \prod_{j \in J} \|\mathbf{b}_j^\star\|,$$

where J ranges over all subsets of $\{0, \ldots, d - 1\}$ *of size k.*

We will now apply this to the dense sublattice Λ' generated by the n short vectors out of the $d = 2n$ dimensions of the NTRU lattice. This gives a (log) left-hand side of $\log \text{vol} (\Lambda') \leq n \log R$, where $R = \|(\mathbf{f}, \mathbf{g})\| \approx \sqrt{d}\sigma$ (and in fact we can argue that $\log \text{vol} (\Lambda') \approx n \log R$). For the right-hand side, the minimum is reached by the n last indices $J = \{n, n + 1, \ldots, 2n - 1\}$.

Pictorially, the usual one-dimensional argument forbids the last Gram–Schmidt vector to go above R; if the heuristically predicted shape contradicts this rule, then the shortest vector must have been detected somehow. The multi-dimensional version of Pataki and Tural instead forbids the black-hashed region to have a surface larger than the grey-filled region in Figure 2.9.

We make our prediction under the Z-shape model, denoting $s = \lg \alpha_\beta$ the slope of the middle section, between indices $n - z$ and $n + z$. The inaccuracies of this model discussed in Section 2.4.3 should be asymptotically negligible, as we will be interested in regimes for which $\beta = o(z)$. The picture also makes it easy to compute the right-hand side of the inequality. It is given by the surface of a right-angled triangle of height $h = \frac{1}{2} \lg q$. Its surface is given by $S = \frac{1}{2}hz = \frac{1}{2}h^2/s = (\lg q)^2/(8 \lg \alpha_\beta)$. We therefore predict that the Pataki-Tural inequality would be violated when $nR = S$ that is: $\lg \alpha_\beta = \lg^2 q/(8nR)$. Noting that $\lg \alpha_\beta = \Theta\left(\frac{\lg \beta}{\beta}\right)$, we conclude that the lattice reduction is going to detect the dense sublattice when

$$\frac{\beta}{\lg \beta} = \Theta\left(\frac{nR}{\lg^2 q}\right) .$$

The required block size is therefore $\beta = \tilde{\Theta}(n/\lg^2 q)$ as it was for the subfield attack, however a more careful analysis of the hidden constants [324] reveals that going to the subfield is slightly unfavourable.

Concrete Behaviour
Although we kept the above development asymptotic for simplicity, it is not hard to keep track of the hidden constants – or even to run simulations – and to predict precisely when the Pataki–Tural lemma would be violated. However, even such a methodology would only lead to an upper bound on the cost of this attack and not an estimate. Indeed, this methodology would essentially correspond to the one of Section 2.5.2 for LWE-uSVP; it is based on an im-possibility argument, but it does not explain or predict the phenomenon, unlike the 2016 estimate.

We therefore emphasise this gap as our last and foremost open problem: give a more detailed explanation of how BKZ detects the hidden sublattice,

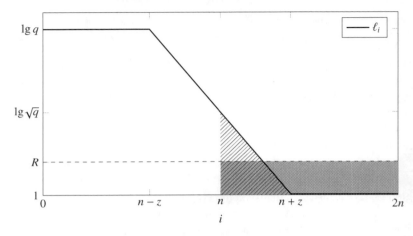

Figure 2.9 The Pataki–Tural constraint on reduced NTRU bases.

leading to a heuristic estimate on when the phenomenon happens, confirmed by extensive experiments. A possible answer may be found by extending the probabilistic analysis of [148], this time accounting for more than the n shortest vectors $(x^i \cdot f, x^i \cdot g)$ for $0 \leq i < n$. Indeed, one could instead consider all the vectors $(p \cdot f, p \cdot g)$ for elements p up to a certain length. These vectors are longer and therefore the probability of finding a given one of them is smaller. Yet, it might be that, in some regimes of parameters, their number outgrows this decrease in probability. When considering multiple vectors from the same dense sublattice the events of finding each of them may not be independent, which might require some care when modelling.

Acknowledgments We are indebted to Damien Stehlé, Rachel Player and Benjamin Wesolowski for their careful review of this chapter. Part of this work was done while the authors were visiting the Simons Institute for the Theory of Computing. The research of Albrecht was supported by EPSRC grants EP/S020330/1 and EP/S02087X/1. The research of Albrecht and Ducas was supported by the European Union Horizon 2020 Research and Innovation Program Grant 780701 (Project PROMETHEUS).

3

History of Integer Factorisation

Samuel S. Wagstaff, Jr

The integer factorisation problem is well known as the mathematical foundation of the public-key cryptosystem RSA. When RSA was introduced 40 years ago by Rivest, Shamir and Adleman, the largest number one could factor was still tiny. Spurred on by the practical deployment of RSA, more and more factoring algorithms were developed and fine-tuned, from the elliptic-curve method to the number field sieve. This chapter treats algorithmic developments in factoring integers and their effect on cryptography, not just on RSA. For a long time, factoring records followed a regular pattern. In this chapter, an overview of factoring progress through the years is given, with a bold prediction of what might come to be.

This chapter covers some of the material of the author's book [611], in abbreviated form. See that work for a fuller treatment, proofs and more examples.

3.1 The Dark Ages: Before RSA

Many years ago, people published tables of primes and of prime factors of integers. The first such table was created by Eratosthenes more than 2500 years ago. Cataldi published a table of factors of numbers up to 800 in 1603. In 1811, Chernac went up to 1 020 000. D. N. Lehmer [355] published the last factor table (to 10 017 000) more than a century ago. He also published a table of all primes up to 10 006 721. No one will publish any more tables of factors or primes up to some limit because one can compute them in seconds using a sieve.

3.1.1 Trial Division

The basic trial-division algorithm is old and slow. For hundreds of years, factorers who used trial division found ways to omit trial divisors that could not possibly divide the candidate number. The algorithm need not try composite numbers as divisors. A simple way to skip some composites is to alternate adding 2 and 4 to a trial divisor to form the next one. This trick skips multiples of 2 and 3. The method, called a wheel, can be extended to skip multiples of 2, 3 and 5 by adding the differences between consecutive residue classes relatively prime to 30.

Another approach is to compute a table of primes first and divide the candidate number only by these numbers. A sieve can be used to build a table of primes between two limits quickly. See Section 3.4.1 for sieves.

Quadratic residues can be used to speed trial division by skipping some primes that cannot be divisors. Euler, Gauss and others used this trick hundreds of years ago. Let us try to factor n and assume we know a non-square quadratic residue r modulo n. Then r must also be quadratic residue modulo any prime factor p of n. If r is not a square, the law of quadratic reciprocity restricts p to only one-half of the possible residue classes modulo $4|r|$. Where does one find small non-square quadratic residues for use in the technique described above? One way to construct them is to compute $x^2 \bmod n$, where x is slightly larger than an integer multiple of $\sqrt{n}$. This idea led to the quadratic sieve algorithm. Another source of small quadratic residues is the continued fraction expansion of $\sqrt{n}$, as we shall see in Section 3.3.

3.1.2 Fermat's Difference of Squares Method

Fermat's difference of squares method (see Algorithm 3.1) may be used to factor an odd number n by expressing n as a difference of two squares, $x^2 - y^2$, with the pair x, y different from $(n + 1)/2$, $(n - 1)/2$ (which gives $x + y = n$ and $x - y = 1$). Any other representation of n as $x^2 - y^2$ produces a non-trivial factorisation $n = (x - y)(x + y)$. Here we have $x \geq \sqrt{n}$.

When the **while** loop terminates we have $r = x^2 - n$, where $x = (t - 1)/2$, and also r is a square: $r = y^2$. Then $n = x^2 - y^2$ and n is factored.

In most cases this algorithm will be much slower than trial division. However, the algorithm works quite well when n has a divisor within $O\left(\sqrt[4]{n}\right)$ of $\sqrt{n}$. It is the reason why one must not choose the two primes for RSA too close together.

Algorithm 3.1 Fermat's difference of squares factoring algorithm.

Input: An odd composite positive integer n.

Output: p and q which are the factors of n.

1: $x \leftarrow \lfloor \sqrt{n} \rfloor$

2: $t \leftarrow 2x + 1$

3: $r \leftarrow x^2 - n$

4: **while** r is not a square **do**

5:      $r \leftarrow r + t$

6:      $t \leftarrow t + 2$

7: $x \leftarrow (t - 1)/2$

8: $y \leftarrow \sqrt{r}$

9: **return** $p = x - y$ and $q = x + y$ are the factors of n.

3.1.3 Pollard's Methods

In the 1970s, Pollard invented two factoring algorithms: the rho method and the $p - 1$ method. Both methods are better than trial division at finding small factors of a large number.

Let n be composite and let p be an unknown prime factor of n, which we hope to find. Pollard [473] suggested choosing a random function f from the set $\{0, 1, \ldots, n - 1\}$ into itself, picking a random starting number s in the set and then iterating f:

$$s, f(s), f(f(s)), f(f(f(s))), \ldots.$$

If we reduce these numbers modulo the unknown prime p, we get a sequence of integers in the smaller set $\{0, 1, \ldots, p - 1\}$. Because of the birthday paradox in probability, a number in the smaller set will be repeated after about $\sqrt{p}$ iterations of f. If u, v are iterates of f with $u \equiv v \pmod{p}$, then it is likely that the greatest common divisor $\gcd(u - v, n) = p$ because p divides $u - v$ and n, and probably no other prime factor of n divides $u - v$. Here is a simple version of the algorithm. The main loop iterates the function in two ways, with one iteration per step for A and two iterations per step for B. Then it computes $\gcd(A - B, n)$ and stops when this gcd first exceeds 1. The Pollard rho method is a Monte Carlo probabilistic algorithm.

The Pollard rho method (see Algorithm 3.2) takes about $O\left(\sqrt{p}\right)$ steps to discover the prime factor p of n. It is the reason why the primes for an RSA public modulus must have at least 75 decimal digits (for 128-bit security).

In 1974, Pollard [475] proposed the $p-1$ method. Fermat's theorem says that $a^{p-1} \equiv 1 \pmod{p}$ when p is a prime not dividing a. Pollard's $p - 1$ method (see

---

Algorithm 3.2 Pollard rho factorisation method.

Input: A composite number n to factor.

Output: A proper factor g of n, or else 'give up'.

1: Choose a random b in $1 \le b \le n - 3$
2: Choose a random s in $0 \le s \le n - 1$
3: $A \leftarrow s$; $B \leftarrow s$
4: Define a function $f(x) = (x^2 + b) \bmod n$
5: $g \leftarrow 1$
6: **while** $g = 1$ **do**
7:     $A \leftarrow f(A)$
8:     $B \leftarrow f(f(B))$
9:     $g \leftarrow \gcd(A - B, n)$
10: **if** $g < n$ **then**
11:     **write** g 'is a proper factor of' n
12: **else**
13:     either give up or try again with new s and/or b

Algorithm 3.3 Simple Pollard $p - 1$ factorisation method.

Input: A composite positive integer n to factor and a bound B.

Output: A proper factor g of n, or else give up.

1: $a \leftarrow 2$
2: **for** $i = 1, \ldots, B$ **do**
3:     $a \leftarrow a^i \bmod n$
4:     $g \leftarrow \gcd(a - 1, n)$
5: **if** $1 < g < n$ **then**
6:     **write** 'g divides n'
7: **else**
8:     give up

Algorithm 3.3) chooses a large integer L with many divisors of the form $p - 1$ to try many potential prime factors p of n at once. Since we cannot compute $a^L \bmod p$ because p is an unknown factor of n, we compute $a^L \bmod n$.

If the largest prime factor of $p - 1$ is $\le B$, then $p - 1$ will divide L when L is the product of all primes $\le B$, each repeated an appropriate number of times. One simple choice for L is $B!$. Usually, B is large and L is enormous. We need not compute L. At step i, we compute $\leftarrow a^i \bmod n$. Here is a basic form of the algorithm.

It is best to compute the gcd operation once every few thousand iterations of the **for** loop rather than just once at the end. When this is done, the **for** loop continues in case $g = 1$. If $g = 1$ at the end, one can either give up or try a second stage.

Baillie found the 16-digit prime divisor $p = 1\,256\,132\,134\,125\,569$ of the Fermat number $F_{12} = 2^{4096} + 1$ using Pollard $p - 1$ with $B = 30\,000\,000$. He succeeded since the largest prime factor of $p - 1$ is less than B:

$$p - 1 = 2^{14} \cdot 7^2 \cdot 53 \cdot 29\,521\,841.$$

The algorithm has a second stage, which chooses a second bound $B_2 > B$ and looks for a factor p of n for which the largest prime factor of $p - 1$ is $\leq B_2$ and the second largest prime factor of $p - 1$ is $\leq B$.

The Pollard $p - 1$ algorithm shows that if p and q are the factors of an RSA public-key, then each of $p - 1$ and $q - 1$ must have a prime factor of at least 35 decimal digits (see also Section 6.2.3).

3.1.4 Primality Testing

Suppose we desire the prime factorisation of a positive integer n. The first step is to learn whether n is prime. If it is prime, no factoring is needed. If n is composite and factorisation finds that $n = ab$ with $1 < a \leq b < n$, then we must determine whether a and b are prime to decide whether we are done. Some methods of proving that an integer is prime, such as Theorem 3.1, require factoring a different integer.

Primality testing was as hard as factoring 100 years ago. Mathematicians would try for a while to factor a large number; if they could not factor it, they might conjecture it is prime. Later, some other mathematician might factor it or prove it is prime.

About a century ago, mathematicians discovered fast tests that reported whether an integer is definitely composite or 'probably prime'. At about the same time, theorems were developed that gave a rigorous proof that a proba-ble prime number is prime, but some factoring might be needed to complete the proof. These tests were refined until 30 years ago, people found a very fast test for primeness that has never been proved correct, but that has never failed either. The theorems also were improved until, 20 years ago, one could prove with modest effort that virtually every 1000-digit probable prime really is prime. Some of these primality tests are probabilistic, meaning that: the al-gorithm chooses random numbers as it runs, most choices for these numbers will lead to a short running time, some random choices could make the pro-gram run for a long time, and the program always gives a correct answer when it finishes. In 2002, a deterministic general polynomial-time algorithm for primality testing was found. Although it runs in deterministic polynomial time and always gives the correct answer, this test is still slower for large numbers than some probabilistic primality tests. The remainder of this section gives highlights of the development of primality testing during the past century. See Pomerance [480] for a common theme of some of the prime proving methods below.

Theorem 3.1 (Lehmer). *Let m be an odd positive integer. Then m is prime if and only if there exists an integer a such that $a^{m-1} \equiv 1$ (mod m), but for every prime q that divides $m - 1$, $a^{(m-1)/q} \not\equiv 1$ (mod m).*

In order to use Theorem 3.1 to prove that m is prime, we must factor $m - 1$ completely. If we cannot factor $m - 1$ but are willing to allow a tiny chance of error, there are fast algorithms to decide whether m is prime or composite.

Call an integer $m > 1$ a probable prime to base a if $a^{m-1} \equiv 1$ (mod m). By Fermat's theorem, every odd prime m is a probable prime. A composite probable prime to base a is called a pseudoprime to base a. For integers $a > 1$, most probable primes to base a are prime and very few of them are pseudoprimes.

A better test on large numbers is to use a strong probable prime test. A strong probable prime to base a is an odd positive integer m with this property: if we write $m-1 = 2^e f$ with f odd, then either $a^f \equiv 1$ (mod m) or $a^{f \cdot 2^c} \equiv -1$ (mod m) for some c in $0 \leq c < e$. A strong pseudoprime is a composite strong probable prime. Every prime is a strong probable prime to every base it does not divide. Also, every strong probable prime m is a probable prime to the same base. There are infinitely many strong pseudoprimes to every base.

It is known that if m is prime and $\equiv 3$ or $7 \bmod 10$, then m divides the Fibonacci number u_{m+1}. Baillie, Pomerance, Selfridge and Wagstaff [28, 481] conjectured that no composite integer m, whose last decimal digit is 3 or 7, is a strong probable prime to base 2 and also divides the Fibonacci number u_{m+1}. (They made a similar, but slightly more complicated, conjecture for numbers with last digit 1 or 9. See [28] and [481] for details.) No one has ever proved or disproved this conjecture. It has been used millions of times to construct large primes with not a single failure. The authors of [481] offer US\$620 to the first person who either proves that this test is always correct or exhibits a composite number that the test says is probably prime. Heuristic arguments suggest that there are composite numbers that the test says are probably prime but that the least such examples have hundreds or even thousands of decimal digits. This probable prime test takes $O\big((\log m)^3\big)$ steps to test m, that is, polynomial time. This primality test has become the (ANSI) standard method of selecting primes for an RSA public key. It is used in the Secure Sockets Layer in computer networks.

In 2002, Agrawal, Kayal and Saxena [9] found the first general deterministic polynomial-time primality algorithm. Their original version had running time $O\left((\log m)^{12}\right)$, but others [248, 442] soon reduced this to about $O\left((\log m)^6\right)$, better, but still much slower than the probabilistic test described above. See Agrawal, Kayal and Saxena [9] and articles that refer to it for details.

3.2 The Enlightenment: RSA

People have concealed messages through cryptography for more than 2000 years. Until recently, the sender and receiver had to meet before the secret communication and decide how to hide their future secret messages. They would choose an algorithm, fixed for a long time, and a secret key, changed often.

Some businesses need secret codes to communicate securely with their offices. Some people want cryptography for their personal secrets. In 1975, the National Bureau of Standards announced a cipher, the Digital Encryption Standard, DES, which they approved for use by individuals and businesses.

As computer networks were being built in the early 1970s, people began to think about electronic communication like email. Computer scientists began to ponder how one could 'sign' a digital document so that the recipient could be certain who wrote it.

Others thought about how to create a system in which people who had never met could communicate securely. Symmetric-key ciphers like DES require users to exchange keys securely by meeting before their secret communication happens. All ciphers known then were of this type.

Whit Diffie and Marty Hellman pondered these matters and found some brilliant solutions in a 1976 paper [163].

First, they defined one-way functions, easy to compute forwards, but nearly impossible to invert. They invented a method of choosing a secret key based on the one-way function $f(x) = b^x \bmod p$. Given large numbers b, p and y, with prime p, it is almost impossible to find an x with $f(x) = y$. Their method allows two users who have never met to choose a common secret key, for DES, say, while eavesdroppers listen to their communication.

This key-exchange method provides secure communication between two users connected on the Internet. However, it is subject to the man-in-the-middle attack in which a wiretapper hijacks a computer between the two parties trying to communicate and executes the protocol with each of them separately. After that action, the hijacker can read or change encrypted messages between the two parties as she decrypts a message from one and re-enciphers it to pass on to the other. Diffie and Hellman [163] suggested a second innovation to solve this problem. They split the key into two parts, one public and one private. Bob would reveal his public enciphering key and the enciphering algorithm. But he would tell no one his private deciphering key. It would be impossible to find the deciphering key from the enciphering key. This cipher is called a public-key cipher. To send Bob a secret message, Alice would fetch his public key and use it to encipher her message. Once it was enciphered, only Bob could decipher it because only Bob knows his private key. If Alice wanted to communicate with Bob by DES, say, she could choose a random DES key, encipher the key with Bob's public key and send it to Bob. When Bob got this message, he would decipher it and then use the DES key to communicate with Alice. No eavesdropper could learn the DES key.

However, a public-key cipher lacks authenticity. Anyone could write a message to Bob, sign it with Alice's name, encipher it using Bob's public key, and send it to Bob. Bob would not know whether it actually came from Alice. This way Eve could send a random DES key to Bob, saying that it came from Alice. Then she could communicate with Bob privately, pretending to be Alice. The third innovation of Diffie and Hellman in [163] was the notion of a digital signature. Alice could create her own public and private keys, just as Bob did above. She would 'sign' a message by applying the deciphering algorithm with her secret key to the plaintext message. Then she could encipher this signed message with Bob's public key and send it to him. Bob would be certain that the message came from Alice when he deciphered it using his private key, enciphered the result with Alice's public key and obtained a meaningful text. This signature verification shows that the message came from Alice because only someone who knows her private key could have constructed a message with these properties.

Diffie and Hellman did not suggest any enciphering and deciphering algorithms for public-key cryptography in their article. Later, Rivest, Shamir and Adleman (RSA) read their paper and found an algorithm [501] for doing this based on the presumed difficulty of factoring large integers. The RSA cipher gives a simple formula for enciphering a message. The public key is the product of two primes large enough so that their product could not be factored. Their formula for deciphering the ciphertext uses the two large primes. They created a trapdoor one-way function, that is, a function f that cannot be inverted unless one knows a secret, in which case it is easy to find x given $f(x)$. The secret that unlocks the RSA cipher is the prime factors of the public key. Since an obvious attack on this cipher is to factor the public key, the publication of their paper created tremendous interest in the problem of factoring large integers.

Here is the simplest form of the RSA cipher. In order to use RSA to receive secret messages, Alice chooses two large secret primes p and q and computes $n = pq$. Alice also chooses an exponent e relatively prime to $\varphi(n) = (p-1)(q-1)$ and computes a number d so that $ed \equiv 1 \pmod{\varphi(n)}$. Alice publishes n and e as her public key and keeps p, q and d secret. Plaintext and ciphertext are positive integers less than n. When Bob wishes to send a secret message M to Alice he finds her public key. He enciphers M as $C = M^e \bmod n$ and sends C to Alice. Alice deciphers the message by computing $M = C^d \bmod n$. The original message M is recovered because of Euler's theorem, which says that $M^{\varphi(n)+1} \equiv M \pmod{n}$ when $\gcd(M, n) = 1$. (For the case $\gcd(M, n) \neq 1$ Fermat's theorem and the Chinese remainder theorem combine to prove correctness.) Only Alice can decipher C because only she knows d. Someone who could factor n to discover p and q could compute d the same way Alice computed it.

Martin Gardner [207] wrote an article describing the papers of Diffie, Hellman, Rivest, Shamir and Adleman. In it, RSA offered a challenge message encoded with a 129-digit product of two secret primes. This article was shocking because it revealed to the general public a powerful cipher that even the government couldn't break. The challenge message in this article was deciphered in 1993–1994 by Derek Atkins, Michael Graff, Arjen Lenstra and Paul Leyland [24] who factored the 129-digit number.

It is clear that if you can factor the RSA modulus n, then you can decipher any ciphertext. But it is less clear, and no one has ever proved, that if you can decipher arbitrary ciphertext in RSA, then you can factor n. Two ciphers were invented that do enjoy this equivalence. Rabin and Williams each devised public-key ciphers with the property that one can prove that breaking the cipher is equivalent to factoring a large integer n. In Rabin's [489] system, Alice chooses two large primes p and q with $p \equiv q \equiv 3 \pmod 4$. She publishes the product $n = pq$ and keeps p and q secret. When Bob wants to send a message M in $0 < M < n$ to Alice, he enciphers it as $C = M^2 \bmod n$. When Alice receives C, which is a quadratic residue modulo n, she uses her knowledge of p and q to find all four square roots of C modulo n. If M is ordinary plaintext, then only one square root will make sense and M is that one. If M is a binary string or otherwise indistinguishable from the other three square roots of $M^2 \bmod n$, then Bob must indicate which square root is M.

Rabin [489] proved that if one has an algorithm to decipher any message M enciphered with his cipher in a reasonable time, then there is an algorithm to factor the modulus n in a reasonable time.

Williams [623] constructed a similar cipher using primes $p \equiv 3 \pmod 8$ and

$q \equiv 7 \pmod 8$, and $n = pq$. His cipher eliminates the ambiguity in deciphering and also has the property that breaking the cipher is equivalent to factoring n.

3.3 The Renaissance: Continued Fractions

Although the factoring method in this section is slow compared with the fastest-known ones, its ideas led to them. The Continued Fraction Algorithm, CFRAC, was the first factoring algorithm with subexponential time complexity.

3.3.1 Basic Facts about Continued Fractions

A simple continued fraction is an expression of the form

$$x = q_0 + \cfrac{1}{q_1 + \cfrac{1}{q_2 + \cfrac{1}{q_3 + \cdots}}}. \tag{3.1}$$

The q_i are required to be integers for all i, and also positive when $i > 0$. A simple continued fraction may be finite,

$$x_k = q_0 + \cfrac{1}{q_1 + \cfrac{1}{q_2 + \cfrac{1}{q_3 + \cdots + \cfrac{1}{q_k}}}}, \tag{3.2}$$

which we write as $[q_0; q_1, q_2, q_3, \ldots, q_k]$.

Consider the infinite continued fraction from Eq. (3.1). If it is truncated at q_k, then it has the value x_k from Eq. (3.2), clearly a rational number.

Given a finite continued fraction as given in Eq. (3.2) we can find its value $x_k = A_k/B_k$ as a rational number in lowest terms working backwards clearing the denominators starting from $q_{k-1} + 1/q_k$. Here is a way of finding this rational number working forwards.

Theorem 3.2. *The rational number* $A_k/B_k = [q_0; q_1, q_2, \ldots, q_k]$ *is determined from* $q_0, \ldots, q_k$ *by* $A_{-1} = 1$, $B_{-1} = 0$, $A_0 = q_0$, $B_0 = 1$ *and*

$$A_i = q_i A_{i-1} + A_{i-2} \tag{3.3}$$
$$B_i = q_i B_{i-1} + B_{i-2} \quad \text{for } i = 1, 2, \ldots, k.$$

A continued fraction is finite if and only if it represents a rational number. If n is a positive integer and not a square, then $x = \sqrt{n}$ is irrational and its continued fraction is infinite but periodic.

Theorem 3.3 (Galois). *Let n be a positive integer and not a square. Then the continued fraction for $\sqrt{n}$ has the form*

$$\sqrt{n} = [q_0; \overline{q_1, q_2, \ldots, q_{p-1}, 2q_0}],$$

where the overbar marks the period.

Example 3.4.

$$\sqrt{44} = [6; \overline{1, 1, 1, 2, 1, 1, 1, 12}]$$
$$\sqrt{77} = [8; \overline{1, 3, 2, 3, 1, 16}]$$
$$\sqrt{85} = [9; \overline{4, 1, 1, 4, 18}].$$

When factoring a large integer n by CFRAC the algorithm will compute only a small portion of the continued fraction for $\sqrt{n}$. There is a simple iteration that computes the q_i in the continued fraction for $\sqrt{n}$ using only integer arithmetic. Several other integers are computed during the iteration. One of them is an integer Q_i, which satisfies $A_{i-1}^2 - nB_{i-1}^2 = (-1)^i Q_i$ and $0 < Q_i < 2\sqrt{n}$. If we consider the equation as a congruence modulo n, we have $A_{i-1}^2 \equiv (-1)^i Q_i \pmod{n}$. In other words, the continued fraction iteration produces a sequence $\{(-1)^i Q_i\}$ of quadratic residues modulo n whose absolute values are $< 2\sqrt{n}$, very small indeed. We want small quadratic residues because they are easier to compute, easier to factor, and more likely to be smooth. (An integer is B-smooth if all of its prime factors are smaller than the real number B.)

One can prove that $x_i = (P_i + \sqrt{n})/Q_i$ for $i \geq 0$, where

$$P_i = \begin{cases} 0 & \text{if } i = 0, \\ q_0 & \text{if } i = 1, \\ q_{i-1}Q_{i-1} - P_{i-1} & \text{if } i \geq 2, \end{cases} \tag{3.4}$$

and

$$Q_i = \begin{cases} 1 & \text{if } i = 0, \\ n - q_0^2 & \text{if } i = 1, \\ Q_{i-2} + (P_{i-1} - P_i)q_{i-1} & \text{if } i \geq 2. \end{cases} \tag{3.5}$$

i	q_i	P_i	Q_i	A_i	B_i
−1	−	−	−	1	0
0	8	0	1	8	1
1	1	8	13	9	1
2	3	5	4	35	4
3	2	7	7	79	9
4	3	7	4	272	31
5	1	5	13	351	40
6	16	8	1	5 888	671
7	1	8	13	6 239	711
8	3	5	4	24 605	2 804
9	2	7	7	55 449	6 319
10	3	7	4	190 952	21 761
11	1	5	13	246 401	28 080
12	16	8	1	4 133 368	471 041

Table 3.1 *Continued fraction expansion for* $\sqrt{77}$.

The q_i can be computed using

$$q_i = \lfloor x_i \rfloor = \begin{cases} \left\lfloor \sqrt{n} \right\rfloor & \text{if } i = 0, \\[2ex] \left\lfloor \dfrac{q_0 + P_i}{Q_i} \right\rfloor = \left\lfloor \dfrac{\sqrt{n} + P_i}{Q_i} \right\rfloor & \text{if } i > 0. \end{cases} \tag{3.6}$$

Define A_i and B_i as in Theorem 3.2 so that $[q_0; q_1, q_2, \ldots, q_i] = A_i/B_i$ for $i \geq 0$.

Example 3.5. Table 3.1 gives the sequences for the continued fraction of $\sqrt{77} = 8.774\,964\,387\,392\,123$.

3.3.2 A General Plan for Factoring

The following theorem is ancient.

Theorem 3.6. *If n is a composite positive integer, x and y are integers, and $x^2 \equiv y^2 \pmod{n}$, but $x \not\equiv \pm y \pmod{n}$, then $\gcd(x - y, n)$ and $\gcd(x + y, n)$ are proper factors of n.*

Later we will tell how to find x and y with $x^2 \equiv y^2 \pmod{n}$. However, it is difficult to ensure that $x \not\equiv \pm y \pmod{n}$, so we ignore this condition. The next theorem tells how several modern factoring algorithms finish.

Theorem 3.7. *If n is an odd positive integer having at least two different prime factors, and if integers x and y are chosen randomly subject to $x^2 \equiv y^2 \pmod{n}$, then, with probability ≥ 0.5, $\gcd(x - y, n)$ is a proper factor of n.*

One can compute in probabilistic polynomial time a square root of any quadratic residue r modulo n, provided the factors of n are known. In fact, computing square roots modulo n is polynomial-time equivalent to factoring n.

Corollary. *Let n have at least two different odd prime factors. If there is a (probabilistic) polynomial time algorithm $\mathcal{A}$ to find a solution x to $x^2 \equiv r \pmod{n}$ for any quadratic residue r modulo n, then there is a probabilistic polynomial time algorithm $\mathcal{B}$ to find a factor of n.*

The general plan of several factoring algorithms is to generate (some) pairs of integers x, y with $x^2 \equiv y^2 \pmod{n}$, and hope that $\gcd(x - y, n)$ is a proper factor of n. Theorem 3.7 says that we will not be disappointed often. It says that each such pair gives at least a 50 per cent chance to factor n. If n has more than two (different) prime factors, then at least one of the greatest common divisor and its co-factor will be composite and we will have more factoring to do. In the fastest modern factoring algorithms it may take a long time to produce the first pair x, y, but after it is found many more random pairs are produced quickly, and these will likely yield all prime factors of n.

3.3.3 The Continued Fraction Factoring Algorithm

The Continued Fraction Factoring Algorithm, CFRAC, of Morrison and Brillhart [444], uses the fact that the Q_i are more likely to be smooth than numbers near $n/2$ because they are small. The algorithm uses the continued fraction expansion for $\sqrt{n}$ to generate the sequences $\{P_i\}$, $\{Q_i\}$, $\{q_i\}$ and $\{A_i \bmod n\}$ via Eqs. (3.4), (3.5), (3.6) and (3.3), and tries to factor each Q_i by trial division. Morrison and Brillhart restricted the primes in the trial division to those below some fixed bound B, called the factor base. CFRAC saves the B-smooth Q_i, together with the corresponding A_{i-1}, representing the relation $A_{i-1}^2 \equiv (-1)^i Q_i \pmod{n}$. When enough relations have been collected, Gaussian elimination is used to find linear dependencies (modulo 2) among the exponent vectors of the relations. We have enough relations when there are more of them than primes in the factor base. Each linear dependency produces a congruence $x^2 \equiv y^2 \pmod{n}$ and a chance to factor n by Theorem 3.7.

Assuming two plausible hypotheses, Pomerance [478] proved that the time complexity of CFRAC is $L(n)^{\sqrt{2}}$, where $L(x) = \exp\left(\sqrt{(\ln x)\ln\ln x}\right)$.

Let me say more about the linear algebra step. Suppose there are K primes in the factor base. Call them $p_1, p_2, \ldots, p_K$. (These are the primes $p \leq B$ for which n is a quadratic residue modulo p. They comprise about half of the primes $\leq B$.) The goal is to find a set S of i for which the product

$\prod_{i \in S} (-1)^i Q_i$ is the square of an integer. Since a square must be positive, the 'prime' $p_0 = -1$ is added to the factor base. For each i for which $(-1)^i Q_i$ is B-smooth, write $(-1)^i Q_i = \prod_{j=0}^{K} p_j^{e_{ij}}$. When $(-1)^i Q_i$ is B-smooth, define the vector $\mathbf{v}_i = (e_{i0}, e_{i1}, \ldots, e_{iK})$. Note that when $(-1)^i Q_i$ and $(-1)^k Q_k$ are multiplied, the corresponding vectors $\mathbf{v}_i$, $\mathbf{v}_j$ are added. A product such as $\prod_{i \in S} (-1)^i Q_i$ is a square if and only if all entries in the vector sum $\sum_{i \in S} \mathbf{v}_i$ are even numbers. Form a matrix with $K + 1$ columns whose rows are the vectors $\mathbf{v}_i$ (reduced modulo 2) for which $(-1)^i Q_i$ is B-smooth. If there are more rows than columns in this matrix, Gaussian elimination will find non-trivial dependencies among the rows modulo 2. Let S be the set of i for which the row $\mathbf{v}_i$ is in the dependency. Each non-trivial dependency, say, $\sum_{i \in S} \mathbf{v}_i = \mathbf{0}$, gives a product $\prod_{i \in S} (-1)^i Q_i$ which is a square, say, y^2. Let $x = \prod_{i \in S} A_{i-1} \bmod n$. Then $x^2 \equiv y^2 \pmod{n}$, an instance of Theorem 3.7.

The idea of combining several relations $x_i^2 \equiv r_i \bmod n$ to form a congruence $x^2 \equiv y^2 \pmod{n}$ and get a chance to factor n goes back at least to Kraitchik [343] and Lehmer and Powers [354] around 1930.

Here are the most important new ideas in [444]:

- the fixed factor base,
- using linear algebra modulo 2 to combine relations to form a square, and
- using large primes to augment the factor base.

Here is how the third idea works. When you try to factor Q_i using the primes in the factor base, you often fail because a co-factor > 1 remains. It is easy to test the co-factor for primality. If it is prime, save the relation. If another relation with the same large prime is found, then one can create a useful relation by multiplying them. Suppose $A_{i-1}^2 \equiv (-1)^i Q_i \pmod{n}$ and $A_{j-1}^2 \equiv (-1)^j Q_j \pmod{n}$. Their product, $(A_{i-1} A_{j-1})^2 \equiv (-1)^{i+j} Q_i Q_j \pmod{n}$, will have one prime not in the factor base appearing on the right side, but this prime will be squared, so it will not appear in the matrix, and the product relation can be used to find a congruence $x^2 \equiv y^2 \pmod{n}$. If the number of possible large primes is t, then repeated large primes will begin to appear roughly when the number of relations with large primes reaches $1.18 \sqrt{t}$ by the birthday paradox. Using large primes this way does not change the theoretical time complexity of CFRAC, but it does have practical value. If B and the upper limit on the size of large primes are chosen well, then most of the relations will come from pairing repeated large primes.

In the 1980s, Smith and Wagstaff [482, 552, 612] fabricated an Extended-Precision Operand Computer for factoring large integers by the CFRAC. It had a 128-bit wide main processor to generate the A_i and Q_i, and 16 remaindering units to divide a Q_i by 16 different primes p_j in parallel, finding only the remainders. As each Q_i was generated, it was loaded into a wide shift register. Sixteen trial divisors p_j were loaded into registers of simple arithmetic-logic units. While the main processor generated the next Q_i and A_i, the current Q_i was shifted out of its register, one bit at a time, and broadcast to the remaindering units, which reported when a remainder was 0. The main processor used the reports to determine whether Q_i was smooth enough for the relation to be saved. A personal computer connected to the main processor stored the smooth relations. The linear algebra was done on a larger computer.

3.4 The Reformation: A Quadratic Sieve

This section adds a new idea to CFRAC: factor the right-hand sides of many relations together with an efficient algorithm called a sieve rather than separately with slow trial division.

3.4.1 The Sieve of Eratosthenes

The sieve of Eratosthenes finds the primes below some limit J. It writes the numbers $1, 2, \ldots, J$. Then it crosses out the number 1, which is not prime. After that, let p be the first number not crossed out. Cross out all multiples of p greater than p. Repeat these steps, replacing p by the next number not yet crossed out, as long as $p \leq \sqrt{J}$. When p reaches $\sqrt{J}$ all numbers crossed out are composite (or 1) and all numbers not crossed out are prime.

A computer program would use an array $P[]$ to represent the numbers 1 to J. Let '1' mean that the number is not crossed out and '0' mean that the number is crossed out. The algorithm starts by marking 1 as 'crossed out' and the other numbers as 'not crossed out'. For each prime p, cross out all multiples of the prime p. Then the first number not yet crossed out is the next prime p. At the end, the value of $P[i]$ is 1 if i is prime and 0 if i is 1 or composite. The sieve of Eratosthenes computes all primes $\leq J$ in $O(J \log \log J)$ steps.

A variation of this sieve factors the numbers in the range of a polynomial $f(x)$ with integer coefficients, but it only finds the prime factors of each $f(x)$ that lie in a finite set $\mathcal{P}$ of primes. The polynomial $f(x)$ is fixed. This sieve algorithm (see Algorithm 3.4) is the heart of the quadratic and number field sieve factoring algorithms. For each i, a linked list $L[i]$ holds the distinct prime factors of $f(i)$.

The output $L[i]$ lists only the distinct prime factors of $f(i)$ in $\mathcal{P}$. A modification of this algorithm also finds the number of repetitions of each

---

Algorithm 3.4 Sieve to factor the range of a polynomial.

Input: Integers $J > I > 1$ and a finite set $\mathcal{P}$ of primes.
Output: For $I \leq i \leq J$, $L[i]$ lists the factors in $\mathcal{P}$ of $f(i)$.

1: **for** $i = I, \ldots, J$ **do**
2:     $L[i] \leftarrow$ **empty**
3: **for** each $p \in \mathcal{P}$ **do**
4:     Find the roots $r_1, \ldots, r_d$ of $f(x) \equiv 0 \pmod{p}$
5:     **for** $j = 1, \ldots, d$ **do**
6:         $i \leftarrow$ the least integer $\geq I$ and $\equiv r_j \pmod{p}$
7:         **while** $i \leq J$ **do**
8:             append p to $L[i]$
9:             $i \leftarrow i + p$

---

prime factor of $f(i)$ without ever forming that number, which may be huge.

3.4.2 The Quadratic Sieve Factoring Algorithm

The quadratic sieve factoring algorithm, QS, and the CFRAC differ only in the method of producing relations $x^2 \equiv q \pmod{n}$ with q factored completely. The CFRAC forms x and q from the continued fraction expansion of $\sqrt{n}$ and factors q by slow trial division. The QS produces x and q using a quadratic polynomial $q = f(x)$ and factors the q with a sieve, much faster than trial division. The quadratic polynomial $f(x)$ is chosen so that the q will be as small as possible. Most of these q will exceed $2\sqrt{n}$, but not by a large factor, so that they are almost as likely to be smooth as the q in CFRAC.

Let $f(x) = x^2 - n$ and $s = \lceil \sqrt{n} \rceil$. The QS factors some of the numbers

$$f(s), f(s+1), f(s+2), \ldots$$

by Algorithm 3.4. If there are K primes in the factor base and we find $R > K$ B-smooth numbers $f(x)$, then there will be R relations involving K primes and linear algebra will produce at least $R - K$ congruences $x^2 \equiv y^2 \pmod{n}$, each of which has probability at least $1/2$ of factoring n, by Theorem 3.7.

Sieve by using the fast Algorithm 3.4 to find the B-smooth numbers among $f(s), f(s+1), f(s+2), \ldots$. The factor base $\mathcal{P}$ consists of the primes $p < B$ for which the Legendre symbol $(n/p) \neq -1$. Write the numbers $f(s+i)$ for i in some interval $a \leq i < b$ of convenient length. The first interval will have $a = s$. Subsequent intervals will begin with a equal to the endpoint b of the previous interval. For each prime $p < B$, remove all factors of p from those $f(s + i)$ that p divides. Because $f(x) = x^2 - n$, p divides $f(x)$ precisely when $x^2 \equiv n$ (mod p). The solutions x to this congruence lie in the union of two arithmetic progressions with common difference p. If the roots of $x^2 \equiv n$ (mod p) are x_1 and x_2, then the arithmetic progressions begin with the first numbers $\equiv x_1$ and x_2 (mod p) which are $\geq a$. The sieve is much faster than trial division. Pomerance [478] proved that the time complexity of the QS is $L(n) = \exp\left(\sqrt{(\ln n) \ln \ln n}\right)$.

Several variations speed the practical QS algorithm, although they do not change its theoretical complexity. They include:

- using large primes (larger than B), as in CFRAC,
- multiple polynomials (not just $f(x) = x^2 - n$) [18, 547],
- self-initialising polynomials (amortise the root-finding) [18].

The front page of the *New York Times* newspaper of 12 October 1988, had an article titled, 'A Most Ferocious Math Problem Tamed'. In it, journalist Malcom W. Browne described how Lenstra and Manasse factored a hard 100-digit number by the QS.

In 1994, the QS was used to factor the 129-digit RSA challenge number mentioned in Section 3.2. Atkins, Graff, Lenstra and Leyland [24] used a variation of the QS with two large primes [372] allowed in each relation. This means that each relation may have zero, one or two primes larger than the greatest one in the factor base. When the relations are harvested at the end of the sieve, after the primes in the factor base have been removed from $f(s)$, if the remaining co-factor is composite, then some effort is made to factor this number, often with the elliptic-curve method of Section 3.6.2. If it can be factored easily, then the relation is saved. Using two large primes complicates the linear algebra only slightly.

In the 1990s, George Sassoon organised all the personal computers on the Isle of Mull, Scotland, into a factoring group. They used the multiple polynomial version of the QS, with each machine sieving a different set of polynomials. As these computers were not interconnected, the relations were collected on floppy disks and carried to one machine for the linear algebra. They factored more than a dozen numbers this way, including one with 101-digits.

In another large factoring endeavour, many people used multiple personal computers to distribute the sieving of the quadratic or number field sieve, and then emailed the relations to one large machine used for the linear algebra. See Lenstra and Manasse [371] for factoring by email.

3.5 The Revolution: A Number Field Sieve

Pollard [474] invented the number field sieve (NFS) by suggesting raising the degree of the polynomial in the QS, although only for numbers with special form. He factored the Fermat number $F_7 = 2^{2^7} + 1$ (which had been factored earlier by Morrison and Brillhart with CFRAC) using the cubic polynomial $2x^3 + 2$ on a small computer. Manasse and the Lenstra brothers [381] soon extended Pollard's ideas to higher-degree polynomials, but still only for numbers of the form $r^e - s$, for small integers r and $|s|$. Their goal was to factor F_9, the smallest Fermat number with no known prime factor. They hoped to use the special form of $F_9 = 2^{512} - (-1)$ to make the numbers that had to be smooth smaller than those needed for the QS. After they factored F_9 in 1990, they and others extended the NFS to general numbers. See Lenstra and Lenstra [360] for more details of the early history of this factoring algorithm. See Pomerance [479] for a summary of the algorithm. Crandall and Pomerance [144] describe the modern algorithm.

Recall that the QS produces many relations $x_i^2 \equiv q_i \pmod{n}$ with q_i factored. After we have enough relations, we match the prime factors of the q_i and create a subset of the q_i whose product is square. In this way, we find congruences $x^2 \equiv y^2 \pmod{n}$ which may factor n by Theorem 3.7.

Now drop the requirement that the left side of a relation must be square. Instead seek relations $r_i \equiv q_i \pmod{n}$ in which both r_i and q_i have been factored completely. Use linear algebra to match the prime factors of r_i and the prime factors of q_i and select a subset of the relations for which both the product of the r_i and the product of the q_i are square. This is a fine idea, but too slow to be practical. The main difficulty is that at least one of $|r_i|$, $|q_i|$ must exceed $n/2$, so it has little chance of being smooth.

The NFS solves this problem by letting the numbers on one side of each relation be algebraic integers from an algebraic number field. The idea is to match the irreducible factors so that each occurs an even number of times and hope the product of the algebraic integers in the selected subset of the relations might be a square in the algebraic number field.

3.5.1 Number Fields

See the books [172, 272, 284, 288] for more about number fields. An algebraic number is the zero of a polynomial with integer coefficients. If the polynomial is monic, then the algebraic number is called an algebraic integer. An algebraic number field is a field that contains only algebraic numbers. The smallest algebraic number field containing the algebraic number α is written $\mathbb{Q}(\alpha)$.

The set of all algebraic integers in $\mathbb{Q}(\alpha)$ is written $\mathbb{Z}(\alpha)$. This set forms a commutative ring with unity. A unit in $\mathbb{Z}(\alpha)$ is an element having a multiplicative inverse in $\mathbb{Z}(\alpha)$. A non-zero, non-unit element γ of $\mathbb{Z}(\alpha)$ is irreducible if it can be factored in $\mathbb{Z}(\alpha)$ only as $\gamma = u\beta$ where u is a unit. When $\gamma = u\beta$, where u is a unit, β is called an associate of γ (and γ is an associate of β). An algebraic integer γ has unique factorisation (in $\mathbb{Z}(\alpha)$) if any two factorisations of γ into the product of irreducible elements and units are the same except for replacing irreducibles by their associates and using different units.

The polynomial of lowest degree having an algebraic number α as a zero must be irreducible, that is, it does not factor into the product of two polynomials of lower degree. If an algebraic number α is a zero of the irreducible polynomial $f(x) \in \mathbb{Z}[x]$, then the conjugates of α are all of the zeros of $f(x)$. The norm $N(\alpha)$ of α is the product of all of the conjugates of α including α. The norm of an algebraic integer is a rational integer. The norm function is multiplicative: $N(\alpha\beta) = N(\alpha)N(\beta)$. Thus if $\beta = \gamma^2$ for some $\gamma \in \mathbb{Z}(\alpha)$, then $N(\beta)$ is the square of the integer $N(\gamma)$. If the algebraic integer α is a zero of the irreducible polynomial $f(x) = x^d + c_{d-1}x^{d-1} + \cdots + c_1 x + c_0$ and a and b are integers, then the norm of $a - b\alpha$ is $N(a - b\alpha) = F(a, b)$, where F is the homogeneous polynomial

$$F(x, y) = x^d + c_{d-1}x^{d-1}y + \cdots + c_1 xy^{d-1} + c_0 y^d = y^d f(x/y). \qquad (3.7)$$

3.5.2 The Number Field Sieve

We now return to the NFS. Recall that we are letting the numbers on one side of each relation be algebraic integers from an algebraic number field. Then we proposed to match the irreducible factors so that each occurs an even number of times and thus form two squares of integers congruent modulo the number n to factor.

The first problem is writing a congruence modulo n with an algebraic integer on one side. This problem is solved by using a homomorphism h from the algebraic integers $\mathbb{Z}(\alpha)$ to Z_n, the integers modulo n. Suppose we have many algebraic integers θ_i, each factored into irreducibles, and also every $h(\theta_i)$ factored into the product of primes. Then we may match the irreducibles and match the primes to choose a subset of the θ_i whose product is a square γ^2 in $\mathbb{Z}(\alpha)$ and so that the product of the $h(\theta_i)$ is a square y^2 in the integers. Let $x = h(\gamma)$, a residue class modulo n. We have

$$x^2 = (h(\gamma))^2 = h(\gamma^2) = h\left(\prod_{i \in S} \theta_i\right) = \prod_{i \in S} h(\theta_i) \equiv y^2 \pmod{n},$$

which may factor n by Theorem 3.7.

Now we explain how to choose the algebraic number field and construct the homomorphism. We want to have an irreducible monic polynomial

$$f(x) = x^d + c_{d-1}x^{d-1} + \cdots + c_1 x + c_0$$

with integer coefficients. Let α be a zero of f in the complex numbers $\mathbb{C}$. The algebraic number field will be $\mathbb{Q}(\alpha)$. Let $\mathbb{Z}[\alpha]$ be the set of all $\sum_{j=0}^{d-1} a_j \alpha^j$, where the a_j are integers. This is a ring contained in the ring $\mathbb{Z}(\alpha)$ of integers of $\mathbb{Q}(\alpha)$. We also need an integer m for which $f(m) \equiv 0 \pmod{n}$. The homomorphism from $\mathbb{Z}[\alpha]$ to Z_n will be defined by setting $h(\alpha) = m \pmod{n}$, that is,

$$h\left(\sum_{j=0}^{d-1} a_j \alpha^j\right) \equiv \sum_{j=0}^{d-1} a_j m^j \pmod{n}.$$

The numbers θ will all have the form $a - b\alpha$. We seek a set S of pairs (a, b) of integers such that

$$\prod_{(a,b)\in S} (a - bm) \text{ is a square in } \mathbb{Z} \tag{3.8}$$

and

$$\prod_{(a,b)\in S} (a - b\alpha) \text{ is a square in } \mathbb{Z}[\alpha]. \tag{3.9}$$

Let the integer y be a square root of the first product. Let $\gamma \in \mathbb{Z}[\alpha]$ be a square root of the second product. We have $h(\gamma^2) \equiv y^2 \pmod{n}$, since $h(a - b\alpha) \equiv a - bm \pmod{n}$. Let $x = h(\gamma)$. Then $x^2 \equiv y^2 \pmod{n}$, which will factor n with probability at least $1/2$, by Theorem 3.7.

The degree d of the polynomial $f(x)$ is 4, 5 or 6 for numbers we currently factor. In addition to being irreducible and having a known zero m modulo n, we want $f(x)$ to have 'small' coefficients compared to n. The next two sections give ways in which we might satisfy all these conditions.

3.5.3 The Special Number Field Sieve

The requirements on $f(x)$ are easily met in the special number field sieve, which factors numbers $n = r^e - s$, where r and $|s|$ are small positive integers. Suppose we decide (based on experience) that d is the proper degree for the polynomial. Let k be the least positive integer for which $kd \geq e$. Let $t = sr^{kd-e}$. Let $f(x)$ be the polynomial $x^d - t$. Let $m = r^k$. Then $f(m) = r^{kd} - sr^{kd-e} = r^{kd-e}n \equiv 0 \pmod{n}$.

Example 3.8. Let us factor $n = 7^{346} + 1$. Let $d = 6$, $m = 7^{58}$ and $f(x) = x^6 + 49$. then $f(m) = (7^{58})^6 + 49 = 7^{348} + 7^2 = 7^2(7^{346} + 1) = 49n \equiv 0 \pmod{n}$.

Kleinjung, Bos and Lenstra [329] factored many large Mersenne numbers $2^p - 1$ with the Special NFS. These factorisations are the current record for factoring integers by the NFS.

3.5.4 The General Number Field Sieve

In the general number field sieve, n lacks the special form just considered, so there is no obvious polynomial. One standard approach to finding a good polynomial (of degree 5, say) to factor n is to let m be an integer slightly larger than $n^{1/5}$. Write $n = \sum_{i=0}^{5} d_i m^i$ in base m with digits d_i in the interval $0 \le d_i < m$, small compared to n. Let the polynomial be $f(x) = \sum_{i=0}^{5} d_i x^i$. See the article [114] by Buhler, Lenstra and Pomerance for the origins of the general number field sieve. Montgomery and Murphy [440] give better ways to choose a polynomial for the GNFS. Of course, the GNFS would be used to factor an RSA public key because it would not have the special form $r^e - s$.

3.5.5 The Number Field Sieve Again

We now return to both forms of the NFS. The program will have two sieves, one for $a - bm$ and one for $a - b\alpha$. The sieve on $a - bm$ is simple: for each fixed $0 < b < M$ we try to factor the numbers $a - bm$ for $-M < a < M$ by the sieve algorithm (Algorithm 3.4).

The goal of the sieve on the numbers $a - b\alpha$ is to allow us to choose a set S of pairs (a, b) so that the product in Eq. (3.9) is a square. Rather than try to factor the algebraic integers $a - b\alpha$, let us work with their norms. If the product in Eq. (3.9) is a square, then its norm is a square, and its norm is the product of all $N(a - b\alpha)$ with $(a, b) \in S$. Since the norms are rational integers, rather than algebraic integers, it is easy to match the prime factors of norms to form squares. Furthermore, the norm of $a - b\alpha$ is a polynomial, $F(a, b)$ in Eq. (3.7), and therefore is something we can factor with the sieve algorithm in Section 3.4. For each fixed b between 0 and M, sieve the polynomial $F(x, b) = N(x - b\alpha)$ for x between $-M$ and M to find smooth values of $N(a - b\alpha)$.

Whenever both $a - bm$ and $N(a - b\alpha)$ are smooth, save the pair (a, b) to represent the relation $h(a - b\alpha) \equiv a - bm \pmod{n}$. After we have found many relations, use linear algebra to construct sets S of pairs (a, b) for which the product of $a - bm$ is a square and the product of the norms of $a - b\alpha$ is a square.

Several problems arise, even in this simple description of the NFS algorithm. For example, the fact that $N(\theta)$ is square need not imply θ is square. One problem is that the norm function does not distinguish among associates. Another is the lack of unique factorisation in most number fields. A third problem is computing the square root of an algebraic number. All of these problems can be solved. See Crandall and Pomerance [144] for the details.

One can show that the NFS is faster than the QS when n is large and the degree d is chosen well. A careful analysis shows that the time complexity of the NFS is $\exp\left(c(\ln n)^{1/3}(\ln\ln n)^{2/3}\right)$ for some constant $c > 0$. The constant c is smaller for the Special NFS than for the General NFS because the coefficients can be made smaller. Lenstra and co-workers [117] factored several RSA challenge numbers using the General NFS.

Shamir [532] (see also Lenstra and Shamir [373]) proposed an optoelectronic device called TWINKLE to execute the number field sieve algorithm. Each prime in the factor base is represented by a light-emitting diode. The LED for prime p glows once every p time units; its intensity is proportional to $\log p$. A sensor views the total intensity of all LEDs and saves the current value of a time unit counter whenever the intensity exceeds a threshold. It locates B-smooth numbers by adding the logarithms of their prime factors.

3.6 An Exquisite Diversion: Elliptic Curves

In 1985, H. W. Lenstra, Jr [388] invented a factoring method using elliptic curves. Lenstra freed Pollard's $p - 1$ factoring algorithm from its dependency on a single number, namely $p-1$, being smooth. He replaced the multiplicative group of integers modulo p in that algorithm with an elliptic curve modulo p. Now the integer that must be smooth is the size of the elliptic curve modulo p. This number is about the same size as p. Soon after this discovery, Miller [433] and Koblitz [340] made the same group replacement in many cryptographic algorithms. This change allows one to use smaller p and construct fast cryptographic algorithms with the same level of security as slower ones. Now elliptic curves are an important tool in cryptography.

3.6.1 Basic Properties of Elliptic Curves

An elliptic curve is the graph of an equation like $y^2 = x^3 + ax^2 + bx + c$ with one extra point ∞. For simplicity we use the Weierstrass form $y^2 = x^3 + ax + b$. For cryptography, a, b, x and y lie in a finite field. For factoring, they are integers modulo p or n. We begin by allowing them to be real numbers so that the elliptic curve is a curve in the plane with an added point ∞.

We will define a way to 'add' points of the set

$$E_{a,b} = \{(x,y) : x, y \in K, \ y^2 = x^3 + ax + b\} \cup \{\infty\}.$$

If $P = (x, y)$ lies on the graph of $y^2 = x^3 + ax + b$, define $-P = (x, -y)$, that is, $-P$ is P reflected in the x-axis. Also define $-\infty = \infty$.

Given two points P and Q, on the graph but not on the same vertical line, define $P + Q = R$, where $-R$ is the third point on the straight line through P and Q. If P and Q are distinct points on the graph and on the same vertical line, then they must have the form $(x, \pm y)$, that is, $Q = -P$, and we define $P + Q = P + (-P) = \infty$. Also define $P + \infty = \infty + P = P$ for any element P of the elliptic curve (including $P = \infty$). To add a point $P \neq \infty$ to itself, draw the tangent line to the graph at P. If the tangent line is vertical, then $P = (x, 0)$ and we define $P + P = \infty$. If the tangent line is not vertical, then it intersects the graph in exactly one more point R, and we define $P + P = -R$. (If P is a point of inflection, then $R = P$.)

Theorem 3.9. *An elliptic curve $E_{a,b}$ with the addition operation $+$ forms an abelian group with identity ∞. The inverse of P is $-P$.*

We will need formulas for computing the coordinates of $P + Q$ in terms of those of P and Q. If one of P, Q is ∞, then we have already defined $P + Q$. Let $P = (x_1, y_1)$ and $Q = (x_2, y_2)$ be on the graph of $y^2 = x^3 + ax + b$. If $x_1 = x_2$ and $y_1 = -y_2$, then $P = -Q$ and $P + Q = \infty$. Otherwise, let s be the slope defined as follows. When $P \neq Q$, let $s = (y_2 - y_1)/(x_2 - x_1)$ be the slope of the line through P and Q. When $P = Q$, let $s = (3x_1^2 + a)/(2y_1)$ be the slope of the tangent line to the graph at P. Then $P + Q = (x_3, y_3)$, where $x_3 = s^2 - x_1 - x_2$ and $y_3 = s(x_1 - x_3) - y_1$.

Now consider elliptic curves modulo a prime p. The formulas for adding points show that if a and b and the coordinates of points P and Q on the elliptic curve $E_{a,b}$ are all rational numbers, then the coordinates of $P+Q$ will be rational numbers (unless $P+Q = \infty$). Therefore, if a and b and the coordinates of points P and Q on the elliptic curve $E_{a,b}$ are integers modulo p, then the coordinates of $P + Q$ will be integers modulo p, unless $P + Q = \infty$, provided that any division needed in the slope calculation is by a number relatively prime to p. Of course, the graph is just a set of pairs of numbers modulo p, not a curve in the plane.

When i is a positive integer and P is a point on an elliptic curve, let iP mean P added to itself i times. It is easy to compute iP when i is large. Algorithm 3.5 takes about $\log_2 i$ point additions on the elliptic curve.

Theorem 3.10. *Let p be an odd prime. Let (r/p) denote the Legendre symbol. The number $M_{p,a,b}$ of points on the elliptic curve $y^2 \equiv x^3 + ax + b \pmod{p}$ satisfies $M_{p,a,b} = p + 1 + \sum_{x=0}^{p-1} \left((x^3 + ax + b)/p \right)$.*

Since the Legendre symbol in Theorem 3.10 has the value $+1$ about as often

---

Algorithm 3.5 Fast point multiplication.

Input: A point P on $E_{a,b}$ modulo m and an integer $i \geq 0$.
Output: $Q = iP$

1: $Q \leftarrow \infty$
2: $R \leftarrow P$
3: **while** $i > 0$ **do**
4:     **if** i is odd **then**
5:         $Q \leftarrow (Q + R) \bmod m$
6:     $R \leftarrow (R + R) \bmod m$
7:     $i \leftarrow \lfloor i/2 \rfloor$
8: **return** Q

---

as the value -1, we expect the number of points on a random elliptic curve modulo p to be close to $p + 1$. Hasse proved that this is so.

Theorem 3.11 (Hasse). *Let the elliptic curve $E_{a,b}$ modulo a prime p have $M_{p,a,b}$ points. Then $p + 1 - 2\sqrt{p} \leq M_{p,a,b} \leq p + 1 + 2\sqrt{p}$.*

The range of possible values for $M_{p,a,b}$ is called the Hasse interval.

3.6.2 Factoring with Elliptic Curves

In 1985, H. W. Lenstra, Jr [388] invented the elliptic-curve method (ECM), a factoring algorithm using elliptic curves (see Algorithm 3.6). Let R_p denote the multiplicative group of integers modulo a prime p. Recall that Pollard's $p - 1$ factoring algorithm in Section 3.1.3 performs a calculation in the integers modulo n that hides a calculation in R_p. The factor p of n is discovered when the size $p - 1$ of the group R_p divides $L = B!$, but p is not found when $p - 1$ has a prime divisor larger than B. Lenstra replaced R_p with an elliptic-curve group $E_{a,b}$ modulo p. By Hasse's theorem, the two groups have roughly the same size, namely, approximately p. Lenstra's algorithm discovers p when $M_{p,a,b}$ divides $L = B!$. It fails to find p when $M_{p,a,b}$ has a prime factor larger than B. There is only one group R_p, but lots of elliptic-curve groups $E_{a,b}$ modulo p. If the size of R_p has a prime factor $> B$, we are stuck. But if the size of $E_{a,b}$ modulo p has a prime factor $> B$, we just change a and b and try a new elliptic curve. Each curve gives an independent chance to find the prime factor p.

Compare this algorithm with Pollard's $p - 1$ algorithm. The two algorithms work exactly the same way, except that Pollard's $p - 1$ algorithm raises a to the power i while the elliptic-curve method multiplies P by i. The former algorithm

---

Algorithm 3.6 Simple elliptic-curve factorisation method.

Input: A composite positive integer n to factor and a bound B.

Output: A proper factor p of n, or else give up.

1: Find the primes $p_1 = 2, p_2, \ldots, p_k \leq B$

2: Choose a random elliptic curve $E_{a,b}$ modulo n and a random point $P \neq \infty$
   on it

3: $g := \gcd(4a^3 + 27b^2, n)$

4: **if** $g = n$ **then**

5:      choose a new curve and point P

6: **if** $g > 1$ **then**

7:      report the factor g of n and stop

8: **for** $i = 1, \ldots, k$ **do**

9:      $P \leftarrow iP$ or else find a factor g of n

10: Give up or try another random elliptic curve.

---

explicitly computes a greatest common divisor with n whereas the latter algorithm hides this operation in the slope calculation of the elliptic-curve point addition. The quantity $4a^3 + 27b^2$ in the algorithm is the discriminant of the cubic polynomial, which tells whether it has a repeated zero. For technical reasons, the algorithm excludes polynomials with a repeated zero.

A good way to choose the elliptic-curve parameters is to choose a random a modulo n and a random point $P = (x_1, y_1)$ modulo n and then let $b = (y_1^2 - x_1^3 - ax_1) \bmod n$. In other words, b has the correct value modulo n so that the point P is on $E_{a,b}$. Lenstra [388] proved that when many curves and points are chosen this way, the sizes $M_{p,a,b}$ of the curves modulo a prime p are well distributed in the Hasse interval.

Whenever two points are added during the computation of iP, the coordinates are reduced modulo n. Imagine that the coordinates are also reduced modulo p, an unknown prime divisor of n. If the size $M_{p,a,b}$ of the elliptic curve modulo p divides $L = i!$, then $LP = \infty$ in the elliptic curve modulo p. Because $P \neq \infty$, somewhere during the calculation we must have $P_1 + P_2 = \infty$ for two points $P_1, P_2 \neq \infty$, working with coordinates modulo p. This means that P_1 and P_2 will have the same x-coordinate modulo p. But we are computing modulo n, and the x-coordinates of the two points will probably not be the same modulo n. In the slope computation, we will try to invert a number not relatively prime to n, so we will factor n instead.

Theorem 3.12. *Let n be a positive integer with an unknown prime factor p. Let B be the optimal bound for finding p by the elliptic-curve algorithm. Assume* *that a random elliptic curve $E_{a,b}$ modulo p has a B-smooth size with probability u^{-u}, where $u = (\ln p)/\ln B$, as Dickman's theorem [157] would predict. Define $L(x) = \exp\left(\sqrt{(\ln x)\ln\ln x}\right)$. Then $B = L(p)^{\sqrt{2}/2}$. The expected total number of point additions performed when the elliptic-curve algorithm is used to discover p is $L(p)^{\sqrt{2}}$. The expected total work needed to discover one prime factor of n is at most $L(n)$ point additions.*

The ECM has a second stage, just like the Pollard $p - 1$ algorithm. The second stage finds a factor p of n when the largest prime factor of $M_{p,a,b}$ is less than B_2 and all the other prime factors of $M_{p,a,b}$ are less than B.

Efficient modern versions of the ECM discover prime factors up to about 20 digits in a few seconds and those up to about 40 digits in a few hours. Luck is required to discover factors having more than 60 digits. See Silverman and Wagstaff [546] for some practical aspects of ECM. See Zimmermann and Dodson [634] for more about ECM.

3.6.3 Factoring Helping Elliptic Curves Used in Cryptography

Almost all of this chapter concerns the use of factoring to break the RSA public-key cipher. This section, however, uses factoring integers to build efficient cryptographic protocols. There are many uses of elliptic curves in cryptography. Here we mention two ways factoring helps to construct elliptic curves that are well suited for use in cryptography.

One application is how to choose an elliptic curve with simple formulas for adding points. Elliptic curves used in cryptography are groups whose points are defined by an equation over a finite field. The size of a finite field is a prime power $q = p^k$. The size of q is typically about 256 or 512 bits for cryptography (see Table 11.2). Some of the most useful elliptic curves have size equal to $q\pm1$. The points (other than ∞) of an elliptic curve over $\mathbb{F}_q$ are pairs of numbers in $\mathbb{F}_q$. The formulas for adding two points use arithmetic in $\mathbb{F}_q$. When q is prime, the arithmetic is in the integers modulo q and it is slow, especially for devices with limited computing power. But if p is a small prime, then arithmetic in $\mathbb{F}_{p^k}$ can be done in steps with numbers no larger than p, so it is fast on a slow machine. Elliptic curves often lie over $\mathbb{F}_{p^k}$ with $k > 1$. The size of such a curve might be $p^k + 1$. The factorisation of the size is needed to ensure that the elliptic-curve discrete logarithm problem (ECDLP) is hard. To make ECDLP hard, $p^k + 1$ should have at least one large prime factor. The ECDLP is to find x so that $Q = xP$, where P and Q are two points on an elliptic curve. Many cryptographic algorithms depend on the ECDLP being intractable.

Another use of factoring is for efficient computation of pairings. Let $E_{a,b}$

be an elliptic curve over $\mathbb{F}_q$. A pairing on $E_{a,b}$ is a function that takes pairs of points of $E_{a,b}$ to the nth roots of unity μ_n in an algebraic closure of $\mathbb{F}_q$ satisfying several mathematical properties. The first use of a pairing in cryptography was in an attack on the ECDLP in 1993. The attack by Menezes, Okamoto and Vanstone (MOV) [421] changes the problem $Q = xP$ on an elliptic curve over $\mathbb{F}_1$ to an equivalent problem $a = b^x$ in $\mathbb{F}_{q^k}$ with $k \leq 6$ for the targeted supersingular curves; the latter problem is easier to solve. Since 2000, many constructive uses of pairings have been found in cryptographic protocols, including three-way key agreement, identity-based encryption and short signatures. These protocols could be broken if the ECDLP were easy.

The naive computation of a pairing by its definition is too slow. There are faster ways to compute pairings for supersingular elliptic curves, whose sizes are numbers $p^k \pm 1$. Unfortunately, supersingular elliptic curves are the ones for which the pairing-based MOV attack on the ECDLP works best. Consequently, the parameter choice for these curves is a delicate balance between ease of computing the pairing and the need to keep the ECDLP hard. This decision requires knowledge of the factorisation of the elliptic-curve size, which has the form $p^n \pm 1$.

See Washington [619] for definitions of the pairings. Boneh and Franklin [80] designed an identity-based encryption system using pairings, which allows someone to use your email address as your public key. See Trappe and Washington [591] for a simple description of the Weil pairing and identity-based encryption. See Estibals [184] for ways of using factors of $2^n - 1$ and $3^n - 1$ to construct elliptic curves in which pairings are easy to compute, but the ECDLP is intractable.

3.7 The Future: How Hard Can Factoring Be?

In this section we discuss factoring from a theoretical computer-science point of view and then present some ways to factor RSA public keys. We end with suggestions for new methods of factoring and make some predictions about the future of factoring.

3.7.1 Theoretical Aspects of Factoring

It would be great if we could find an algorithm for factoring n that we could prove always works and runs in a specified time. Of course, trial division described in Section 3.1.1 has this property and factors n in $O\left(n^{1/2}\right)$ steps. There is an algorithm of Pollard [475] and Strassen [576] (see Section 5.5 of [144]) that uses fast polynomial evaluation to factor n in $O\left(n^{1/4+\varepsilon}\right)$ steps. These algorithms are rigorous and deterministic, that is, they do not choose random numbers and one can prove before they begin that they will succeed in a specified number of steps. Shanks [534] invented a fast algorithm for factoring n by computing the class number of primitive binary quadratic forms of discriminant n. This complicated algorithm runs in $O\left(n^{1/4+\varepsilon}\right)$ steps, but can be modified to factor n in $O\left(n^{1/5+\varepsilon}\right)$ steps, assuming the extended Riemann hypothesis. Algorithms like Shanks's $O\left(n^{1/5+\varepsilon}\right)$ one, CFRAC, QS and NFS are fast and deterministic, but the proofs of their running times depend on heuristic hypotheses, so they are not rigorous.

A probabilistic algorithm[1] for factoring integers uses random numbers and may or may not factor the number, depending on the random choices it makes. For a given input, it may perform different steps each time it is invoked and may produce different factorisations or none at all. For example, the elliptic-curve method of Section 3.6.2 chooses a random elliptic curve and a random point on it, and then works deterministically to compute a large multiple of the point. For some choices it will factor the input number, perhaps finding different factors (say, if n has two prime factors of about the same size).

A probabilistic algorithm may or may not be rigorous. The elliptic-curve method is probabilistic and not rigorous because it assumes without proof that, with a certain probability, there is an elliptic curve having B-smooth size in the Hasse interval. The ability to make random choices is a powerful tool that can make a probabilistic algorithm faster than a deterministic algorithm for the same job. In order to say that a probabilistic algorithm for factoring is rigorous, there must be a proof without unproved assumptions that it will factor the input number n with positive probability in a specified number of steps. It turns out that there is a rigorous probabilistic algorithm for factoring integers with a subexponential running time. Dixon [166] invented such an algorithm in 1981.

Dixon's algorithm for factoring n has two parameters, u and v, to be specified later. It begins by choosing a set S of u random integers between 1 and n. The remainder of the algorithm is deterministic. Call the algorithm A_S. Let the factor base consist of all primes $< v$. For each $z \in S$, try to factor $w = (z^2 \bmod n)$ using only the primes in the factor base. If you succeed in factoring w completely, save the relation $z^2 \equiv$ (the product of the factors of w) (mod n). After the factoring is finished, combine the relations using linear algebra modulo 2 as in the QS. The method used to factor the numbers w does not matter; even trial division would be fast enough for the theorem.

---

[1] Technically, a Las Vegas probabilistic algorithm.

Let $L(n) = \exp\left(\sqrt{(\ln n)\ln\ln n}\right)$. Dixon [166] proved this theorem about the set of algorithms $\{A_S : S \subseteq [1,n], \text{size}(S) = u\}$.

Theorem 3.13 (Dixon). *Let n be an odd integer with at least two different prime factors. Let $v = L(n)^{\sqrt{2}}$ and $u = \lceil v^2 \rceil$. Then the average number of steps taken in the execution of algorithm A_S is $L(n)^{3\sqrt{2}}$. The probability that algorithm A_S fails to factor n is proportional to $L(n)^{-\sqrt{2}}$, uniformly in n.*

The theorem says that, when n is large, almost all of the algorithms A_S with correct parameters u and v will factor n successfully and that all A_S run in subexponential time $L(n)^{3\sqrt{2}}$.

Although they have rigorous proofs of their time complexities, the algorithms mentioned above are much slower than the QS and NFS.

The RSA cipher and several other cryptographic algorithms depend on factoring integers being a hard problem. Can we prove that factoring integers is hard?

One must phrase this question carefully. Suppose n has a million decimal digits, and the low-order digit is 6. Then obviously 2 is a prime factor of n. One can trivially 'factor' n as $2 \cdot (n/2)$.

Here is one way to ask the question. As a function of n, what is the minimum, taken over all integer factoring algorithms, of the maximum, taken over all integers M between 2 and n, of the number of steps the algorithm takes to factor M? The integer factoring algorithm has input M and outputs a list of all prime factors of M. Integer factoring would be a polynomial-time problem if the answer were $O((\log n)^c)$ for some constant c.

The model of computation, that is, what constitutes a 'step', probably matters a lot in answering the question. Suppose we decide that a single arithmetic operation $(+, -, \times, \div)$ is one step. Assume that integers are stored in binary notation and that each register (memory location) may hold one integer of any size. Shamir [531] proved that, in this model, one can factor n in $O(\log n)$ steps. Shamir's result shows that we can factor n in a polynomial (in $\log n$) number of steps provided that an arithmetic operation with numbers as large as $n!$ counts as one step. This shows why we need to consider the complexity of arithmetic with large numbers when analysing number theoretic algorithms.

3.7.2 Factor an RSA Public Key

In this section and the following section we consider various ways to factor a large integer with the help of special information available about it. Some of the special information might include its use as a public key for the RSA cipher.

Theorem 3.14. *There is a polynomial time algorithm for factoring a composite integer n, given a base a to which n is a pseudoprime but not a strong pseudoprime.*

If we could find such a base quickly, then we would have a fast method of factoring integers.

As explained by May in his chapter, there are ways to factor an RSA public modulus n when some information is known about it or the parameters are chosen poorly. None of these tricks actually threaten the RSA cipher because, when the cipher is used properly, the needed information about the modulus is not leaked. However, it turns out that, in practice, proper use can be elusive; see the many mishaps presented by Heninger in her chapter.

A. K. Lenstra *et al.* [386] collected several million 1024-bit RSA public keys from X.509 certificates and PGP on the Internet. Most of these keys appeared valid. They computed the gcd of each pair of these n and were able to factor a few of them because they shared a single secret prime factor. This finding suggests poor seeds for random-number generators, that is, two (or more) users used the same seed to generate a prime for their RSA key.

There are other attacks on the RSA cipher besides factoring n. For example, if Alice happens to encipher the same message M via RSA using different enciphering exponents e_1 and e_2, but the same RSA modulus n, then an attacker can recover M from the two ciphertexts without factoring n. Boneh [74] describes all the attacks below and many more.

Theorem 3.15. *If n is the product of two different primes p and q, then one can factor n in polynomial time, given n and $\varphi(n)$.*

Proof We have $\varphi(n) = (p - 1)(q - 1) = n - (p + q) + 1$ and $n = pq$. Thus, $n + 1 - \varphi(n) = p + q = p + n/p$ or $p^2 - (n + 1 - \varphi(n))p + n = 0$. This quadratic equation in the unknown p has the two solutions p and q, which may be computed easily by the quadratic formula.   □

Theorem 3.15 shows why one must not reveal $\varphi(n)$ when n is an RSA public key. One must also not reveal the deciphering exponent d when n is an RSA public key. Not only could one decipher all ciphertext knowing d, but one can factor n, too. Of course, the enciphering exponent e is public. This theorem appears in the original RSA paper [501].

Theorem 3.16. *There is a probabilistic polynomial time algorithm to factor n, given an integer n which is the product of two unknown primes and given two integers e, d, between 1 and n with ed ≡ 1 (mod $\varphi(n)$).*

There is a deterministic method for factoring n, given e and d, but it is more complicated. As we saw above, we can factor n given n and $\varphi(n)$ by solving a quadratic equation. May [407] showed how to find $\varphi(n)$ from n, e and d in deterministic polynomial time. His result depends on a theorem of Coppersmith telling how to find small solutions to a polynomial congruence modulo n. Coppersmith's theorem relies in turn on the LLL [380] algorithm for finding a reduced basis for a lattice. As this work leads to several other attacks on RSA, we describe them in the following section.

Here is one more way to factor n when a parameter is chosen badly. One can factor n when d is unknown, but satisfies $d \leq \sqrt[4]{n}/3$, according to Wiener [621].

Theorem 3.17 (Wiener). *There is a polynomial time algorithm which can factor n, given n and e, provided $e < \varphi(n)$ and $n = pq$, where p and q are (unknown) primes with $q < p < 2q$, and there is an (unknown) integer $d < n^{1/4}/3$ satisfying $ed \equiv 1 \pmod{\varphi(n)}$.*

Finally, we give an example of factoring an RSA modulus n by exploiting a hardware failure during signature generation. We explained in Section 3.2 how Alice can set up RSA as a public-key cipher to receive encrypted messages from Bob. Alice can sign a message M to Bob as $S = M^d \bmod n$, where d is her private RSA deciphering exponent. When Bob receives S he can verify Alice's signature by getting her public RSA key n, e, and computing $M = S^e \bmod n$. The fact that this M is meaningful text shows that it came from Alice because only Alice could construct such a signature S. Now she can accelerate this calculation of S by a factor of four using her knowledge of the factors p and q of n. She computes $S_p = S \bmod p$ and $S_q = S \bmod q$ by fast exponentiation with smaller numbers and then combines these values with the Chinese remainder theorem to obtain S.

Suppose that a hardware error complements one bit of S_p during this calculation, but S_q is computed correctly, so that she sends Bob an incorrect signature S'. Suppose also that an eavesdropper Eve obtains S' and M. Eve can factor Alice's public modulus n, given only S', n, e and M, where e is Alice's public encryption exponent. We have $M = S^e \bmod n$. Since the error occurred in computing S_p, we have $M \equiv (S')^e \pmod{q}$ but $M \not\equiv (S')^e \pmod{p}$. Therefore, $\gcd(M - (S')^e, n) = q$. This attack is due to Lenstra and his co-workers [306]. See Boneh [74] for another similar attack, in which Eve has S and S', but not M.

3.7.3 Factoring With Lattices

In this section we give a brief overview of lattices and applications to factoring an RSA public modulus. The definitions and theorems in this section are somewhat vague. See [380] for the true definitions and theorems.

Lattices have many uses in cryptography. They may be used to define cryptosystems and to attack RSA and other ciphers. Several attacks on the RSA public-key cipher with poor parameter choices use lattices and the LLL algorithm. We describe here only the ways lattices can be used to factor an RSA modulus.

Definition 3.18. The lattice generated by linearly independent vectors $\mathbf{v}_1, \ldots, \mathbf{v}_r$ with integer coordinates is the set of all linear combinations $a_1 \mathbf{v}_1 + \cdots + a_r \mathbf{v}_r$ with integers a_i.

A basis for a lattice L is a set of linearly independent vectors $\mathbf{v}_1, \ldots, \mathbf{v}_r$ with integer coordinates such that any vector in L can be written as a linear combination $a_1 \mathbf{v}_1 + \cdots + a_r \mathbf{v}_r$ of the basis vectors with integer coefficients a_i.

Every lattice has a basis. Every basis of a given lattice L has the same size r, called the rank of L. The rank r is the same as the dimension of the vector space (a subspace of $\mathbb{R}^n$) spanned by any basis of L. A lattice has full rank if $r = n$.

Recall that if $\mathbf{v}_1, \ldots, \mathbf{v}_r$ and $\mathbf{w}_1, \ldots, \mathbf{w}_r$ are two bases for the same vector space, then each $\mathbf{w}_i$ can be written as a linear combination of the $\mathbf{v}_j$.

Likewise, if $\mathbf{v}_1, \ldots, \mathbf{v}_r$ and $\mathbf{w}_1, \ldots, \mathbf{w}_r$ are two bases for the same lattice, then each $\mathbf{w}_i$ can be written as a linear combination of the $\mathbf{v}_j$ with integer coefficients. Let B be the $r \times n$ matrix whose rows are $\mathbf{v}_1, \ldots, \mathbf{v}_r$. The 'size' of a lattice L is a real number $\det(L)$ defined in terms of any basis. If L has full rank, then B is square and $\det(L) = |\det(B)|$, the latter 'det' being the usual determinant. This is the only case we will use below.

There exists a vector $\mathbf{v}_1$ in a lattice L with minimum positive norm, the shortest vector. Let $\lambda_1(L)$ be the norm of this shortest $\mathbf{v}_1$. Every vector $\mathbf{w} \in L$ which is linearly dependent on $\mathbf{v}_1$ must be $\mathbf{w} = t\mathbf{v}_1$ for some integer t. At least two vectors have this minimum positive length since the norm of $-\mathbf{v}$ equals the norm of $\mathbf{v}$.

For integer $k \geq 1$, let $\lambda_k(L)$ be the smallest positive real number so that there is at least one set of k linearly independent vectors of L, with each vector having length $\leq \lambda_k(L)$. This defines a sequence

$$\lambda_1(L) \leq \lambda_2(L) \leq \lambda_3(L) \leq \cdots .$$

Note that we count lengths, not vectors, in this definition.

A lattice is often presented by giving a basis for it. This basis sometimes consists of very long, nearly parallel vectors. Sometimes it is more useful to have a basis with shorter vectors that are closer to being orthogonal to each other. The process of finding such a basis from a poor one is called reducing the lattice or lattice reduction.

The nicest basis $\{v_1, v_2, \ldots, v_r\}$ for L would have the length of v_i be $\lambda_i(L)$ for each i. It turns out to be $\mathcal{NP}$-hard to find a nicest basis.

The Gram–Schmidt process does not work for lattices because the $m_{i,j}$ are usually not integers, so the new basis vectors are not integer linear combinations of the original vectors. There are several definitions of reduced basis and they are not equivalent. Lenstra, Lenstra and Lovász [380] give one definition. Its vectors approximate the shortest possible vectors for a basis of a given lattice, and the angle between any two reduced basis vectors is not allowed to be too small. They give a polynomial time algorithm, related to the Gram–Schmidt process, but more complicated, that computes a reduced basis from a given one. They prove that their algorithm, called the LLL algorithm runs in polynomial time and constructs a reduced basis.

We will use only the following special application of the LLL algorithm.

Theorem 3.19 (LLL [380]). *Let L be a lattice spanned by $\mathbf{v}_1, \ldots, \mathbf{v}_r$. With input $\mathbf{v}_1, \ldots, \mathbf{v}_r$, the LLL algorithm finds in polynomial time a vector $\mathbf{b}_1$ with length*

$$\|\mathbf{b}_1\| \leq 2^{(r-1)/4} \det(L)^{1/r}.$$

Now we describe some attacks on RSA using LLL.

Recall the notation of RSA: $n = pq$ is the product of two large primes and is hard to factor. Choose e with $\gcd(e, \varphi(n)) = 1$, where $\varphi(n) = (p-1)(q-1)$. Via the Extended Euclidean Algorithm, find d with $ed \equiv 1 \pmod{\varphi(n)}$. Discard p and q. The public key is n, e and the private key is d. Encipher plaintext M as $C = M^e \bmod n$. Decipher ciphertext C as $M = C^d \bmod n$.

Consider these three problems about the RSA cipher.

(1) The RSA problem: given n, e and C, find M.
(2) Compute d: given n and e, find d.
(3) Factor n: given n, find p and q.

Clearly, if we can solve (3), then we can solve (2); and if we can solve (2), then we can solve (1). In fact, (3) is equivalent to (2). It is not known whether (3) is equivalent to (1). The next theorem was proved by May [407]. Its proof uses the LLL algorithm [380].

Theorem 3.20 (May). *Let $n = pq$, where p and q are two primes of the same*

bit-length. Let positive integers e and d satisfy ed ≡ 1 (mod $\varphi(n)$) and ed ≤ n^2.
There is a deterministic polynomial time algorithm which factors n given the
input n, e, d.

Theorem 3.20 is a beautiful theoretical result. It says that knowing the RSA deciphering exponent is deterministically polynomial time equivalent to knowing the factors of n. Of course, if an RSA deciphering exponent were accidentally leaked, one could factor n easily in probabilistic polynomial time by the method of Theorem 3.16.

Now assume we have some partial information about p or q or d because they are limited somehow, perhaps by a poor choice of parameters or a faulty random number generator.

Coppersmith [137] proved that if $f(x)$ is a monic polynomial and b is an unknown factor of a given integer n, then one can find all 'small' solutions x to $f(x) \equiv 0$ (mod b) quickly via the LLL lattice reduction algorithm.

Theorem 3.20 is one application of his method. Here is another application. We can factor an RSA modulus given a few more than half of the high-order bits of p (as a number $\tilde{p} \approx p$).

Theorem 3.21 (Coppersmith [137]). *One can factor n = pq, where p > q, in polynomial time, given n and an integer $\tilde{p}$ with $|p - \tilde{p}| < n^{5/28}/2$.*

Similar theorems allow one to factor $n = pq$ given the high-order half of the bits of q or the low-order bits of either p or q.

Coppersmith [136, 138] also proved a theorem that lets one find small roots of a polynomial with two variables in polynomial time using the LLL algorithm. This result gives a faster algorithm than Theorem 3.21 for factoring $n = pq$ when the high- or low-order half of the bits of either p or q are known. The bivariate polynomial theorem also gives polynomial time algorithms for factoring an RSA modulus n when some bits of d are known. Here is an example of these results.

Theorem 3.22 (Boneh and Durfee [75]). *There is a polynomial time algorithm to factor n = pq, given n and e, provided that there exists an integer $d < n^{0.292}$ such that ed ≡ 1 (mod $\varphi(n)$).*

The lesson of Theorems 3.17 and 3.22 is that the deciphering exponent d in RSA should not be too small, or else an attacker will be able to factor n. If you are choosing RSA keys and notice that $d < n^{2/3}$, say, then you should choose new e and d. So far as I know, there is no risk in letting e be small; even $e = 3$ appears to be safe.

m	NFac	Year	Method	Who
5	2	1732	Trial division	Euler
6	2	1855	Trial division and $p \equiv$ 1 (mod 2^{m+2})	Clausen and Landry
7	2	1970	CFRAC	Morrison and Brillhart
8	2	1980	Pollard rho	Brent and Pollard
9	2	1990	NFS	Lenstra, Manasse, *et al.*
10	4	1995	ECM	Brent
11	5	1988	ECM	Brent

Table 3.2 *The history of factoring Fermat numbers.*

3.7.4 The Future of Factoring

One way of predicting the future is by extrapolating from the past. Consider the problem of completely factoring Fermat numbers, and consider the history of how and when each Fermat number was factored.

The Fermat numbers $F_m = 2^{2^m} + 1$ for $5 \leq m \leq 32$ are composite. These numbers have been completely factored for $5 \leq m \leq 11$. Table 3.2, adapted from [611], gives the year and method by which the factorisation of these seven numbers was finished. The abbreviation NFac denotes the number of prime factors of F_m.

Can one predict from this table when more Fermat numbers will be factored? The year increases with m except for the last one. This is because F_{11} has four factors small enough to discover easily by ECM, while F_9 and F_{10} have penultimate factors too big to find easily in 1989. In what year would you guess F_{12} will be completely factored? In 1996, Richard Guy bet John Conway that at least one more Fermat number would be factored completely by his 100th birthday in 2016. Richard lost that bet. Six small factors of F_{12} are known. The remaining co-factor is composite with 1133 decimal digits. This number is too large to factor by QS or NFS. The only way it will be factored soon by a known method is if it has just one very large prime factor and we can discover its small prime factors by ECM. Six different algorithms were used to factor the seven numbers. Perhaps someone will invent a new algorithm to finish F_{12}.

A new factoring algorithm was invented about every five years from 1970 to 1995, as shown in Table 3.3 taken from [611]. Each of these algorithms did something that previous algorithms could not do.

Why have we found no new faster factoring algorithms since 1995? Many have tried to find new ones. People have fashioned variations of QS, ECM and NFS that lead to faster programs for these algorithms. The improved

Year	Method	Who
1970	CFRAC	Morrison and Brillhart
1975	Pollard $p-1$ and rho	Pollard
1980	Quadratic sieve	Pomerance
1985	ECM	H. W. Lenstra, Jr
1990	SNFS	Pollard and Lenstra
1995	GNFS	Pollard and Lenstra

Table 3.3 *The evolution of factoring methods.*

algorithms are much faster than the first versions, but they still have the same asymptotic time complexity as the first version. Have we already discovered the fastest integer-factoring algorithms? I doubt it.

We need a new factoring algorithm. The time complexities of the fastest known algorithms have the form

$$\exp\left(c(\ln n)^t (\ln \ln n)^{1-t}\right), \tag{3.10}$$

for some constants $c > 0$ and $0 < t < 1$. For QS and ECM, $t = 1/2$; for NFS, $t = 1/3$. The time complexity has this shape because these algorithms have to find one or more smooth numbers, and the density of such numbers is stated in Dickman's [157] theorem. Any new factoring algorithm that works by finding smooth numbers probably would have time complexity of the form (3.10). A polynomial time factoring algorithm likely would not rely on smooth numbers.

We suggest ideas for discovering new, faster ways to factor large numbers.

(1) Combine the number field sieve and lattices. Sieve the NFS polynomial on the points of a lattice in the plane. C.-P. Schnorr [514, 516] has taken the first steps in this direction.
(2) The fastest general algorithms, the QS and GNFS, solve $x^2 \equiv y^2 \pmod{n}$ and use Theorem 3.7. Find a way to construct congruent squares modulo n faster than by using smooth numbers to build relations.
(3) Can you quickly find a $1 < b < n$ so that n is a pseudoprime to base b, but not a strong pseudoprime to base b, and apply Theorem 3.14?
(4) Factoring algorithms exist with running time (3.10) with $t = 1/2$ and $1/3$. Find one with $t = 1/4$. Note that this t is the reciprocal of 1 plus the number of parameters that you can adjust to 'tune' the algorithm. CFRAC, QS and NFS have factor base size as one parameter; NFS adds the degree of the polynomial as a second parameter. ECM has the parameter B.
(5) Find an algorithm to answer this question quickly. Given integers $1 < B < n$, does n have a factor $p \le B$? If you had a fast algorithm to answer this

yes/no question, then you could use it $\log_2 n$ times in a binary search for the least prime factor of n. Note that when $\lfloor \sqrt{n} \rfloor \leq B < n$, the question just asks whether n is composite, and Agrawal, Kayal and Saxena [9] answered that question in polynomial time in 2002. See Section 3.1.4.

(6) Build a quantum computer and use Shor's method. In 1994, Shor [538] gave an algorithm for factoring integers in polynomial time using a quantum computer. Basically, Shor's quantum computer computes a Fourier transform with a very sharp peak at $\varphi(n)$. When you observe this transform you will learn $\varphi(n)$. Then use Theorem 3.15.

(7) Simulate Shor's method on a conventional computer. Simulate each qubit by a real variable between 0 and 1. Use non-linear optimisation methods (as in [155]) to locate the peak in the transform to find $\varphi(n)$.

(8) If none of the above works, then try to prove that factoring is hard. Remember Shamir's $O(\log n)$-step algorithm of Section 3.7.1. Can you even prove that there is a constant $c > 0$ such that when factoring some numbers n on a computer with fixed word size, at least $c \log^2 n$ steps are required?

If quantum computers can be made to work, then factoring integers will become easy. A secure RSA public modulus has more than 2000 bits (see also Table 11.2). A quantum computer using Shor's method to factor such a number would require a few million entangled qubits. At present, we can build quantum computers with no more than a few dozen entangled qubits. One problem with building larger quantum computers is decoherence, in which ambient energies observe the entangled qubits before the computation ends. I cannot guess whether this problem will be solved.

Let me go out on a limb and predict that the number of qubits in quantum computers will increase gradually. I also predict that mathematicians will invent slightly faster, but not polynomial time, factoring algorithms. As these new methods are programmed and as quantum computers gradually improve, there will be a slow increase over the next century in the size of integers that humans can factor. This is what has happened during the past 50 years with the Cunningham Project [111], which factors numbers of the form $b^n \pm 1$. The smallest interesting number of this form that remains to be factored has increased smoothly from 30 decimal digits 50 years ago to about 200 digits now, the result of faster algorithms and faster computers.

4

Lattice-Based Integer Factorisation: An Introduction to Coppersmith's Method

Alexander May

'I love this Coppersmith stuff, taking polynomials and shifting them around, it is so real! Way better than this cryptography meta-reduction non-sense.'
– Arjen Lenstra, personal communication at Asiacrypt 2007

Abstract

Coppersmith's method is used to find unknowns in polynomial equations. In this chapter, our polynomial equations stem from cryptanalysis problems related to the RSA cryptosystem and integer factorisation with the cryptographic secrets modelled as unknowns, but Coppersmith's method can be applied in a much more general context (see, e.g., [77, 128], just to mention a couple).

The beauty of Coppersmith's method comes from exploring exponentially sized search spaces, while preserving polynomial time using the famous Lenstra–Lenstra–Lovász (LLL) lattice reduction algorithm [380]. In Coppersmith-type literature, researchers are mainly focussing on maximising the limits of the search space using an asymptotic lattice dimension analysis. This makes it quite cumbersome to follow a Coppersmith-type analysis for non-experts.

In this chapter, we follow a different approach to make the method more accessible to newcomers. We review some of the most famous applications of Coppersmith's approach, such as RSA attacks with known parts of the message, Håstad's broadcast attack, factoring with known bits, certification of RSA keys, factorisation via the RSA secret key, and RSA small secret-key attacks. Instead of focussing on the method's full asymptotic extent, we provide small lattice bases that illustrate the applications, which can be easily and efficiently implemented. We hope that this approach results in a larger target audience, including security engineers, lecturers and students, and allows them to experiment with and study the beauty of Coppersmith's LLL-based method and its applications.

4.1 Introduction to Coppersmith's Method

In 1996, Coppersmith proposed two lattice-based methods for finding small roots of polynomial equations, one method for polynomial equations over the integers [137] and one for modular polynomial equations [136]. In the following, we mainly focus on the latter modular technique, but in Section 4.5 we also discuss the relation to the integer technique.

Coppersmith's method is inspired by lattice-techniques from Håstad [262] and Girault, Toffin and Vallée [222] but contrarily to these methods, Coppersmith's method has the benefit that it provides provable guarantees to find all roots smaller than a certain bound X in polynomial time in the bit-length of, e.g., the modulus of the polynomial equation. The polynomial run time is especially fascinating for two reasons.

First, the root bound X is in general of exponential size in the bit-length of the modulus. That is, we are able to find all roots in an interval of exponential size. Put differently, we are able to scan a search space as large as a polynomial fraction of the modulus in polynomial time.

Second, the use of LLL reduction [380] is sufficient, whereas former methods [262] often relied on finding shortest lattice vectors, which is in general a hard problem [10]. Hence, the efficiency of Coppersmith's methods is completely inherited from the efficieny of the famous Lenstra–Lenstra–Lovász algorithm [380, 551]. In fact, using stronger lattice reduction instead of LLL would not increase the bound X up to which roots can be found.

Applications The first applications of Coppersmith's method given in the original works [136, 137, 138] already impressively demonstrated the technique's power. Using the modular technique [136], Coppersmith showed that one can invert the RSA function $x \mapsto x^3 \bmod N$ in polynomial time given only a $\frac{2}{3}$ fraction of the bits of x (see Section 4.3.1). This has crucial implications for proving RSA security [78, 543], because under the assumption that inverting the RSA function is hard the Coppersmith result implies that even recovering a $2/3$-fraction must be hard.

By using the integer technique [137], Coppersmith showed that, for an RSA modulus $N = pq$, having primes p, q of the same bit-size, recovering half of the bits of p leads to polynomial factorisation (see Section 4.3.3, and Lenstra [367] for an application). This result was used in many practical settings. In 2012,

Lenstra *et al.* [386] showed that improper practical use of randomness for generating p, q is a severe real-world issue, by factoring many public RSA moduli by just doing gcd computations. This gcd-attack was later extended to nearby gcds using Coppersmith's attack [52, 451] and is known as 'ROCA', short for the return of Coppermith's attack (see also Section 6.2.3 for settings where ROCA arises.)

On the constructive side, the 'factoring with known bits' method can be used to certify correctness of RSA public keys (N, e), in the sense that one can efficiently prove that e does not divide $\varphi(N)$, see Section 4.3.3 and [307].

Certainly, the most prominent cryptanalysis application of Coppersmith's method is the improvement of Wiener's famous attack on RSA secret exponents $d < N^{\frac{1}{4}}$ [622] to $d < N^{0.292}$ by Boneh and Durfee [79] (see Section 4.4.2). Similar attacks have been designed for RSA secret exponents with small Chinese Remainder Theorem (CRT) representation $(d_p, d_q) = (d \bmod p - 1, d \bmod q - 1)$ (see Section 4.4.3 and [66, 290, 406, 583]). In 2017, Takayasu, Lu and Peng showed that one can efficiently factor N if $d_p, d_q \leq N^{0.122}$, thereby significantly improving over the previously best bound $d_p, d_q \leq N^{0.073}$ from [290].

We conclude our chapter in Section 4.5 by discussing some open questions related to the optimisation of Coppersmith's method, and by giving an outlook for the promising Coppersmith-type research direction for polynomial systems. This direction was recently used in the amazing cryptanaltic result of Xu, Sarkar, Lu, Wang and Pan [627] in 2019 that fully breaks the so-called modular inversion hidden number problem [83].

4.2 Useful Coppersmith-Type Theorems

Assume that we receive as input a polynomial $f(x)$ of degree δ over the ring $\mathbb{Z}_N$ for some integer N of unknown factorisation, and we want to find all roots of $f(x)$ in a certain interval. The description length of f, N is $\Theta(\delta \log N)$. Without loss of generality, we may assume that f is monic, otherwise we multiply by the inverse coeffient of the leading monomial x^δ modulo N. If this inverse does not exist, then we can factor N and proceed recursively.

4.2.1 Idea of Coppersmith's Method

Coppersmith's method constructs from $f(x)$ a polynomial $g(x)$ of usually larger degree such that every small modular root x_0 of f, i.e.,

$$f(x_0) = 0 \bmod N \text{ with } |x_0| < X \, ,$$

is also a root of g over $\mathbb{Z}$. Thus, we reduce modular univariate modular root finding to integer univariate root finding, for which we have standard methods.

Let us fix some integer $m \in \mathbb{Z}$. We construct g as an integer linear combination of multiples of

$$h_{i,j} = x^j N^i f^{m-i}(x).$$

Notice that every root x_0 of f satisfies $h_{i,j}(x_0) = 0 \bmod N^m$. Hence if g is an integer linear combination of the $h_{i,j}$s then we have $g(x_0) = 0 \bmod N^m$ as well.

Let us identify the polynomials $h_{i,j}(x)$ with their coefficient vectors. The integer linear combinations of these vectors form an integer lattice L. A key observation is that small vectors in L correspond to linear combinations $g(x)$ with small coefficients. Moreover, if $g(x)$ has small coefficients, and is evaluated at small points x_0 with $|x_0| \leq X$ only, then the result must also be (somewhat) small. More precisely, assume that $g(x_0)$ is in absolute value smaller than N^m for all $|x_0| \leq X$. Then we have for all x_0 that

$$g(x_0) = 0 \bmod N^m \text{ and } |g(x_0)| < N^m.$$

These two equations together imply that $g(x_0) = 0$, since the only multiple of N^m smaller in absolute value then N^m is $0 \cdot N^m = 0$. This implies that $g(x)$ has the desired roots over the integers!

Notice that our so-far used criterion $|g(x_0)| < N^m$ for all $|x_0| \leq X$ is not efficiently checkable, since we do not know any root x_0, only their upper bound X. However, if $g(x)$ has sufficiently small coefficients, this criterion should automatically be fulfilled. The following lemma makes this intuition precise. The lemma is usually contributed to Howgrave-Graham [278], but already appeared in the work of Håstad [262]. It provides an easily testable upper bound on the norm of g's coefficient vector that is sufficient to guarantee $|g(x_0)| < N^m$.

To this end let us introduce some useful notion. Let $g(x) = \sum_{i=0}^{n} c_i x^i$ be a univariate polynomial with coefficient vector $(c_0, c_1, \ldots, c_n)$. Then the polynomial $g(xX)$ has coefficient vector $\mathbf{v} = (c_0, c_1 X, \ldots, c_n X^n)$, and we denote by $\|g(xX)\|$ the Euclidean norm of $\mathbf{v}$.

Lemma 4.1 (Håstad/Howgrave-Graham). *Let $g(x)$ be a univariate polynomial with n monomials. Let m, X be positive integers. Suppose that*

1. $g(x_0) = 0 \bmod N^m$ where $|x_0| \leq X$,
2. $\|g(xX)\| < \frac{N^m}{\sqrt{n}}$.

Then $g(x_0) = 0$ holds over the integers.

Proof   Property 2 implies

$$|g(x_0)| = \left| \sum_i c_i x_0^i \right| \le \sum_i |c_i x_0^i|$$

$$\le \sum_i |c_i| X^i \le \sqrt{n} \, \|g(xX)\| < N^m.$$

By property 1 we know that $g(x_0)$ is a multiple of N^m, and therefore $g(x_0) = 0$.   □

Whereas in Coppersmith's method we construct polynomials $g(x)$ that automatically satisfy property 1 of Lemma 4.1, the norm property 2 of Lemma 4.1 guarantees that $g(x)$ has the same small roots as $f(x)$ over the integers (but it may have additional roots). Of course, our goal is to maximise X, i.e., the range of small roots that we can efficiently recover.

4.2.2 The Crucial Role of Lenstra–Lenstra–Lovász Reduction

Let L be the lattice defined by the coefficient vectors of $h_{i,j}(x)$. The following celebrated theorem relates the length of a shortest vector in an LLL-reduced basis of L to the lattice determinant $\det(L)$, which is an invariant of L.

Theorem 4.2 (Lenstra, Lenstra, Lovász). *Let $L \in \mathbb{Z}^n$ be a lattice spanned by $B = \{\mathbf{b}_1, \ldots, \mathbf{b}_n\}$. The LLL algorithm outputs a lattice vector $\mathbf{v} \in L$ satisfying*

$$\|\mathbf{v}\| \le 2^{\frac{n-1}{4}} \det(L)^{\frac{1}{n}}$$

in time $O\left(n^6 \log^3 B_{max}\right)$, *where* $B_{max} = \max_{i,j}\{|(\mathbf{b}_i)_j|\}$ *is the largest basis entry.*

Neumaier and Stehlé [452] showed that the same output quality as in Theorem 4.2 can be achieved in run time

$$O\left(n^{4+\epsilon} \log^{1+\epsilon} B_{max}\right) \text{ for any constant } \epsilon > 0. \tag{4.1}$$

Notice that by Minkowski's first theorem [249], every lattice L contains a non-zero vector $\mathbf{v}' \le \sqrt{n} \det(L)^{\frac{1}{n}}$. Hence, the smallest vector found by LLL reduction may be longer by an exponential factor (in the dimension n) than the shortest lattice vector. This is, however, perfectly fine for Coppersmith's method. Recall that a vector $\mathbf{v}$ in a Coppermith-type lattice corresponds to some polynomial $g(x)$, for which we have to satify the second property of Lemma 4.1: $\|g(xX)\| < \frac{N^m}{\sqrt{n}}$. Thus, the LLL output vector $\mathbf{v}$ is short enough if

$$\|\mathbf{v}\| \le 2^{\frac{n-1}{4}} \det(L)^{\frac{1}{n}} < \frac{N^m}{\sqrt{n}}.$$

We will see that in Coppersmith's method we have $\det(L) = N^{\Theta(\mathsf{m})}$ with $\mathsf{m} \approx \log N$. Thus, for sufficiently large N, the terms $2^{\frac{n-1}{4}}$ and $\sqrt{n}$ can be neglected, which leads to the simplified so-called enabling condition

$$\det(L) \leq N^{mn}, \tag{4.2}$$

that plays a crucial role in all constructions using Coppersmith's method. The enabling condition is used to optimise X. Namely, we have to define a collection of polynomials $h_{i,j}$ such that their coefficient vectors span a lattice L with $\det(L)$ as small as possible.

4.2.3 Coppersmith-Type Theorems

We are now ready to formulate Coppersmith's theorem for univariate polynomials.

Theorem 4.3. *Let N be an integer of unknown factorisation. Let $f(x)$ be a univariate monic polynomial of constant degree δ. Then we can find all solutions x_0 of the equation*

$$f(x) = 0 \bmod N \quad with \quad |x_0| \leq N^{\frac{1}{\delta}}$$

in time $O\left(\log^{6+\epsilon} N\right)$ for any $\epsilon > 0$.

Proof   Let us just briefly sketch the proof; a full proof can be found in [408]. Choose $\mathsf{m} \approx \frac{\log N}{\delta}$ and define the collection of polynomials

$$h_{i,j}(x) = x^j N^i f(x)^{\mathsf{m}-i} \text{ for } 0 \leq i < \mathsf{m}, 0 \leq j < \delta.$$

It is not hard to see that the coefficient vectors of $h_{i,j}(xX)$ form an $n = \mathsf{m}\delta \approx \log N$-dimensional lattice basis B with $\det(L) = \det(B) \approx N^{\frac{\delta \mathsf{m}^2}{2}} X^{\frac{n^2}{2}}$. Hence the enabling condition from Eq. (4.2) translates to

$$N^{\frac{\delta \mathsf{m}^2}{2}} X^{\frac{n^2}{2}} \leq N^{mn}.$$

Using $n = \mathsf{m}\delta$ this is equivalent to $X^{\delta^2 \mathsf{m}^2} \leq N^{\delta \mathsf{m}^2}$, from which we easily derive the desired root bound $X \leq N^{\frac{1}{\delta}}$.

It remains to show the run time. We work in an $n \approx \log N$-dimensional lattice with largest entries of bit-size $\log B_{max} = O\mathsf{m} \log N = O\left(\log^2 N\right)$. Using the runtime of the Neumeier–Stehlé LLL variant of Eq. (4.1) lattice reduction runs in time $O\left(\log^{6+\epsilon} N\right)$.   □

Extending Coppersmith's Bound As already pointed out in Coppersmith's original work, any small root bound X can be be extended to cX for some real number c at the expense of an additional run time factor of c. It is an important open question whether this can be improved, or whether such a linear run time factor is unavoidable.

Theorem 4.4. *Let N be an integer of unknown factorisation and $c \geq 1$. Let $f(x)$ be a univariate monic polynomial of constant degree δ. Then we can find all solutions x_0 of the equation*

$$f(x) = 0 \bmod N \quad \text{with} \quad |x_0| \leq cN^{\frac{1}{\delta}}$$

in time $O\left(c \log^{6+\epsilon} N\right)$ for any $\epsilon > 0$.

Proof Split the interval $[-cN^{\frac{1}{\delta}}, cN^{\frac{1}{\delta}}]$ in c sub-intervals of size each $2N^{\frac{1}{\delta}}$, centered at some x_i. For each sub-interval with centre x_i, we apply Theorem 4.3 to find all roots within this sub-interval. $\qquad\square$

Theorem 4.4 has immediate consequences for cryptographic attacks (see Section 4.3), as it easily allows to extend Coppersmith-type root bounds.

Coppersmith's Method for an Unknown Divisor Somewhat surprisingly, one can also extend Coppersmith's method to find roots of $f(x)$ modulo b where $b \geq N^{\beta}$ is an unknown divisor of N. In principle, we keep the same strategy as before and simply work modulo b instead of N. For example, $h_{i,j} = x^j N^i f(x)^{m-i}$ now has small roots modulo b^m. Moreover, the simplified enabling condition from Eq. (4.2) becomes

$$\det(L) \leq N^{\beta mn}. \tag{4.3}$$

Working out the details yields the following theorem that was already stated in Coppersmith [136, 138] and Howgrave-Graham [278] for polynomial degree $\delta = 1$, and first appeared in its full form in [408].

Theorem 4.5. *Let N be an integer of unknown factorisation, which has a divisor $b \geq N^{\beta}$, $0 < \beta \leq 1$. Let $c \geq 1$, and let $f(x)$ be a univariate monic polynomial of constant degree δ. Then we can find all solutions x_0 of the equation*

$$f(x) = 0 \bmod b \quad \text{with} \quad |x_0| \leq cN^{\frac{\beta^2}{\delta}}$$

in time $O\left(c \log^{6+\epsilon} N\right)$ for any $\epsilon > 0$.

Notice that Theorem 4.4 is a special case of Theorem 4.5 with $\beta = 1$. For $\beta < 1$, the small root bound decreases polynomially with exponent β^2.

4.3 Applications in the Univariate Case

4.3.1 Inverting the RSA Function with Partial Knowledge

Small Messages Assume we work with RSA with small public encryption exponent e. As a simple example take $e = 3$, and let $c = m^3 \bmod N$ be an RSA ciphertext with $m < N^{\frac{1}{3}}$. Then $m^3 < N$ and therefore $c = m^3$ holds over the integers. Thus, we may compute $m = c^{\frac{1}{3}}$ over $\mathbb{Z}$ by standard efficient root finding methods such as Newton iteration.

Let us now formulate the same problem in terms of Coppersmith's method. Define the polynomial $f(x) = x^3 - c \bmod N$. Then an application of Theorem 4.3 yields that we find all roots $m < N^{\frac{1}{3}}$ in polynomial time. In fact, there is nothing to do, since Coppersmith's method constructs a polynomial equation with the same roots over the integers, but $f(x)$ already has the desired root over the integers.

The crucial advantage of Coppersmith's method is that it also covers the inhomogeneous case, in which the pre-image m of the RSA function is not small, but we know it up to an additive term of size at most $N^{\frac{1}{3}}$ (or $N^{\frac{1}{e}}$ in general). This inhomogeneous case is addressed in the following.

Stereotyped Messages Let us start with an illustrative example solvable in small lattice dimension. Let $c = m^3 \bmod N$, where we know an approximation m_1 of the message up to an additive error of size at most $m - m_1 < N^{\frac{5}{21}}$, i.e.,

$$m_0 = m - m_1 \text{ for some unknown } m_0 < N^{\frac{5}{21}}.$$

One may think of m_1 as a $\frac{16}{21}$-fraction of the most significant bits. Our goal is to recover the $\frac{5}{21}$-bit fraction m_0 efficiently with Coppersmith's method.

Example: Stereotyped Messages

Given: $c = m^3 \bmod N$ and some m_1 satisfying $m - m_1 < N^{\frac{5}{21}}$

Polynomial: $f(x) = (x + m_1)^3 - c \bmod N$ with root $x_0 = m - m_1 < N^{\frac{5}{21}}$

Parameters: degree $\delta = 3$

Lattice Basis Define $m = 2$, i.e., all polynomials have root x_0 modulo N^2. Define the collection of seven polynomials

$$N^2, \ N^2 x, \ N^2 x^2, \ N f(x), \ x N f(x), \ x^2 N f(x), \ f^2(x).$$

Let $X = N^{\frac{5}{21}}$ and $h_1(x), \ldots, h_7(x)$ denote the above collection. The coefficient vectors of $h_i(xX)$, $1 \le i \le 7$, define the following lattice basis

$$B = \begin{pmatrix} N^2 & & & & & & \\ & N^2X & & & & & \\ & & N^2X^2 & & & & \\ N(m_1^3-c) & 3Nm_1^2X & 3Nm_1X^2 & NX^3 & & & \\ & N(m_1^3-c)X & 3Nm_1^2X^2 & 3Nm_1X^3 & NX^4 & & \\ & & N(m_1^3-c)X^2 & 3Nm_1^2X^3 & 3Nm_1X^4 & NX^5 & \\ (m_1^3-c)^2 & 6(m_1^3-c)m_1^2X & (6(m_1^3-c)m_1+9m_1^4)X^2 & (20m_1^3-2c)X^3 & 15m_1^2X^4 & 6m_1X^5 & X^6 \end{pmatrix}.$$

Lattice basis B spans an $n = 7$-dimensional lattice L with $\det(L) = |\det(B)| = N^9X^{21}$. Using the enabling condition from Eq. (4.2), we obtain

$$N^9X^{21} \leq N^{2\cdot7} \text{ which implies } X \leq N^{\frac{5}{21}}.$$

Using the fast Neumaier–Stehlé variant of LLL reduction from Eq. (4.1), we recover the lower $\frac{5}{21} \approx 0.238$-fraction of m in almost linear time $O\left(\log^{1+\epsilon} N\right)$.

An application of Theorem 4.4 yields a superior bound for the bits that one can recover at the cost of an increased running time.

Theorem 4.6. *Let $c' = m^e \bmod N$ with constant e. Assume we know some m_1 satisying $m - m_1 < cN^{\frac{1}{e}}$ for some $c \geq 1$. Then m can be found in time $O\left(c\log^{6+\epsilon} N\right)$.*

Proof   Theorem 4.4 yields that in time $O\left(c\log^{6+\epsilon} N\right)$ we can recover all roots $x_0 = m - m_1$ of size

$$|x_0| \leq cN^{\frac{1}{\delta}} = cN^{\frac{1}{e}}.$$

□

Theorem 4.6 implies that, e.g., for RSA exponent $e = 3$ given a $\frac{2}{3}$-fraction of the pre-image m we find the remaining $\frac{1}{3}$-fraction in polynomial time. Maybe somewhat surprisingly, this has important applications for proving RSA security [78, 543]. Namely, if we assume that it is hard to invert the RSA function $m \mapsto m^e \bmod N$, then by Theorem 4.6 it must already be hard to recover an $\frac{e-1}{e}$-fraction of m.

4.3.2 Systems of Univariate Polynomial Equations

Polynomially Related Messages   The following is known as Håstad's RSA broadcast attack [262]. Assume that we obtain three textbook RSA ciphertexts c_1, c_2, c_3 under three different RSA public keys $(N_1, 3), (N_2, 3), (N_3, 3)$ for the same message m (without randomised padding function [44]). In other words, we obtain the following system of congruences

$$c_1 = m^3 \bmod N_1$$
$$c_2 = m^3 \bmod N_2$$
$$c_3 = m^3 \bmod N_3.$$

Let $N = \min_i\{N_i\}$, and let us assume $m < N$, which implies $m^3 < N^3$. Let us further assume without loss of generality that the public moduli N_i are pairwise co-prime. By the Chinese Remainder Theorem (CRT) we can easily compute some

$$c = \mathrm{CRT}(c_1, c_2, c_3) = m^3 \bmod N_1 N_2 N_3.$$

Since $m^3 < N^3 < N_1 N_2 N_3$, we know that $c = m^3$ holds over the integers. Thus, we can simply compute $m = c^{\frac{1}{3}}$.

Now assume that we obtain the following system of congruences

$$c_1 = m^2 \bmod N_1$$
$$c_2 = m^3 \bmod N_2 \qquad\qquad (4.4)$$
$$c_3 = m^3 \bmod N_3.$$

Strictly speaking $c_1 = m^2 \bmod N_1$ is not a textbook RSA ciphertext (since squaring is not a bijection), but we may ignore that, since m is uniquely defined by the above system. Because the right-hand side of the equations in Eq. (4.4) take different values, we may no longer simply compute $\mathrm{CRT}(c_1, c_2, c_3)$ as in Håstad's broadcast attack.

Instead, we define the following degree-6 polynomials

$$f_1(x) = (x^2 - c_1)^3, \; f_2(x) = (x^3 - c_2)^2, \; f_3(x) = (x^3 - c_3)^2$$

that all share the same root $x_0 = m$ modulo N_1^3, N_2^2, respectively N_3^2. Via the Chinese Remainder Theorem we now compute a degree-6 polynomial $f = \mathrm{CRT}(f_1, f_2, f_3)$ that has the root $x_0 = m$ modulo $N_1^3 N_2^2 N_3^2$.

Example: Polynomially Related Messages

Given: equations from (4.4). Let $M = N_1^3 N_2^2 N_3^2$ and $N = \min_i\{N_i\}$

Polynomial: $f(x)$ of degree 6 with root $x_0 = m$ modulo M, where $m < N$

Parameters: degree $\delta = 6$

Lattice Basis Let $m = 6$, i.e., all polynomials have root x_0 modulo M^6. We define the following collection of 37 polynomials

$$
\begin{array}{cccccc}
M^6, & M^6 x, & M^6 x^2, & M^6 x^3, & M^6 x^4, & M^6 x^5, \\
M^5 f(x), & M^4 x f(x), & M^5 x^2 f(x), & M^5 x^3 f(x), & M^5 x^4 f(x), & M^5 x^5 f(x), \\
M^4 f^2(x), & M^4 x f^2(x), & M^4 x^2 f^2(x), & M^4 x^3 f^2(x), & M^4 x^4 f^2(x), & M^4 x^5 f^2(x), \\
M^3 f^3(x), & M^3 x f^3(x), & M^3 x^2 f^3(x), & M^3 x^3 f^3(x), & M^3 x^4 f^3(x), & M^3 x^5 f^3(x), \\
M^2 f^4(x), & M^2 x f^4(x), & M^2 x^2 f^4(x), & M^2 x^3 f^4(x), & M^2 x^4 f^4(x), & M^2 x^5 f^4(x), \\
M f^5(x), & M x f^5(x), & M x^2 f^5(x), & M x^3 f^5(x), & M x^4 f^5(x), & M x^5 f^5(x), \\
f^6(x). & & & & &
\end{array}
$$

Let $h_1(x), \ldots, h_{37}(x)$ be the above collection, and let $X = N$. Then the coefficient vectors of $h_1(xX), \ldots, h_{37}(xX)$ define a lattice basis B of some lattice L with $\dim(L) = 37$ and

$$
\det(L) = |\det(B)| = M^{6 \sum_{i=1}^{6} i} X^{\sum_{i=1}^{36} i} = M^{126} X^{666}.
$$

Thus, the enabling condition from Eq. (4.2) yields

$$
\det(L) \le M^{6 \cdot 37} \text{ which is equivalent to } X \le M^{\frac{16}{111}}.
$$

Since $M = N_1^3 N_2^2 N_3^2 > N^8$, we have $M^{\frac{16}{111}} > N^{\frac{128}{111}} > N$. Thus, we find in time $O\left(\log^3 N\right)$ via LLL-reduction (or time $O\left(\log N^{1+\epsilon}\right)$ via Eq. 4.1) the unique $m < N = \min_i \{N_i\}$.

Using Theorem 4.3 we can show a slightly stronger and more general result.

Theorem 4.7 ([409]). *Let $N_1, \ldots, N_k$ be pairwise co-prime RSA-moduli and let $m < \min_i \{N_i\}$. Let $g_i(x)$, $i = 1, \ldots, k$, be polynomials of degree δ_i. Assume we are given*

$$
c_i = (g_i(m))^{e_i} \bmod N_i \text{ satisfying } \sum_{i=1}^{k} \frac{1}{\delta_i e_i} \ge 1.
$$

Then m can be computed in time $O\left(\log^{6+\epsilon} N\right)$ for all $\epsilon > 0$.

Proof Without loss of generality we assume that all $g_i(m)$ are monic. Otherwise, we multiply by the inverse of their leading coefficient. If this inverse computation fails, we obtain the factorisation of some N_i, which in turns enables us to compute m.

We define $\delta = \text{lcm}_i \{\delta_i e_i\}$ as the least common multiple. Furthermore, we define the degree-δ polynomials

$$
f_i(x) = (g_i(x)^{e_i} - c_i)^{\frac{\delta}{\delta_i e_i}} \text{ with root } f_i(m) = 0 \bmod N_i^{\frac{\delta}{\delta_i e_i}}.
$$

Let $M = \prod_{i=1}^{k} N_i^{\frac{\delta}{\delta_i e_i}}$ be the product of all moduli. Via Chinese Remaindering we compute the polynomial

$$f(x) = CRT(f_1, \ldots, f_k) = \sum_{i=1}^{k} b_i f_i(x) \bmod M, \text{ where } b_i \bmod N_j = \begin{cases} 1 \text{ if } i = j \\ 0 \text{ else} \end{cases}.$$

An application of Theorem 4.3 shows that we can find in time $O\left(log^{6+\epsilon}N\right)$ all roots x_0 smaller than

$$|x_0| \leq M^{\frac{1}{\delta}}.$$

Using our condition $1 \leq \sum_{i=1}^{k} \frac{1}{\delta_i e_i}$, we know that our desired root m satisfies

$$m < \min_i\{N_i\} \leq \left(\min_i\{N_i\}\right)^{\sum_{i=1}^{k} \frac{1}{\delta_i e_i}} \leq \prod_{i=1}^{k} N_i^{\frac{1}{\delta_i e_i}} = M^{\frac{1}{\delta}}.$$

Thus, we recover the desired root via Coppersmith's method. $\qquad\square$

An important special case of Theorem 4.7 is when all $g_i(x) = x$ and therefore $\delta_i = 1$.

Theorem 4.8 ([409]). *Let $N_1, \ldots, N_k$ be pairwise co-prime RSA-moduli and let $m < \min_i\{N_i\}$. Assume we are given*

$$c_i = m^{e_i} \bmod N_i \text{ satisfying } \sum_{i=1}^{k} \frac{1}{e_i} \geq 1.$$

Then m can be computed in time $O\left(log^{6+\epsilon} N\right)$ for all $\epsilon > 0$.

Let us apply Theorem 4.8 to the equation system of Eq. (4.4). Since $e_1 = 2, e_2 = e_3 = 3$ we have $\frac{1}{2} + \frac{1}{3} + \frac{1}{3} = \frac{7}{6} \geq 1$. Thus, we can find m in time $O\left(log^{6+\epsilon} N\right)$. Recall that the analysis at the beginning of the section even allowed run time $O\left(log^{1+\epsilon} N\right)$. The reason is that $\sum_i \frac{1}{e_i}$ is significantly larger than 1, which in turn allows for a smaller lattice basis, and therefore also for superior run time.

4.3.3 Factoring with Known Bits and RSA Key Certification

Let $N = pq$ be an RSA modulus, without loss of generality $p > q$ and therefore $p > N^{\frac{1}{2}}$. Assume that we know a good approximation p_1 of p up to an additive term of size at most $N^{\frac{1}{5}}$, i.e., we know some p_1 such that

$$p_0 = p - p_1 \text{ for some unknown } p_0 < N^{\frac{1}{5}}.$$

One may think of p_1 as a $\frac{4}{5}$-fraction of the most significant bits of p. Our task is to recover the $\frac{1}{5}$ least significant bits p_0.

Example: Factoring with Known Bits

Given: $N = pq$, $p > N^{\frac{1}{2}}$ and some p_1 satisfying $p - p_1 < N^{\frac{1}{5}}$

Polynomial: $f(x) = x + p_1 \bmod p$ with root $x_0 = p - p_1 < N^{\frac{1}{5}}$

Parameters: degree $\delta = 1$, divisor size $\beta = \frac{1}{2}$

Lattice Basis Let $m = 2$, i.e., all polynomials have root x_0 modulo p^2. We define the collection of five polynomials

$$N^2, \ Nf(x), \ f^2(x), \ xf^2(x), \ x^2 f^2(x).$$

Let $X = N^{\frac{1}{5}}$ and $h_1(x), \ldots, h_5(x)$ denote the above collection. Then the coefficient vectors of $h_i(xX)$, $i = 1, \ldots, 5$, form the lattice basis

$$B = \begin{pmatrix} N^2 & & & & \\ Np_1 & NX & & & \\ p_1^2 & 2p_1X & X^2 & & \\ p_1^2X & 2p_1X^2 & X^3 & \\ p_1^2X^2 & 2p_1X^3 & X^4 \end{pmatrix}. \tag{4.5}$$

Notice that $n = \dim(L) = 5$ and $\det(L) = N^3 X^{10}$. Thus the enabling condition from Eq. (4.3) yields

$$N^3 X^{10} \leq N^{\beta m n} = N^5 \text{ which implies } X \leq N^{\frac{1}{5}}.$$

The LLL reduction runs on lattice basis B in time $O\left(\log^3 N\right)$ (and in $O\left(\log^{1+\epsilon} N\right)$ using Eq. (4.1)).

Using Theorem 4.5, we may increase the bound $N^{\frac{1}{5}}$ at the cost of an increased run time.

Theorem 4.9. *Let N be composite with divisor $p > N^\beta$. Assume we are given p_1 satisfying $p - p_1 < cN^{\beta^2}$ for some $c \geq 1$. Then p can be found in time $O\left(c \log^{6+\epsilon} N\right)$ for any $\epsilon > 0$.*

Proof Theorem 4.5 implies that in time $O\left(c \log^{6+\epsilon} N\right)$ we find all $x_0 = p - p_1$ of size

$$|x_0| \leq cN^{\frac{\beta^2}{\delta}} = cN^{\beta^2}.$$

$\square$

Let $N = pq$ be an RSA modulus with $N^{\frac{1}{2}} < p < 2N^{\frac{1}{2}}$, i.e., $\beta = \frac{1}{2}$. Theorem 4.9 implies that we can factor N in polynomial time given half of the bits of p, i.e., $p - p_1 < N^{\frac{1}{4}} < p^{\frac{1}{2}}$.

Moreover, if we set $p_1 = 0$ and $c = 2N^{\frac{1}{4}}$ then Theorem 4.9 gives us a deterministic factorisation algorithm for RSA moduli with running time $\tilde{O}\left(N^{\frac{1}{4}}\right)$, where the $\tilde{O}$-notion suppresses polylogarithmic factors.

Certifying RSA For the correctness of many RSA-based protocols, we have to guarantee that the RSA parameters (N, e) are properly chosen. This means in particular that the RSA function $m \mapsto m^e \bmod N$ is a bijection, which is true if and only if

$$\gcd(e, \varphi(N)) = 1.$$

However, the co-primality of e and the Euler totient function $\varphi(N)$ is in general not publically checkable, since the knowledge of $\varphi(N)$ implies the factorisation of N (at least for RSA moduli).

However, let us for the moment assume that N is an RSA modulus, and that $e \geq \sqrt{2}N^{\frac{3}{10}}$ is prime. Primality of e can easily be checked in polynomial time [431]. Because $N = pq$ we have $\varphi(N) = (p - 1)(q - 1)$. By primality of e we conclude that $\gcd(e, \varphi(N)) \in \{1, e\}$. Therefore, (N, e) does not define a bijective RSA function if and only if $\gcd(e, \varphi(N)) = e$, in which case we have $e|p - 1$ or $e|q - 1$. Without loss of generality, let us assume $e|p - 1$, i.e., there exists some unknown $k \in \mathbb{N}$ with

$$p = ek + 1. \tag{4.6}$$

Without loss of generality we may also assume $\gcd(e, N) = 1$. On input e, N the extended Euclidean algorithm outputs Bézout coefficients $r, s \in \mathbb{Z}$ satisfying $er + Ns = 1$. Multiplying Eq. (4.6) by r yields

$$pr = ekr + r = (1 - Ns)k + r.$$

Using $N = pq$, we obtain

$$p(r + qsk) = k + r.$$

Thus, the polynomial $f(x) = x + r$ satisfies $f(x_0) = 0 \bmod p$ with root $x_0 = k = \frac{p-1}{e}$. Notice that $r = e^{-1} \bmod N$ plays the role of p_1 from the previous 'factoring with known bits' approach. Namely, r is an approximation of some (unknown) multiple of p. Because N is an RSA modulus, we know that $\frac{1}{2}p < q$. Multiplication by p gives $p < \sqrt{2N}$. Thus, we obtain

$$x_0 = k = \frac{p-1}{e} < \frac{\sqrt{2N}}{\sqrt{2}N^{\frac{3}{10}}} = N^{\frac{1}{5}}. \tag{4.7}$$

> **Example: Certification of RSA Parameters**
> **Given:** $N = pq$ and some prime $e \geq \sqrt{2}N^{\frac{3}{10}}$ with $e|p-1$
> **Polynomial:** $f(x) = x + (e^{-1} \bmod N) \bmod p$ with root $x_0 = \frac{p-1}{e} < N^{\frac{1}{5}}$
> **Parameters:** degree $\delta = 1$, divisor size $\beta = \frac{1}{2}$

Lattice Basis Setting $p_1 = e^{-1} \bmod N$, we can directly use lattice basis B from Eq. (4.5). Thus, we may in time $O\left(\log^3 N\right)$ via LLL (or $\log^{1+\epsilon} N$ via Eq. (4.1)) compute the factorisation of N if $e|p-1$. If Coppersmith's method fails to factor N, then by the provable guarantees of Coppersmith's method we certified that $\gcd(e, \varphi(N)) = 1$, and therefore that the RSA function is indeed a bijection.

Using Theorem 4.5, our certification procedure holds for even smaller public exponents $e \geq N^{\frac{1}{4}}$.

Theorem 4.10. *Given an RSA modulus $N = pq$ and some prime $e \geq N^{\frac{1}{4}}$. Then we can decide in time $O\left(\log^{6+\epsilon} N\right)$ for any $\epsilon > 0$ whether* $\gcd(e, \varphi(N)) = 1$.

Proof We have $x_0 = k = \frac{p-1}{e} \leq \frac{\sqrt{2N}}{N^{\frac{1}{4}}} < 2N^{\frac{1}{4}}$. Thus, an application of Theorem 4.9 yields the desired result.   □

Notice that Theorem 4.10 assumes that we start with some RSA modulus $N = pq$ that is a product of two primes p, q of equal bit-size. In fact, the proof of Theorem 4.10 builds on this property. Somewhat surprisingly, we can certify RSA parameters (N, e) for any N of unknown factorisation, even for products of more than two primes.

Theorem 4.11 ([307]). *Given a composite N and some prime $e \geq N^{\frac{1}{4}}$. Then we can decide in time $O\left(\log^{6+\epsilon} N\right)$ for any $\epsilon > 0$ whether* $\gcd(e, \varphi(N)) = 1$.

Proof Sketch It is not hard to see that in the case $\gcd(e, \varphi(N)) > 1$, we have $e|p-1$ for some prime divisor of p of N. But, as opposed to the RSA case from Theorem 4.10, we now only have a trivial upper bound $p < N$ on p. This implies that an application of Coppersmith's method finds p in time $O\left(\log^{6+\epsilon} N\right)$, provided that $x_0 = k = \frac{p-1}{e} \leq \frac{N}{N^{\frac{1}{4}}} = N^{\frac{3}{4}}$.

Assume that $p \geq N^{\sqrt{3/4}}$. Then an application of Theorem 4.5 with $\beta = \sqrt{3/4}, \delta = 1$ yields that we find x_0 if

$$|x_0| \leq N^{\frac{\beta^2}{\delta}} = N^{\frac{3}{4}},$$

as desired. In summary, we find p if and only if $p \in [N^{\sqrt{3/4}}, N]$. Otherwise we know that $p < N^{\sqrt{3/4}} \approx N^{0.866}$. Therefore, we obtain a new upper bound on p, and may apply Coppersmith's method iteratively, finding a yet decreased lower bound on p, etc.

The proof in [307] shows that, after $O(\log N)$ iterative applications of Coppersmith's method, we succeed to cover the whole interval $[e, N]$ of all possible candidates for p. If all of these iterations fail to factor N, we certified that $\gcd(e, \varphi(N)) = 1$. □

Computing the RSA Secret Key Implies Factoring Let (N, e) be an RSA public key. From the factorisation of N one can derive the Euler totient function $\varphi(N)$, and compute the RSA secret key as $e^{-1} \bmod \varphi(N)$. Thus, factoring implies computation of the RSA secret key.

A natural question is whether the converse is also true. If we are able to compute $d = e^{-1} \bmod \varphi(N)$, does this lead to an efficient factorisation algorithm? This question was already addressed in the original RSA paper [501], where the authors mentioned a probabilistic factorisation algorithm due to Miller [431]. It remained open whether there also exists a deterministic factorisation algorithm.

Let us first assume that $e \leq \frac{1}{3} N^{\frac{1}{2}}$. We will see that in this case we have a completely elementary factorisation algorithm. Assume that we are able to compute the secret key d with $2 < d < \varphi(N)$. Then there exists some unknown $k \in \mathbb{N}$ such that

$$ed = 1 + k\varphi(N).$$

Let $\tilde{k} = \frac{ed-1}{N} < k$. Then

$$k - \tilde{k} = \frac{ed - 1}{\varphi(N)} - \frac{ed - 1}{N} = \frac{(ed - 1)(N - \varphi(N))}{\varphi(N)N} > 0.$$

Since p, q are of equal bit-size, we have

$$N - \varphi(N) = p + q - 1 < 3N^{\frac{1}{2}} \text{ and } \frac{1}{2}N < \varphi(N) < N.$$

Since $e \leq \frac{1}{3}N^{\frac{1}{2}}$, we obtain

$$0 < k - \tilde{k} < \frac{\frac{1}{3}N^{\frac{1}{2}}\varphi(N) \cdot 3N^{\frac{1}{2}}}{\varphi(N)N} = 1.$$

This implies that $k = \lceil \tilde{k} \rceil$. Knowledge of k gives us $\varphi(N) = \frac{ed-1}{k}$. Solving the two equations $\varphi(N) = (p - 1)(q - 1)$ and $N = pq$ eventually yields the factorisation p, q.

What happens if $ed > N^{\frac{3}{2}}$? Let us model this case as a Coppersmith-type problem. Notice that $M = ed - 1$ is a multiple of the (unknown) $\varphi(N)$, and the polynomial

$$f(x) = N - x \bmod \varphi(N) \text{ has a small root } x_0 = p + q - 1 < 3N^{\frac{1}{2}}.$$

Notice here the analogy to 'factoring with known bits': M plays the role of N, $\varphi(N)$ plays the role of p, but this time we know by construction a good approximation N of $\varphi(N)$.

Example: Factoring via RSA Secret Key

Given: $N = pq$ and e, d satisfying $ed = 1 \bmod \varphi(N)$, $ed < (9N)^{\frac{5}{3}}$

Polynomial: $f(x) = N - x \bmod \varphi(N)$ with root $x_0 = p + q - 1 < 3N^{\frac{1}{2}}$

Parameters: degree $\delta = 1$, divisor size $\beta = \log_{ed-1} \varphi(N) \geq \frac{3}{5}$

Lattice Basis Set $m = 2$, then all polynomials have root x_0 modulo $\varphi(N)^2$. Set $M = ed - 1 = (9N)^{\alpha}$ for some $\alpha < \frac{5}{3}$. Let $\beta = \frac{1}{\alpha}$ and $X = 3N^{\frac{1}{2}}$. We define the collection of four polynomials

$$h_1(x) = M^2, \ h_2(x) = Mf(x), \ h_3(x) = f^2(x), \ h_4(x) = xf^2(x).$$

The integer linear combinations of the coefficient vectors of $h_1(xX), \ldots, h_4(xX)$ form a lattice L with basis

$$B = \begin{pmatrix} M^2 & & & \\ MN & -MX & & \\ N^2 & -2NX & X^2 & \\ N^2X & -2NX^2 & X^3 \end{pmatrix}.$$

We have $n = \dim(L) = 4$ and $\det(L) = |\det(B)| = M^3X^6$. Thus the enabling condition from Eq. (4.3) becomes $M^3X^6 \leq M^{\beta mn} = M^{\frac{8}{\alpha}}$. Using $M = ed - 1 = (9N)^{\alpha}$, this implies

$$X \leq M^{\frac{4}{3\alpha} - \frac{1}{2}} = (9N)^{\frac{4}{3} - \frac{\alpha}{2}}.$$

Because $\alpha < \frac{5}{3}$, the exponent satisfies $\frac{4}{3} - \frac{\alpha}{2} > \frac{1}{2}$. Therefore, we find the root $x_0 = p + q - 1 < 3N^{\frac{1}{2}}$ in time $O\left(\log^{1+\epsilon} N\right)$ via LLL reduction. From x_0 we easily derive the factorisation of N.

An application of Theorem 4.5 yields the following theorem.

Theorem 4.12. *Let $N = pq$ be an RSA modulus with key pair $2 < e, d < \varphi(N)$ satisfying $ed = 1 \bmod \varphi(N)$. On input N, e, d one can compute the factorisation of N in time $O\left(\log^{6+\epsilon} N\right)$ for any $\epsilon > 0$.*

Proof Define $M = ed - 1 = \varphi(N)^{\alpha}$ for some $\alpha < 2$. Let $f(x) = N - x \bmod \varphi(N)$

with root $x_0 = p + q - 1 < 3N^{\frac{1}{2}}$. Let $\delta = 1, \beta = \frac{1}{\alpha}, c = 6$. Via Theorem 4.5 we find x_0 in time $O\left(\log^{6+\epsilon} N\right)$ if

$$|x_0| \leq 6M^{\beta^2} = 6\varphi(N)^{\frac{1}{\alpha}}.$$

Since $\alpha < 2$ and $\varphi(N) > \frac{N}{2}$ we obtain

$$6\varphi(N)^{\frac{1}{\alpha}} > 6\varphi(N)^{\frac{1}{2}} > 3N^{\frac{1}{2}}.$$

Thus, Coppersmith's method succeeds to recover $x_0 = p + q - 1$, from which we derive N's factorisation.   □

4.4 Multivariate Applications: Small Secret Exponent RSA

Throughout this section we assume that e is approximately of size N, denoted $e \approx N$. If we choose some small secret key d, then $e = d^{-1} \bmod \varphi(N)$ should be randomly distributed, i.e., $e \geq \varphi(N)/2 \gg N/4$ with probability $\frac{1}{2}$. In fact, all attacks that we study in this section become even more effective (in the sense of larger achievable root bounds), if e is significantly smaller than N. For the ease of notation, we ignore the case of smaller e.

4.4.1 Wiener's Attack

In 1990, Wiener [622] discovered that, for RSA public keys (N, e) with corresponding secret $d < \frac{1}{3}N^{\frac{1}{4}}$, the factorisation of N can be found via continued fractions in time $O\left(\log^3 N\right)$. We formulate Wiener's result in terms of Coppersmith's method. We have $ed = 1 + k(N + 1 - p - q)$, and therefore the bivariate polynomial

$$f(x, y) = x(N - y) + 1 \bmod e \text{ has root } (x_0, y_0) = (k, p + q - 1).$$

We already know that for RSA moduli with primes p, q of same bit-size we have $y_0 = p + q - 1 < 3N^{\frac{1}{2}}$. Moreover, since $e < \varphi(N)$ we have

$$x_0 = k = \frac{ed - 1}{\varphi(N)} < d.$$

Notice at this point that $e \ll N$ would imply a smaller upper bound on x_0. By setting

$$u = 1 - xy \tag{4.8}$$

we linearise $f(x, y)$ as

$$f(u, x) = u + xN \bmod e \text{ with root } (u_0, x_0) = (1 - x_0 y_0, x_0) = (1 - k(p + q - 1), k).$$

Notice that $|u_0| < k(p + q - 1) < 3dN^{\frac{1}{2}}$.

Example: RSA with Small Secret Exponent d

Given: $N = pq$, e satisfying $ed = 1 \bmod \varphi(N)$ with $d < \frac{1}{3}N^{\frac{1}{4}}$

Polynomial: $f(u, x) = u + xN \bmod e$ with root

$(u_0, x_0) = (1 - k(p + q - 1), k)$

Lattice Basis Set $\mathsf{m} = 1$. Then all polynomials have root (x_0, y_0) modulo e. Define $X = \frac{1}{3}N^{\frac{1}{4}}$ and $U = 3N^{\frac{3}{4}}$. We also define the collection of polynomials

$$h_1(u, x) = ex, \quad h_2(u, x) = f(u, x).$$

The integer linear combinations of the coefficient vectors of $h_1(uU, xX)$ and $h_2(uU, xX)$ form a lattice L with basis

$$B = \begin{pmatrix} eX & \\ NX & U \end{pmatrix}.$$

We have $n = \dim(L) = 2$ and $\det(L) = eUX$. The enabling condition from Eq. (4.2) becomes

$$eUX \le e^2, \quad \text{which implies } X \le \frac{e}{3N^{\frac{3}{4}}}.$$

Assuming $e \approx N$, we obtain $X \le \frac{1}{3}N^{\frac{1}{4}}$. Thus, we recover d in time $O\left(\log^{1+\epsilon} N\right)$.

Notice that Coppersmith's method guarantees only that we find a polynomial $g(u, x) = g_0 u + g_1 x$ from an LLL-reduced shortest lattice vector such that $g(u_0, x_0) = 0$. In our bivariate case, it remains to recover the root (u_0, x_0). We conclude from $g(u_0, x_0) = 0$ that

$$g_0 u_0 = -g_1 x_0.$$

Since $\gcd(u_0, x_0) = 1$, it follows that $|g_0| = x_0$ and $-|g_1| = u_0$.

4.4.2 Boneh–Durfee Attack with Unravelled Linearisation

Let us start with the polynomial $f(u, x) = u + xN$ as in the description of Wiener's attack. Our goal is to improve on Wiener's $N^{\frac{1}{4}}$-bound.

Example: RSA with Small Secret Exponent d

Given: $N = pq$, e satisfying $ed = 1 \bmod \varphi(N)$ with $d < N^{0.256}$

Polynomial: $f(u, x) = u + xN \bmod e$ with root

$(u_0, x_0) = (1 - k(p + q - 1), k)$

Lattice Basis (First Try) Instead of choosing $m = 1$ as in Wiener's attack, we take $m = 2$ and define the following collection of polynomials with root modulo e^2

$$h_1 = e^2 x, \ h_2 = ef(u, x), \ h_3 = e^2 x^2, \ h_4 = exf(u, x), \ h_5 = f^2(u, x).$$

This leads to lattice L with basis

$$B = \begin{pmatrix} e^2 X & & & & \\ eNX & eU & & & \\ & & e^2 X^2 & & \\ & & eNX^2 & eUX & \\ & & N^2 X^2 & 2NUX & U^2 \end{pmatrix}.$$

We obtain enabling condition

$$\det(L) = e^6 U^4 X^4 \leq e^{10}.$$

This is equivalent to the enabling condition $UX \leq e$ from Wiener's attack. It seems that Wiener's bound is stable under increasing m. However, we may also include in our collection the polynomial $h_6 = y f^2(u, x)$. This polynomial has monomials

$$x^2 y, \ uxy, \ u^2 y.$$

Using Eq. (4.8) substitute each occurence of xy by $1 - u$. Such a substitution is called unravelled linearisation [270, 271], and leaves us with monomials

$$x, \ ux, \ u, \ u^2, \ u^2 y,$$

of which only the last monomial $u^2 y$ does not appear in $h_1, \ldots, h_5$.

Unfortunately, the new collection is not yet good enough for improving the $\frac{1}{4}$-bound. However, we succeed with unravelled linearisation when using larger m.

Lattice Basis (Second Try) Let us construct a similar lattice basis with the choice $m = 4$. Let $d < N^\delta$ with $\delta = 0.256$, set $X = N^\delta$ and $U = Y = N^{\frac{1}{2} + \delta}$.

We take the powers of $f^i(u, x)$ for $i = 3, 4$ and in addition the polynomial $y f^4(u, x)$. That is, we take the 10 polynomials

$$e^4 x^3, \quad e^3 x^2 f, \quad e^2 x f^2, \quad e f^3, \quad e^4 x^4,$$
$$e^3 x^3 f, \quad e^2 x^2 f^2, \quad e x f^3, \quad f^4, \quad y f^4.$$

This leads to lattice L with basis $B =$

$$
\begin{pmatrix}
e^4x^3 & & & & & & & & & \\
e^3NX^3 & e^3UX^2 & & & & & & & & \\
e^2N^2X^3 & 2e^2NUX^2 & e^2U^2X & & & & & & & \\
eN^3X^3 & 3eN^2UX^2 & 3eNU^2X & eU^3 & & & & & & \\
& & & & e^4X^4 & & & & & \\
& & & & e^3NX^4 & e^3UX^3 & & & & \\
& & & & e^2N^2X^4 & 2e^2NUX^3 & e^2U^2X^2 & & & \\
& & & & eN^3X^4 & 3eN^2UX^3 & 3eNU^2X^2 & eU^3X & & \\
& & & & N^4X^4 & 4N^3UX^3 & 10N^2U^2X^2 & 4NU^3X & U^4 & \\
N^4X^3 & 4N^3UX^2 & 10N^2U^2X & 4NU^3 & -N^4UX^3 & -4N^3U^2X^2 & -10N^2U^3X & -4NU^4 & U^4Y
\end{pmatrix}.
$$

We obtain $\dim(L) = 10$ and $\det(L) = e^{20}U^{20}X^{16}Y$. For $e \approx N$, our enabling condition becomes

$$e^{20}U^{20}X^{16}Y \le e^{40}, \text{ which implies } N^\delta \le N^{\frac{19}{74}} \approx N^{0.256}.$$

Thus, we find via lattice reduction a polynomial with root (u_0, x_0, y_0) over $\mathbb{Z}$ in time $O(\log^{1+\epsilon} N)$. Using $u = 1 - xy$ from Eq. (4.8) gives us a polynomial in x, y only. In fact, one can show that under the same enabling condition LLL reduction provides two polynomials with the desired root over the integers. We then compute $y_0 = p + q - 1$ from both polynomials via resultant computation. From y_0 we easily obtain the factorisation of N.

In general, one can improve the bound for d to $N^{1-\sqrt{\frac{1}{2}}} \approx N^{0.292}$.

Theorem 4.13 (Boneh–Durfee 1999 [79]). *Let (N, e) be an RSA modulus with $ed = 1 \bmod \varphi(N)$, $e \approx N$ and $d < N^{0.292}$. Then N can be factored in time $O(\log^{6+\epsilon} N)$ for any $\epsilon > 0$.*

Proof For a proof we directly refer to Boneh and Durfee [79]. A slightly simpler proof using the unravelled linearisation strategy from above can be found in [271].   □

If e is significantly smaller than N, then one obtains larger bounds on d in Theorem 4.13. On the other hand, if $e > N^{1.875}$ then the attack of Theorem 4.13 no longer works for any secret d.

Whether the $N^{0.292}$ bound can be improved, or whether it is optimal is one of the major open problems in Coppersmith-type cryptanalysis. Boneh and Durfee [79] argued that their polynomial has a unique solution for all $d < N^{\frac{1}{2}}$. Clearly, it is a necessary requirement for the efficiency of Coppersmith's method that there are not exponentially many solutions, since we have to output all of them. However, even a unique solution does not imply that we can efficiently find it.

In fact, there have been serious efforts to improve the Boneh–Durfee attack [42, 271, 349, 581], indicating that the attack is either optimal, or that significantly new techniques have to be developed for an improvement.

Ernst *et al.* [183] and Takayasu and Kunihiro [579, 580, 582] showed that the Boneh–Durfee attack admits a smooth extension to partial key exposure attacks, where one gets a constant fraction of d's most significant bits. If $d \le N^{0.292}$ then we need no bits at all, while with growing d we need a larger known fraction for recovering d (and the factorisation) in polynomial time. Eventually, for full size d we need all bits of d, coinciding with the result of Theorem 4.12.

4.4.3 Small CRT Exponents

Since Wiener's small d attack in 1990, it was an open problem whether there exist polynomial time attacks for RSA secret keys d with a small Chinese Remainder representation, i.e., $d_p = d \bmod p - 1$ and $d_q = d \bmod q - 1$ are both small. This is a question with practical significance, since for performance reasons almost all real-world RSA implementations compute the RSA decryption function using CRT-exponents (d_p, d_q).

The first polynomial time attack on CRT-exponents [406] was described in 2003, but only worked for RSA with (significanty) imbalanced prime factors p, q. This was improved in 2005 by Bleichenbacher and May [66], but their attack still required (less) imbalanced prime factors or small e. Here, we review the idea of the Bleichenbacher–May attack, since it is the basis for later improvements.

We start with the set of equations

$$ed_p = 1 + k(p - 1)$$
$$ed_q = 1 + \ell(q - 1). \tag{4.9}$$

and rewrite these as

$$ed_p + k - 1 = kp$$
$$ed_q + \ell - 1 = \ell q.$$

Multiplication of both equations leads to the identity

$$e^2 d_p d_q + e(d_p(\ell - 1) + d_q(k - 1)) + k\ell(1 - N) - (k + \ell - 1) = 0 \tag{4.10}$$

with four unknowns d_p, d_q, k, ℓ. Let $d_p, d_q < N^\delta$. Then k (and analogously ℓ) can be upper-bounded as

$$k = \frac{ed_p - 1}{p - 1} \le N^{\frac{1}{2}+\delta}.$$

Let us take Eq. (4.10) modulo e^2 and work with the linear polynomial equation

$$f(x, y, z) = ex + y(1 - N) - z \text{ with root} \tag{4.11}$$
$$(x_0, y_0, z_0) = (d_p(\ell - 1) + d_q(k - 1), k\ell, k + \ell - 1).$$

Choose $X = N^{\frac{1}{2}+2\delta}, Y = N^{1+2\delta}, Z = N^{\frac{1}{2}+\delta}$. We take the collection of three polynomials

$$e^2x, \ e^2y, \ f(x,y,z)$$

that lead to a lattice L with basis

$$\begin{pmatrix} e^2X & & \\ & e^2Y & \\ eX & (1-N)Y & -Z \end{pmatrix}.$$

Our enabling condition gives us

$$e^4XYZ \leq e^6, \text{ resulting in } \delta \leq 0.$$

Thus, the Bleichenbacher–May attack does not succeed for full-size e (but for e that are significantly smaller than N).

However, the identity of Eq. (4.10) was used in 2007 by Jochemsz and May [290] to show an attack on CRT exponents $d_p, d_q \leq N^{0.073}$. Unfortunately, to the best of our knowledge, dimension 30 is the smallest lattice dimension for which the Jochemsz–May attack provides a positive bound, and the attack uses Coppersmith's method over the integers (rather than the modular approach that we used in this work). Therefore, we omit a concrete lattice basis here.

Theorem 4.14 (Jochemsz–May 2007[290]). *Let (N, e) be an RSA modulus with $ed = 1 \mod \varphi(N)$ satisfying $d_p = d \mod p-1, d_q = d \mod q-1 \leq N^{0.073}$. Then the factorisation of N can be found in time $O\left(\log^{6+\epsilon} N\right)$ for any $\epsilon > 0$.*

The Jochemsz–May attack was optimised in [271] using smaller lattice bases that still asymptotically achieve the same bound $N^{0.073}$. This optimisation indicated that there is no possible improvement using solely Eq. (4.10).

However in 2017, Takayasu, Lu and Peng used all three polynomial equations from Eqs. (4.9) and (4.10) to further improve the bound to $N^{0.091}$ [583], and later even to an impressive $N^{0.122}$ [584].

Theorem 4.15 (Takayasu, Lu and Peng [584]). *Let (N, e) be an RSA modulus with $ed = 1 \mod \varphi(N)$ satisfying $d_p = d \mod p-1, d_q = d \mod q-1 \leq N^{0.122}$. Then the factorisation of N can be found in time $O\left(\log^{6+\epsilon} N\right)$ for any $\epsilon > 0$.*

4.5 Open Problems and Further Directions

4.5.1 Optimal Use of Coppersmith's Method

We have seen that Coppersmith's method leads to powerful results, where all roots within an exponentially sized search space can be found in polynomial time. In the previous sections we derived polynomial equations f, defined certain polynomial collections using algebraic shifts of f, and sometimes further optimised using back-substitution via unravelled linearisation.

All this is currently handcrafted, and it is unclear how to find optimal strategies for any of the above steps.

Initial Polynomial Selection The most important step is to define the initial polynomial(s) to work with. The results of Section 4.4.3 illustrate the importance of the polynomial selection. If one takes into account the polynomial from Eq. (4.11) only, then there seems to be no hope of improving upon the $N^{0.073}$ bound from Theorem 4.14. However, if we also take the polynomials from Eq. (4.9) then we get the significantly improved $N^{0.122}$ bound from Theorem 4.15.

In general, we are not aware of any strategy for properly selecting the initial polynomial(s). Different polynomial choices for the same problem lead to different Newton polytopes of the polynomial, which in turn lead to different optimisation in Coppersmith's method, as shown in [69].

In fact, some improved bounds in literature stem (solely) from a more clever polynomial selection.

Shift Selection, Theoretically After polynomial selection of f, one has to find an optimal collection of polynomials. This is usually a major part of the analysis of Coppersmith's method. The only strategy we are aware of is due to Jochemsz and May [289], and is based on f's Newton polytope. A polynomial's Newton polytope is the convex hull of the exponent vectors of f's monomials. For instance, if $f(x, y) = ax^2 + bxy + cy^2 + d$ then the convex hull of its exponent vectors $(2, 0), (1, 1), (0, 2), (0, 0)$ forms a triangle in $\mathbb{Z}^2$.

In a nutshell, the trade-off that has to be optimised in Coppersmith's method is that one has to introduce as many shifts as possible, while keeping the newly introduced monomials as small as possible. This is realised in the Jochemsz–May method by shifting with the points in the interior of the Newton polytope.

However, to obtain optimal bounds one also has to shift in the direction of some smaller unknown. As an example, let X, Y be upper bounds for the root x_0, y_0 of a polynomial $f(x, y)$. If $X \ll Y$ then it is beneficial to additionally shift in the x-direction.

This optimisation is by now a quite handcrafted process. Can we derive an optimal algorithm that on input of a polynomial including (parameterised) upper bounds on the desired root outputs an optimal collection of shifts together with the resulting maximised root bound parameter(s)?

The Ingenuity of LLL Reduction: Shift Selection, Practically From a practical perspective, the optimal selection of polynomials can be delegated to lattice reduction [432]. To this end, we use as many shifts as possible, and look which vectors are chosen by LLL reduction as a linear combination of the shortest vectors. These vectors usually form a proper sublattice that might admit a better root bound.

There are many examples in the literature, where authors reported that experimentally they were able to find larger roots than were predicted by theory. Usually, such a behaviour is explained by a sublattice structure. As an example, in the Boneh–Durfee attack on small secret d one may find experimentally via LLL reduction within the original lattice basis that led to the $N^{0.284}$ bound the sublattice that admits the $N^{0.292}$ bound.

While it is nice that LLL automatically optimises the shift selection, it also leads to some open questions. In order to find better root bounds (asymptotically), we have to find an optimal shift selection (asymptotically). How do we derive an optimal asymptotic shift selection for arbitrary parameter m in Coppersmith's method, from some optimised shift selection for fixed m? Even if we are capable of finding an optimal asymptotic shift selection, how do we analyse the resulting root bound? The standard approach only handles asymptotic determinant calculation for triangular lattice bases, but in general we do not have triangularity (see [79]).

Use of Unravelled Linearisation Sometimes we know additional relations involving the desired root. For example, in the analysis of Boneh–Durfee attack (Section 4.4.2) we used a polynomial $f(u, x)$, where $u = 1 - xy$ was related to x. Using this relation improved the root bound, we called this technique unravelled linearisation.

At the moment, we are not aware of any general strategy that exploits the power of unravelled linearisation in a systematic way.

Coppersmith's Method: Modular or Integer? Usually, we derive a polynomial equation (system) that we somehow artifically transform into a modular polynomial. As an example, take the system of equations from Eq. (4.9) that led to the integer equation from Eq. (4.10). For our analysis, we looked for a polynomial root of Eq. (4.10) modulo e^2, but we could as well have worked modulo e, or modulo N.

Equation (4.10) is an especially interesting example, since all unknowns are linked. If we work modulo e^2, then the unknown $d_p d_q$ vanishes, but d_p, d_q still appear in the unknown coefficient $(kd_p + \ell d_q)$ of e. So intuitively, by working modulo e^2 we lose information from Eq. (4.10). As a consequence, Jochemsz and May chose to work with Coppersmith's method over the integers directly on the polynomial equation from Eq. (4.10), without any modular reduction.

Nevertheless, the currently best bound on CRT-exponents from Takayasu, Lu and Peng (see Theorem 4.15) works with Coppersmith's method modulo e^2. Can we also express the Takayasu–Lu–Peng attack [584] in terms of Coppersmith's method over the integers? Does it lead to the same attack bound, or does it even improve? More general, can we express all modular attacks as integer attacks? A result of Blömer and May [69] gives some indication that there are modular attacks that might not be convertible to integer attacks with the same bound, even for the univariate modular case.

4.5.2 Further Directions in Coppersmith-Type Cryptanalysis

Optimality of RSA Small Secret Exponent Attacks Without any doubt, small secret RSA exponent attacks are the best studied cryptanalysis application of Coppersmith's method, and are considered the most exciting results in this area. Yet, we do not have strong guarantees for the optimality of the bounds.

While the $N^{0.292}$ Boneh–Durfee bound consistently resisted serious efforts for improvements within the last two decades [42, 271, 349, 581], a first positive CRT-bound of $N^{0.073}$ (Theorem 4.14) was derived in 2007, and further improved in 2017 to $N^{0.122}$ (Theorem 4.15). Since the interaction of Eqs. (4.9) and (4.10) is much more involved than for the simple RSA key equation $ed = 1 \bmod \varphi(N)$, the optimality of the $N^{0.122}$-bound is much harder to study. Takayasu, Lu and Peng [584] reported that they experimentally found sublattices that admit practically better bounds than theoretically predicted. However, whether this might lead to better asymptotic bounds remains an open problem.

If we do not develop methods to prove the optimality of an attack directly, including the optimality of polynomial selection and shift selection, then one might at least be able to link attacks. Could we say derive the Boneh–Durfee bound as a special case of the Takayasu–Lu–Peng attack for the setting $d_p = d_q = d$? This would give us some more confidence in the optimality of Takayasu–Lu–Peng.

Systems of Polynomial Equations Even for a single polynomial we do not yet fully understand how to optimise Coppersmith's method. The situation becomes much more challenging when we move to systems of polynomials.

However, information-theoretically speaking, polynomial systems also provide more power to an attacker.

We take as an example (yet again) the RSA secret exponent attack. Although we do not know how to improve the $N^{0.292}$ bound, it was already argued in the original work of Boneh and Durfee [79] that we cannot beat the $N^{0.5}$ bound, because beyond this bound we expect exponentially many roots. Thus, with single polynomials we usually can only find roots within a polynomial fraction of the modulus.

Another example are the RSA stereotyped messages from Section 4.3.1, where we recover in polynomial time an unknown $\frac{1}{e}$-fraction of the messages. The situation changes if we move to polynomial equation systems. From the results for univariate polynomial systems by May and Ritzenhofen [409], we derived in Theorem 4.8 a result that basically states that every RSA equation with public exponent e_i yields a $\frac{1}{e_i}$-fraction of the message, and these fractions add up! Thus, if we have sufficiently many equations, we can fully recover the whole message in polynomial time, not just a fraction of the message.

A natural example, where an arbitrary number of polynomial equations appear are pseudorandom number generators. In 2009, Hermann and May [270] looked at power generators of the form $s_i = s_{i-1}^3 \bmod N$ and showed that these generators can successfully be attacked if one outputs a $(1 - \frac{1}{e})$-fraction of all $\lg N$ bits in each iteration. This implies that the Blum–Blum–Shub generator [72] with $e = 2$ should only output significantly less than half of its bits per iteration.

In 2001, Boneh, Halevi and Howgrave-Graham [83] constructed the so-called Inverse Congruential Pseudo Random Number generator with security based on the Modular Inversion Hidden Number Problem (MIHNP). In the MIHNP one obtains samples of the form

$$(t_i, \mathrm{MSB}_\delta(\alpha + t_i)^{-1} \bmod p)$$

for some random $t_i \in \mathbb{Z}_p$, where MSB_δ reveals the δ most significant bits. The goal in MIHNP is to recover the hidden number $\alpha \in \mathbb{Z}_p$.

In 2019, Xu, Sarkar, Lu, Wang and Pan [627] showed the really impressive result that for any constant fraction $\frac{1}{d}$ of output bits per iteration, given n^d MIHNP samples one can solve MIHNP in polynomial time.

Theorem 4.16 (Xu, Sarkar, Lu, Wang and Pan [627]). *The Modular Inversion Hidden Number Problem can be solved for any constant fraction $\frac{\delta}{\log p}$ in polynomial time.*

Notice that this result has a completely new quality from a cryptanalytic perspective. While other cryptanalysis results using Coppersmith's method work in polynomial time only if especially small parameters are chosen, or if some side-channel information reveals parts of the secret, the result of Xu *et al.* completely breaks MIHNP for any parameter setting. Before [627], it was conjectured in [83] that MIHNP is hard whenever one outputs less than a $\frac{\delta}{\log p} = \frac{1}{3}$-fraction of the bits, a quite typical result using Coppersmith's method.

We strongly believe that a careful understanding of Coppersmith's method applied to systems of polynomial equations might lead to similar ground-breaking cryptanalysis results in future.

Acknowledgement The author thanks Christine van Vredendaal for a thorough proofreading.

5

Computing Discrete Logarithms

Robert Granger and Antoine Joux

5.1 Introduction

Let G be a multiplicatively written finite cyclic group, let $g \in G$ be a generator and let $h \in G$. The discrete logarithm problem (DLP) for (G, g, h) is the computational problem of determining an integer x such that $h = g^x$. Note that the integer x is uniquely determined modulo the group order. Just as for the continuous logarithm function, one also writes $x = \log_g h$ and refers to x as the discrete logarithm of h to the base g.

The DLP has been central to public key cryptography ever since its inception by Diffie and Hellman in 1976 [163], and its study can be traced at least as far back as 1801, when discrete logarithms featured in Gauss's *Disquisitiones Arithmeticae*, referred to there as indices with respect to a primitive root modulo a prime [213, Article 57–60]. Indeed, the multiplicative group $\mathbb{F}_p^\times$ of the field $\mathbb{F}_p$ of integers modulo a prime p is perhaps the most natural example of a group in which the DLP can be posed – which is presumably why Diffie and Hellman used this setting for their famous key agreement protocol – and it is still believed to be hard for well-chosen primes.

In general, if the DLP is hard in a particular group then one can instantiate numerous important cryptographic protocols. So the issue at hand is: how hard is it to compute discrete logarithms in various groups? In this chapter we shall describe some cryptographically relevant DLPs and present some of the key ideas and constructions behind the most efficient algorithms known that solve them. Since the topic encompasses such a large volume of literature, for the finite-field DLP we limit ourselves to a selection of results reflecting recent advances in fixed characteristic finite fields. We start by briefly recalling the so-called generic algorithms, which do not exploit any representational properties of group elements and may thus be applied to any finite cyclic group, and then recall the more-sophisticated approach known as the index calculus method, which may be applied whenever the representation of elements of a group can be imbued with a suitable notion of smoothness. In Section 5.2 we introduce elliptic curves and pairings over finite fields and consider various discrete logarithm algorithms. Then in Section 5.3 we consider some groups in which the DLP is easier than for the strongest elliptic curves, including some families of weak curves. In Section 5.4 we focus on discrete logarithm algorithms for the cryptosystem XTR ($\approx$ ECSTR for Efficient and Compact Subgroup Trace Representation) and algebraic tori when defined over extension fields, and finally in Section 5.5 we present some of the key insights behind the breakthroughs between 2012 and 2014 that led to the downfall of finite fields of fixed characteristic in cryptography.

First, we introduce some useful notation for describing the running time of discrete logarithm algorithms (or equivalently the complexity or hardness of the DLP), which has become customary. Let N be the order of a group G. We define

$$L_N(\alpha, c) := \exp\left((c + o(1))(\log N)^\alpha (\log \log N)^{1-\alpha}\right),$$

where $\alpha \in [0, 1]$, $c > 0$ and log denotes the natural logarithm. When there is no ambiguity we often omit the subscript N, and sometimes write $L(\alpha)$ to mean $L(\alpha, c)$ for some $c > 0$. Observe that $L(0) = (\log N)^{c+o(1)}$, which therefore represents polynomial time, while $L(1) = N^{c+o(1)}$ represents exponential time. If an algorithm has a running time of $L(\alpha)$ for some $0 < \alpha < 1$ it is said to be of sub-exponential complexity.

5.1.1 Generic Algorithms

In the context of the DLP, a generic algorithm is one that applies in the generic group model, in which elements have a randomly selected unique encoding and one can only perform group operations and check for equality of elements (by comparing the encodings), see [542]. In this context, it was shown by Shoup [542] and Nechaev [450] that the DLP has an exponential running time $\Omega(\sqrt{N})$ if N is prime. This result implies that a sub-exponential algorithm must exploit a suitable group representation. We now describe the main examples of generic algorithms for the DLP in any finite cyclic group G, given a generator g and a target element h, where $|G| = N$.

First, if N is composite and its prime factorisation is known, then one can apply the Pohlig–Hellman algorithm [472], which reduces the DLP in G to DLPs in prime-order subgroups. In particular, let $N = p_1^{e_1} \cdots p_r^{e_r}$. By the Chinese remainder theorem it is sufficient to solve the DLP in each of the subgroups of order $\mathbb{Z}/p_i^{e_i}\mathbb{Z}$ for $i = 1, \ldots, r$, and one can project the DLP into each of them by powering g and h by the respective co-factors $N/p_i^{e_i}$. Let $x_i = \log_g h \pmod{p_i^{e_i}}$. If $e_i = 1$ then one needs only to solve a DLP in a prime order subgroup. If $e_i > 1$ then the digits of the p_i-ary expansion of x_i can be computed sequentially, starting from the least-significant digit via projecting and applying a Hensel lifting approach, each time solving a DLP in a subgroup of order p_i. For this reason, groups of large prime order, or those whose order possesses a large prime factor, are used in practice.

Moving on to algorithms for solving the DLP, we start with the time–memory trade-off known as the Baby-Step–Giant-Step (BSGS) method, attributed to Shanks. Let $M = \lceil \sqrt{N} \rceil$. One first computes a table $\{(j, g^j) \mid j \in \{0, \ldots, M-1\}\}$ (the baby steps) and sorts it according to the second component. Letting $k = g^{-M}$, one then computes $h, hk, hk^2, \ldots$ (the giant steps) until a collision $hk^i = g^j$ is found, at which point one knows that $\log_g h = iM + j$. The algorithm requires $O\left(\sqrt{N}\right)$ storage and $O\left(\sqrt{N}\right)$ group operations. Its precise bit complexity depends on the cost of group operations and on the implementation of the search for collisions.

An alternative approach is Pollard's rho method [476]. It requires some heuristic assumptions but preserves the expected $O\left(\sqrt{N}\right)$ running time, while reducing the storage requirement to $O(1)$. The heuristic can be removed at the cost of introducing an extra logarithm factor in the runtime [276]. The core idea is to define pseudorandom sequences (a_i), (b_i) in $\mathbb{Z}/N\mathbb{Z}$ and $(x_i) \in G$ such that $x_i = g^{a_i} h^{b_i}$. To construct the sequence, we iterate a function $f : G \to G$ that allows the tracking of the exponent and behaves in a pseudorandom fashion. A typical choice is to partition G as a disjoint union $G = G_1 \cup G_2 \cup G_3$ and then define f by setting $f(x) = x^2$ when $x \in G_1$, $f(x) = gx$ when $x \in G_2$ and $f(x) = hx$ when $x \in G_3$.

Once f is defined, we construct the sequence (x_i) iteratively, starting from a random x_0 and computing $x_{i+1} = f(x_i)$. Eventually, because G is finite, one must have $x_j = x_j$ for some $i \neq j$. For such a collision one has $\log_g h = \frac{a_j - a_i}{b_i - b_j}$, provided the denominator is invertible modulo N. In fact, the sequence is ultimately periodic and one has $x_j = x_{j+\ell}$ for some $\ell, j_0 > 0$ and every $j \geq j_0$. In this context, one uses a cycle-finding algorithm to find a collision, for instance Floyd's cycle-finding algorithm which discovers a collision of the form $x_i = x_{2i}$.

Because of its ultimate periodicity, the sequence (x_i) has a tail and a cycle, depicted in Figure 5.1. This is why Pollard called it the rho ('ρ') method. For a random function f the expected length of the tail and the cycle is $\sqrt{\pi N/8}$ and therefore the expected time to solve the DLP is $\sqrt{\pi N/2}$. For concrete choices of f, a similar behaviour is seen in practice and, because of the assumption, the

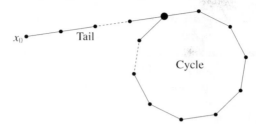

Figure 5.1 Illustration of the shape of Pollard's rho ('ρ') sequences.

algorithm is heuristic. Owing to the negligible storage requirements, the rho method is usually preferred over the BSGS method. For large computations with generic algorithms, the method of choice is often parallel collision search as introduced in [602].

5.1.2 The Index Calculus Method

The index calculus method (ICM) – meaning, rather opaquely, a 'method for calculating the index' – is an approach to solving DLPs that can be far more efficient than generic methods, depending on the group and element representation under consideration, as well as the ingenuity of the mathematician. However, the basic template is the basis for all sub-exponential algorithms and so the use of the definite article is probably justified. The method was first published by Kraitchik in the 1920s in the context of the DLP in prime fields [343, 344], and has been independently discovered many times since; see [359, 411, 460] and the references therein.

We now describe the two stages of the ICM for (G, g, h) abstractly, i.e., without reference to a particular group. First, one must choose a subset $\mathcal{F} \subseteq G$ known as the factor base, such that $\langle \mathcal{F} \rangle = G$, and to which g is usually adjoined if it is not already in $\mathcal{F}$. Informally, the first stage of the ICM is finding the logarithms of all elements in $\mathcal{F}$; this stage is usually divided into two parts, namely, relation generation and linear algebra. The second stage is the individual logarithm stage, i.e., expressing an arbitrary element over $\mathcal{F}$ so as to infer its discrete logarithm.

More formally, let $A = \mathbb{Z}/N\mathbb{Z}$ and consider the surjective group homomorphism

$$\phi: A^{|\mathcal{F}|} \to G, \quad (e_f)_{f \in \mathcal{F}} \mapsto \prod_{f \in \mathcal{F}} f^{e_f}.$$

The aforementioned steps are as follows.

- **Relation generation** Find vectors $(e_f)_{f \in \mathcal{F}}$ in $\ker \phi$, known as relations, which thus generate a subset $R \subseteq \ker \phi$.
- **Linear algebra** Compute a non-zero element $(x_f)_{f \in \mathcal{F}} \in R^{\perp}$, i.e., one satisfying $\sum_{f \in \mathcal{F}} x_f e_f = 0$ for all $(e_f)_{f \in \mathcal{F}} \in R$. Taking the logarithm of the multiplicative relations, we see that the vector of logarithms of the elements of $\mathcal{F}$ (in any basis) form a solution. Assuming that the set of equations is large enough, one does not expect any other solutions.
- **Individual logarithm** Find a preimage $(e_f)_{f \in \mathcal{F}} \in \phi^{-1}(h)$; it then follows that $\log_g h = \sum_{f \in \mathcal{F}} e_f \log_g f$.

Provided that sufficiently many linearly independent relations have been found, the discrete logarithms of elements of $\mathcal{F}$ can be computed, up to a non-zero scalar multiple, which can be normalised by insisting that $\log_g g = 1$.

In order to apply the ICM to a particular group and element representation, one needs to be able to define a suitable factor base. In order to do this one usually requires a natural notion of norm, primes and consequently smoothness, or the ability to impose analogues of these algebraically. An example of the former will be seen in Section 5.5, while an example of the latter will be seen in Section 5.4.

5.2 Elliptic Curves

5.2.1 Elliptic Curves of Finite Fields: A Quick Summary

An elliptic curve is a mathematical object that can be presented through several complementary points of view. When used for cryptographic purposes, the main focus is usually on elliptic curves over finite fields and one often focusses on the following definition.

Definition 5.1. An elliptic curve in Weierstrass form is a smooth projective curve given by a homogeneous equation:

$$Y^2 Z + a_1 XYZ + a_3 YZ^2 = X^3 + a_2 X^2 Z + a_4 XZ^2 + a_6 Z^3.$$

When the coefficients $(a_1, a_2, a_3, a_4, a_6)$ belong to $\mathbb{F}_q$, we say that the curve is defined over $\mathbb{F}_q$.

Let p denote the characteristic of $\mathbb{F}_q$. As soon as $p \geq 5$, it is possible, via a linear change of coordinates, to change the equation into a reduced Weierstrass equation:

$$Y^2 Z = X^3 + aXZ^2 + bZ^3.$$

Most of the time, we consider elliptic curves given by such a reduced equation. Let us briefly recall that the projective plane consists of all classes of non-zero triples (X, Y, Z) obtained from the equivalence relation that identifies (X, Y, Z) and (X', Y', Z') whenever there exists an invertible (i.e., non-zero in the case of $\mathbb{F}_q$) value λ such that $X = \lambda X'$, $Y = \lambda Y'$ and $Z = \lambda Z'$. The equivalence class associated to (X, Y, Z) is usually denoted by $(X : Y : Z)$.

A projective point with $Z = 0$ is said to lie at infinity. On the Weierstrass equation, we see that $Z = 0$ implies $X = 0$. As a consequence, there is a single point at infinity on the elliptic curve defined by that equation, the point with class $(0 : 1 : 0)$. It is simply called the point at infinity on E and written O_E. All other points have $Z \neq 0$; by using the equivalence, they can be written as $(x : y : 1)$.

The pair (x, y) then satisfy the affine equation:

$$y^2 + a_1 xy + a_3 y = x^3 + a_2 x^2 + a_4 x + a_6 \quad \text{or} \quad y^2 = x^3 + ax + b.$$

It is a frequent practice to describe an elliptic curve by such an affine equation, together with the implicit convention that the curve also contains the point O_E at infinity.

The use of the projective form formalises the meaning of O_E; it can also by useful for faster computations. Indeed, it may avoid the need to compute inverses while adding points.

Main Invariant For the curve to be smooth, it should not have any singular points. This can be tested on a reduced Weierstrass equation by computing the discriminant:

$$\Delta = -16(4a^3 + 27b^2).$$

The curve is smooth, if and only if, $\Delta \neq 0$.

Moreover, the reduced Weierstrass form allows many distinct but isomorphic curve equations. To see that, it suffices to consider changes of variables of the form $(x, y) = (u^2 x', u^3 y')$. The change transforms the equation

$$y^2 = x^3 + ax + b$$

into

$$y'^2 = x'^3 + a'x' + b',$$

where $a' = a/u^4$ and $b' = b/u^6$.

The j-invariant of the curve is a simple way to classify isomorphic curves. It is given by:

$$j = -1728 \cdot \frac{64a^3}{\Delta}.$$

Two isomorphic curves have the same j-invariant; furthermore, over an algebraically closed field, two curves with the same j-invariant are isomorphic. However, over finite fields, the situation is more complex. Indeed, from the above formulae we see that $u^2 = a'b/ab'$ and we need to distinguish the case where u^2 is a quadratic residue or not, in the considered finite field.

In the first, u itself exists in the same field and the two curves are isomorphic. In the second, u belongs to a quadratic extension and the curves are said to be quadratic twists. Note that, when $a = a' = 0$ or $b = b' = 0$, the situation is more complex because we can compute only u^4 or u^6 rather than u^2.

To distinguish quadratic twists over a finite field $\mathbb{F}_q$, one also computes the so-called minimal polynomial of Frobenius $X^2 - tX + q$. This is equivalent to point counting and the number of points defined over $\mathbb{F}_q$ is $q + 1 - t$ (including the point at infinity). Two isomorphic curves have the same number of points and, when going from a curve to a quadratic twist, the parameter t changes its sign.

Because point counting can be done efficiently by using the Schoof–Elkies–Atkin method (SEA), introduced by Schoof [522], we can always assume that t is known. This makes elliptic curves quite useful for discrete logarithm-based cryptosystems where it is essential to know the cardinality of the group being used.

Lines and the Group Law

The main interest of elliptic curves, especially for cryptography, is that they can be equipped with a group law (denoted additively) whose neutral element is the point at infinity. This law can be defined geometrically as follows. Take an elliptic curve given by a reduced Weierstrass equation and consider the intersection of an arbitrary line with the curve. Three cases are to be considered: the line at infinity with equation $Z = 0$, a vertical line with equation $x = x_0$ and, finally, other lines with equation $y = \lambda x + \mu$.

In the first case, substituting Z by 0 in the projective equation of E, we find $x^3 = 0$. As a consequence, the intersection is reduced to the point O_E with multiplicity 3. For a vertical line, the affine equation becomes $y^2 = x_0^3 + ax_0 + b$. When the right-hand side is 0 we get as the intersection a double point $(x_0, 0)$. When it is a quadratic residue, the intersection is formed of the two points $\left(x_0, \pm\sqrt{x_0^3 + ax_0 + b}\right)$. When it is a non-residue, the line does not meet the curve on affine points. However, by considering points over a quadratic field extension, we recover the two points of intersection. Furthermore, considering the projective equation of the same line, i.e., $X = x_0 Z$, we see that a vertical line also meets the curve at O_E.

Finally, for a line $y = \lambda x + \mu$, replacing y by this expression in the curve equation, we obtain

$$x^3 - \lambda^2 x^2 + (a - 2\lambda\mu) x + (b - \mu^2) = 0.$$

Counting roots with multiplicities, this polynomial can have 0, 1 or 3 roots. Over a well-chosen field extension, we can always find three roots (counting multiplicities). Let x_1, x_2 and x_3 denote these (not necessarily distinct) roots. For each x_i, we get an intersection point $(x_i, \lambda x_i + \mu)$.

As a consequence, we see that any line intersects the curve E three times (counting multiplicities and the intersection with O_E). The group law on E can be created from the simple rule that the sum of such three points of intersection is always O_E.

From the line at infinity, we find that $3O_E$ is zero; it is therefore natural to choose O_E as the neutral element for the group law. Then, consider a vertical line that meets the curve at (x_0, y_0) and $(x_0, -y_0)$. Because $(x_0, y_0) + (x_0, -y_0) + O_E = O_E$, we see that points that are symmetric about the x-axis are opposites. We thus denote by $-P$ the reflection of P about the x-axis.

Finally, consider a line meeting the curve at P, Q and R, implying that $P + Q + R = O_E$. We can deduce that the sum $P + Q$ is equal to $-R$. From this, we recover the usual addition formulae on an elliptic curve.

More precisely, if P and Q are symmetric about the x-axis, their sum is O_E. Otherwise, we compute the slope λ of the line through P and Q. If the points are equal, the line is tangent to the curve and:

$$\lambda = \frac{2y_P}{3x_P^2 + a}.$$

If they are distinct, we have:

$$\lambda = \frac{y_P - y_Q}{x_P - x_Q}.$$

We then find that $x_{P+Q} = \lambda^2 - (x_P + x_Q)$ and that $y_{P+Q} = -(y_P + \lambda (x_{P+Q} - x_P))$.

Divisors, Functions and Pairings

In truth, this idea of considering the intersection of the curve with a line is a particular case of a more general construction. To explain the more general viewpoint, we need to introduce two essential mathematical objects called divisors and functions of the curve.

Divisors of E A divisor of the curve E is simply a mapping from the set of points on the curve to the integers of $\mathbb{Z}$ that is non-zero only on finitely many points. A frequent representation consists of listing the points where the mapping is non-zero together with the value at each of these points. With this representation, a divisor D is written as a formal finite sum:

$$D = \sum_{i=1}^{n_D} e_i(P_i).$$

Given a divisor D in the above form, we define its degree as $\deg D = \sum_{i=1}^{n_D} e_i$. We also define the support of D as the set of points appearing with a non-zero coefficient in D. The set of divisors can be naturally given a group structure where the addition of two divisors is defined as the sum of the underlying mappings. When considering this group law, we can see that deg is a group homomorphism to $\mathbb{Z}$. Thus, its kernel, called the set of degree-zero divisors, is also a group.

Functions and Function Field of a Curve The concept of functions on a curve will generalise the idea of a line. Given an elliptic curve E and its (reduced) Weierstrass equation, we proceed in two steps. First, we consider the ring of bivariate polynomials in X and Y modulo the curve equation $Y^2 - (X^3 + aX + b)$. For example, in that ring, Y^2 and $X^3 + aX + b$ are representations of the same element. This ring is an integral domain and we can then build its field of fractions. This field is called the function field of the curve. Informally, we are thus considering rational fractions in X and Y modulo the curve's equation.

Let f/g be a fraction representing an element of the function field. Then, for any point P of E, we can compute the value $f(P)$ by replacing in f the variables X and Y by the values of the coordinates of P. In the same way, we can compute $g(P)$. Finally, when $g(P)$ is non-zero, the evaluation of f/g at P is defined as $f(P)/g(P)$. Moreover, if f_1/g_1 and f_2/g_2 represent the same element of the function field and if, in addition, $g_1(P) \neq 0$ and $g_2(P) \neq 0$, then $(f_1/g_1)(P) = (f_2/g_2)(P)$. Indeed, when f_1/g_1 and f_2/g_2 represent the same function, then $f_1 g_2 - f_2 g_1$ is a multiple of the curve equation. Thus, the value of $f_1 g_2 - f_2 g_1$ at P is 0, which implies the equality of evaluations of f_1/g_1 and f_2/g_2 at P. Furthermore, this allows us to define f/g at every point P on E. If $g(P) \neq 0$, we use the value $f(P)/g(P)$ as before. Otherwise, when $g(P) = 0$, we consider $f(P)$. More precisely, if $g(P) = 0$ and $f(P) \neq 0$ we say that $(f/g)(P) = \infty$. In the final case where both $g(P) = 0$ and $f(P) = 0$, it is always possible to find F and G such that $f/g = F/G$ and $(F(P), G(P)) \neq (0, 0)$. We can then use F/G to define the value at P. This definition is valid because (when defined) the value is independent of the choice of representative of the function.

When $(f/g)(P) = 0$, we say that P is a zero of f/g and, when $(f/g)(P) = \infty$, we say that P is a pole of f/g. It is also possible to define multiplicities of zeros and poles.

Evaluation of a function at a point can easily be generalised into an evaluation on a divisor in the following way. If D is given as $D = \sum_i e_i(P_i)$ and F is an element of the function field, with no zero or pole in the support of D, we define $F(D)$ by the following formula

$$F(D) = \prod_i F(P_i)^{e_i},$$

i.e., as the product (with multiplicities) of evaluations on all points in the support of D.

Principal Divisors and the Group Law To every non-zero function F in the function field, we can associate a divisor $\mathrm{Div}(F)$ that regroups the information on its zeros and poles. In the following equation that defines $\mathrm{Div}(F)$, the notation $\mathcal{Z}_F$ stands for the set of its zeros and $\mathcal{P}_F$ for the set of poles. Furthermore, when P is a zero or a pole, m_P denotes its multiplicity. We now define the divisor of F as

$$\mathrm{Div}(F) = \sum_{P \in \mathcal{Z}_F} m_P(P) - \sum_{P \in \mathcal{P}_F} m_P(P).$$

Note that for all $\lambda \in \mathbb{F}_p^\times$ we have $\mathrm{Div}(\lambda F) = \mathrm{Div}(F)$. Indeed, multiplication by a non-zero constant does not change the zeros or poles of a function. Furthermore, $\mathrm{Div}(1)$ is the zero divisor (empty in our notation), $\mathrm{Div}(FG) = \mathrm{Div}(F) + \mathrm{Div}(G)$ and $\mathrm{Div}(1/F) = -\mathrm{Div}(F)$. Thus, the Div operator is a morphism for the multiplicative group of the function field to the additive group of divisors. Divisors in the image of Div are called principal divisors. In addition, it can be shown that the degree of a principal divisor is zero.

As a consequence, principal divisors form a subgroup of degree-zero divisors and we can form the quotient group. This yields back the group law for the elliptic curve. The most relevant property is that for every degree-zero divisor D, there exists a unique point P and function F (up to equivalence) such that D can be written as $(P) - (O_E) + \mathrm{Div}(F)$. It is thus possible to label the elements of the above quotient group by points of E, which equips E with the group law.

Conversely, this correspondence can be used to test whether a given divisor is principal. We first check that the degree is zero then evaluate the expression given the divisor on the curve. More precisely, for $D = \sum_i e_i(P_i)$, we compute $v(D) = \sum_i e_i P_i$. A degree-zero divisor D is principal, if and only if, $v(D)$ is the point at infinity O_E.

Furthermore, if D is a principal divisor, there exists a unique (up to multiplication by a constant) function F such that $\mathrm{Div}(F) = D$.

Weil's Pairing The Weil pairing is a non-degenerate, antisymmetric, bilinear function from the n-torsion subgroup $E[n]$ of the curve E to the nth roots of unity in $\overline{\mathbb{F}}_p$. We recall that the n-torsion $E[n]$ is the set of all points P defined over the algebraic closure $\overline{\mathbb{F}}_p$ such that $nP = O_E$. When n is co-prime to p, we know that $E[n]$ is isomorphic to the group $(\mathbb{Z}/n\mathbb{Z})^2$.

Let P and Q be two n-torsion points. We denote by D_P the divisor $D_P = (P) - (O_E)$ and by D_Q the divisor $(Q) - (O_E)$. We see that nD_P and nD_Q are principal. Let F_P and F_Q be functions such that $\mathrm{Div}(F_P) = nD_P$ and $\mathrm{Div}(F_Q) = nD_Q$. The Weil pairing $e_n(P, Q)$ is then defined as $F_P(D_Q)/F_Q(D_P)$.

When P and Q are defined over a small-degree extension of $\mathbb{F}_p$, there exists an efficient algorithm due to Miller [434] that quickly computes this value $e_n(P, Q)$. Owing to this efficiency, the Weil pairing and its cousin the Tate pairing [193] have been used to construct a large variety of so-called pairing-based cryptographic protocols.

The main parameter that governs the concrete use of a pairing is the embedding degree. Given a curve E with cardinality N_E over $\mathbb{F}_p$ and q a prime divisor of N_E, there is a cyclic subgroup of order q defined over $\mathbb{F}_p$. The embedding degree of this subgroup is the smallest integer k such that $\mathbb{F}_{p^k}$ contains a primitive q-th root of unity. It gives the smallest field $\mathbb{F}_{p^k}$ in which one can compute a non-degenerate pairing involving points of order q.

5.2.2 Discrete Logarithm Algorithms for Families of Weak Curves

Supersingular and Low-Embedding Degree Curves

The pairings we just described can be used as a tool to transport the discrete logarithm problem on an elliptic curve to a discrete logarithm problem in a finite field. This is the MOV method, introduced by Menezes, Okamoto and Vanstone [418]. It initially used the Weil pairing but can also rely on the Tate pairing [193].

The idea, in order to solve $Q = nP$ in a cyclic group of order q, is to find a third point R, also of order q, such that $e_q(P, R) \neq 1$. Then, by linearity of the pairing, we have $e_q(Q, R) = e_q(P, R)^n$. This moves the discrete logarithm to the group of qth roots of unity in the finite field $\mathbb{F}_{p^k}$, where k is the embedding degree.

Because of the sub-exponential algorithms for computing discrete logarithms in $\mathbb{F}_{p^k}^\times$ (discussed briefly in Section 5.4 when p is of cryptographic size), this gives a better than generic algorithm as long as k remains small.

An especially weak case for which the MOV method was initially proposed is the case of supersingular curves that have cardinality $p + 1$ and embedding degree two because $p + 1$ divides $p^2 - 1$.

Descent and Cover Methods

The discrete logarithm techniques that follow apply (to this day) only to curves defined over a finite field $\mathbb{F}_{p^k}$, where the extension degree k has small factors. In particular, we currently do not know how to use them for curves defined over prime fields $\mathbb{F}_p$. Similarly, after studying state-of-the-art methods together with various speed-ups, Galbraith and Gebregiyorgis [200] conclude that logarithms on curves over prime-degree extension fields cannot be computed faster than by generic methods for cryptographic sizes.

The Gaudry–Hess–Smart (GHS) Method This method, described in [202, 212, 273, 419], consists of finding a so-called cover of the curve E defined over $\mathbb{F}_{p^k}$ by a curve H of genus $g > 1$ defined over the smaller field $\mathbb{F}_p$. More precisely, a cover is a surjective map from H to E expressed by rational functions on H. It is particularly useful when the genus g is not too large, ideally when $g = k$. The existence of the cover can be used to transport the discrete logarithm problem from E to the Jacobian of H, where it becomes easier. It turns out that the conditions permitting the existence of a cover are such that the most studied and more vulnerable cases are for $k = 2$ and $k = 3$.

The Gaudry–Semaev Method This method developed by Gaudry [210] and Semaev [529] is an index calculus technique that remains in the curve E. Its basic idea is to write arbitrary points on the curve as a sum of a small number of points with abcissa in the small field $\mathbb{F}_p$. This is done by solving multivariate systems of polynomial equations in k variables, with a degree growing exponentially with k. Again, this is only achievable for small values of k. Gaudry's general method [210] is summarised in Section 5.4.2.

Combining Both, the Joux–Vitse Method For certain curves with $k = 6$, the situation is especially bad. As shown by Joux and Vitse [301], it is possible to combine both attacks. First moving the discrete logarithm to the Jacobian of a genus-three curve of $\mathbb{F}_{p^2}$, where a variant of the Gaudry–Semaev method can be used. Discrete logarithms can then be computed for these specific curves, even for cryptographically meaningful group sizes.

Diem's Asymptotic Analysis In two articles [158, 159], Diem showed that elliptic-curve discrete logarithms over finite fields $\mathbb{F}_{q^n}$ can be solved asymptotically faster than by generic methods. First [158], he achieved an asymptotic complexity of the form $\exp\left(O\left(\max\left(\log q, n^2\right)\right)\right)$ and later [159], he improved it to $\exp\left(O\left(\max\left(\log q, n\log(q)^{1/2}, n^{3/2}\right)\right)\right)$. As far as we know, Diem's methods have not been used in any large-size computation. The main obstruction seems to be the need to solve algebraic systems of equations of large degrees and number of variables, which are not practically accessible with current algebraic methods.

5.3 Some Group Descriptions with Easier Discrete Logarithms

5.3.1 Addition Modulo an Integer

A basic theorem in group theory is that every cyclic group of order N is isomorphic to its structure group $(\mathbb{Z}/N\mathbb{Z}, +)$. Moreover, in this group representation, solving the discrete logarithm problem is trivial. Indeed, because the group law is additive, the discrete logarithm problem is, given a generator of the group, i.e., a number x co-prime with N and a value y, to find n such that $y = nx$ (mod N). This implies $n = y x^{-1}$ (mod N), where x^{-1} is obtained from Euclid's extended GCD algorithm.

A classical question that arises when presenting discrete logarithm-based cryptography to pure mathematicians is related to the remark. Why should the discrete logarithm problem be considered to be hard as it is so easy in the (isomorphic) structure group? In fact, computing discrete logarithms is simply a way to describe explicitly the isomorphism between a cyclic group and the corresponding $(\mathbb{Z}/N\mathbb{Z}, +)$. Furthermore, solving the problem requires not only an explicit expression but also an efficiently computable one.

5.3.2 Matrix Groups

Because the discrete logarithm problem can be defined for any cyclic group, it is quite natural to consider the subgroup of the square matrices over some finite field $\mathbb{F}_p$ generated by an invertible matrix G. In particular, this was suggested by Odoni *et al.* [461].

This problem was studied by Menezes and Wu [420]. It turns out that it can be reduced to discrete logarithms in finite fields. Furthermore, the matrix computations are more expensive for the participants than the corresponding computations in finite fields. As a consequence, this particular instantiation of the discrete logarithm problem does not provide any specific advantage.

We briefly describe here the main idea of the construction. Let A and B be two n by n matrices over $\mathbb{F}_p$. We want to find ℓ such that $B = A^\ell$ knowing that such an integer exists.

We first consider the characteristic polynomial p_A of A. In general, the complete form of the attack depends on the factorisation of p_A. However, the attack is easier to describe when p_A is an irreducible polynomial of degree n. For simplicity of exposition, we limit ourselves to that case. In that case, p_A has n distinct conjugate roots in $\mathbb{F}_{p^n}$ and these roots are eigenvalues of A. Let α denote one of these eigenvalues, the others can be written as α^{p^i} with i in $[1\ldots n-1]$.

As a consequence, A can be diagonalised by writing $A = C^{-1}A_D C$ where A_D is a diagonal matrix whose entries are the values α^{p^i} with i in $[0\ldots n-1]$. We see that $B = C^{-1}A_D^\ell C$, thus $B_D = CBC^{-1}$ is diagonal with entries $\alpha^{\ell p^i}$. Taking the logarithm of the first entry α^ℓ relative to α in $\mathbb{F}_{p^n}$ is thus enough to completely recover ℓ.

5.3.3 Particularly Bad Curves

Singular Curves

A first example of bad curves covers a degenerate case, the case of curves with a zero discriminant. These curves have a singular point and are not valid elliptic curves. However, they can be obtained by reducing modulo a prime p an elliptic curve E with rational coefficients. In that case, we say that E has bad reduction at p.

In this situation, there is a singular point on the curve and we denote by E^{ns} the set of regular points. It is interesting to know that, on this set, the usual geometric construction of an elliptic-curve group law still works and yields a group law on E^{ns}.

We can distinguish two main cases. In the first one, the curve E is given by $y^2 = x^3$ (possibly after a change of variables) in the second by $y^2 = x^3 + Ax^2$ with $A \neq 0$. In both cases, the point $(0,0)$ is singular on E and the set E^{ns} consists of all the other points.

In the first case, i.e., on $y^2 = x^3$, every point of E^{ns} can be written as (ℓ^2, ℓ^3), with $\ell \neq 0$. Moreover, given $P = (x_P, y_P)$ the corresponding value is simply given by $\ell_P = y_P/x_P$. Let P and Q be the two points corresponding to the values

ℓ_P and ℓ_Q. The slope of the line through P and Q is

$$\lambda = \frac{y_Q - y_P}{x_Q - x_P} = \frac{\ell_Q^2 + \ell_P \ell_Q + \ell_P^2}{\ell_Q + \ell_P}.$$

Let $R = (x_R, y_R)$ be the third point of intersection with E. We find that $x_R = \lambda^2 - (x_P + x_Q)$ and $y_R = \lambda(x_R - x_P) + y_p$. Developing the expressions we can write

$$x_R = \frac{\ell_P^2 \ell_Q^2}{\ell_Q^2 + 2\,\ell_P \ell_Q + \ell_P^2} = \left(\frac{\ell_P \ell_Q}{\ell_P + \ell_Q} \right)^2$$

and

$$y_R = -\left(\frac{\ell_P \ell_Q}{\ell_P + \ell_Q} \right)^3.$$

Thus, the sum of P and Q corresponds to the values $\ell_{P+Q} = \frac{\ell_P \ell_Q}{\ell_P + \ell_Q}$. We can remark that

$$\frac{1}{\ell_{P+Q}} = \frac{1}{\ell_P} + \frac{1}{\ell_Q}.$$

Adding the natural convention that the point at infinity has an infinite associated value (with inverse 0), we see that addition on E^{ns} boils down to addition in $\mathbb{F}_p$. The group isomorphism to the structure group is explicit and efficient to compute and the discrete logarithm is thus easy.

For the second case, i.e., the curve $y^2 = x^3 + Ax^2$, we start by writing $A = \alpha^2$. This leads to two subcases depending on whether A is a square in $\mathbb{F}_p$ or not. In the first case, we turn point addition into multiplication in $\mathbb{F}_p^{\times}$, while in the second it becomes multiplication in the subgroup of order $p+1$ of the quadratic extension $\mathbb{F}_{p^2}^{\times}$.

As before, we express the points of the curve E^{ns} as functions of a parameter ℓ_P given by the following formula:

$$\ell_P = \frac{y_P + \alpha\, x_P}{y_P - \alpha\, x_P},$$

with the convention that it is equal to 1 for the point at infinity. The coordinates x_P and y_P can be recovered from ℓ_P by computing:

$$x_P = \frac{4\alpha^2 \ell_P}{(\ell_P - 1)^2} \quad \text{and} \quad y_P = \frac{4\alpha^3 \ell_P(\ell_P + 1)}{(\ell_P - 1)^3}.$$

Finally, we can check that $\ell_{P+Q} = \ell_P \ell_Q$. Thus, the discrete logarithm is transported to the multiplicative group of a finite field $\mathbb{F}_p$ or $\mathbb{F}_{p^2}$. As a consequence, the discrete logarithm becomes much easier than on (general) elliptic curves.

Anomalous Curves

An elliptic curve over a prime field $\mathbb{F}_p$ is said to be anomalous when its trace is equal to 1 or equivalently its cardinality is equal to p. On such curves, the discrete logarithm problem becomes easy. In fact, two distinct methods have been proposed to explain this fact. The first one, by Semaev [528], defines an additive pairing sending a point P to an element of $\mathbb{F}_p$ by defining the function f_P with divisor $p(P) - p(O)$ (which is unique up to a multiplicative constant) and evaluating the ratio of the function f_P and its x-derivative at another (fixed) point.

The second approach, by Smart [550], considers an arbitrary lift of the curve and the point P to the p-adic numbers and uses a multiplication by p and a p-adic elliptic logarithm. All of these computations can be done with low p-adic precision.

For simplicity, we present here a heuristic version of the second method that bypasses the use of p-adic elliptic logarithms. As usual, given the curve E defined modulo p and two non-zero points P and Q, we want to solve the equation $Q = nP \pmod{p}$. Indeed, since p is prime, any non-zero point is a generator of E.

We start by considering an arbitrary Weierstrass equation modulo p^2 that reduces to the equation of E modulo p. We denote by E_2 this lifted curve modulo p^2. Each point of E, including the point at infinity, can be lifted to E_2 in p distinct ways via Hensel's lemma. Thus E_2 contains p^2 points with coordinates modulo p^2. Furthermore, the usual group law construction can be applied to E_2. We thus get an abelian group with order p^2. Its structure is either $(\mathbb{Z}/p\mathbb{Z})^2$ or $\mathbb{Z}/p^2\mathbb{Z}$. Heuristically, for a random lifting, we expect $\mathbb{Z}/p^2\mathbb{Z}$ to occur more frequently.

Let us now assume that E_2 with the elliptic-curve addition is a cyclic group of order p^2. The reduction modulo p of points is a surjective group homomorphism to E, whose kernel is formed of the p-distinct lifting of the point at infinity. We denote by r_E this reduction. The kernel of r_E is thus a subgroup of order p.

Let P_2 and Q_2 be two arbitrary liftings of P and Q. This implies that $r_E(Q_2 - nP_2) = Q - nP = 0$. Thus $Q_2 - nP_2$ is a lift of the point at infinity O_E. As a consequence, $p(Q_2 - nP_2)$ is zero on E_2. Thus, $(pQ_2) = n(pP_2)$. In addition, pP_2 and pQ_2 are in the kernel of r_E.

To see how this leads to a recovery of n, let us study the structure of the kernel of r_E, i.e., of the subgroup formed by all lifts of the point at infinity. To do this, it is useful to consider a weighted projective description of the Weirstrass equation, where the variables X, Y and Z respectively have weights

2, 3 and 1. The Weierstrass equation then has total weight 6 and can be written as:

$$Y^2 + a_1XYZ + a_3YZ^3 = X^3 + a_2X^2Z^2 + a_4XZ^4 + a_6Z^6.$$

Note that each a_i corresponds here to the Z^i term.

In this weighted notation, (X, Y, Z) and (X', Y', Z') represent the same point if and only if there exists an invertible element λ modulo p^2 such that $X' = \lambda^2 X$, $Y' = \lambda^3 Y$ and $Z' = \lambda Z$.

The liftings of the point at infinity are the triple where Z is a multiple of p. Up to equivalence, they can all be written as $(1, 1, \ell p)$. The point $(1, 1, 0)$ is the zero of the group law. For all the other points of this form, we can put them into the equivalent form $(\ell^{-2}, \ell^{-3}, p)$.

If we further assume that $a_1 = 0$, which can be achieved by using a reduced Weierstrass form for E_2, the situation is equivalent to considering the point (ℓ^{-2}, ℓ^{-3}) on the singular curve of equation $y^2 = x^3$. As we saw previously, the logarithm can be obtained by just mapping this point to the slope, we previously called ℓ, in the finite field.

5.4 Discrete Logarithms for XTR and Algebraic Tori

The Diffie–Hellman key exchange protocol [163] and El Gamal encryption and signatures [182] were formulated in the multiplicative group of a prime field $\mathbb{F}_p$. While the fastest algorithms for solving the DLP in $\mathbb{F}_p^\times$ are sub-exponential (for details we refer the reader to the survey article [245]), given a subgroup of prime order l, the fastest-known discrete logarithm algorithms that operate purely within the subgroup are generic. As Schnorr observed, one can therefore base protocols in the subgroup in order to speed up exponentiations, and obtain shorter signatures, for example, provided that the complexity of both attacks is above the required security threshold [515].

However, other than by using the discrete logarithm of a subgroup element relative to a generator, for prime fields there does not seem to be a way to reduce the size of the representation of elements: each requires $\lceil \log_2 p \rceil$ bits. One way to overcome this representational (and operational) inefficiency is to consider instead subgroups of the multiplicative group of extension fields, i.e., $\mathbb{F}_{p^n}^\times$. This idea is the basis of the entirety of Chapter 10, so presently we only briefly mention a couple of important examples.

The cryptosystem LUC [553], developed by Smith and Skinner in 1995, represents elements of the order $p + 1$ subgroup of $\mathbb{F}_{p^2}^\times$ by their trace from $\mathbb{F}_{p^2}$ to $\mathbb{F}_p$. As a result, just one element of $\mathbb{F}_p$ is needed to represent an element, thus providing the optimal compression factor of 2 for the full subgroup. Building upon this idea, in 1999, Brouwer, Pellikaan and Verheul described a compression method for elements of the order $p^2 - p + 1$ subgroup of $\mathbb{F}_{p^6}^\times$, again using the trace function, but this time to $\mathbb{F}_{p^2}$ [112]. This reduces the representation to just two elements of $\mathbb{F}_p$, providing the optimal compression factor of 3 for the full subgroup. Very soon afterwards, Lenstra and Verheul developed the cryptosystem XTR, extending the compression method in [112] by developing a more efficient exponentiation for the trace representation than is available in the usual field representation [361, 377].

Observe that, for both LUC and XTR, the relevant subgroups do not embed into a proper subfield of $\mathbb{F}_{p^n}$, for $n = 2$ and $n = 6$ respectively. Indeed, they are the so-called cyclotomic subgroups of $\mathbb{F}_{p^n}^\times$, which have order $\Phi_n(p)$ where $\Phi_n(x)$ is the n-th cyclotomic polynomial, defined by

$$\Phi_n(x) := \prod_{1 \le k \le n,\ \gcd(k,n)=1} (x - \zeta_n^k),$$

and ζ_n is a primitive (complex) n-th root of unity. Note that the degree of $\Phi_n(x)$ is simply $\phi(n)$, where $\phi(\cdot)$ is the Euler totient function. Furthermore, Because

$$x^n - 1 = \prod_{d|n} \Phi_d(x), \tag{5.1}$$

for each $d|n$ the subgroup of $\mathbb{F}_{p^n}^\times$ of order $\Phi_d(p)$ embeds into $\mathbb{F}_{p^d}$. Hence, for $d < n$ one can solve the DLP in the order $\Phi_d(p)$ subgroup of $\mathbb{F}_{p^n}^\times$ by applying sub-exponential algorithms to $\mathbb{F}_{p^d}^\times$, rather than to $\mathbb{F}_{p^n}^\times$. So the subgroup of order $\Phi_n(p)$ may be regarded as the 'cryptographically strongest' subgroup of $\mathbb{F}_{p^n}^\times$, and this subgroup is always used in cryptographic applications.[1]

Interestingly, the first-listed author of this chapter was informed in 2005 by the second-listed editor of this book that the real purpose of XTR was to stimulate research into the DLP in finite fields of small extension degree, and possibly in the cyclotomic subgroups [237], which was confirmed by Lenstra when the conversation was raised in 2013 [235]. For the former possibility, generally referred to as the medium prime case, research has progressed steadily, with several $L_{p^n}(1/3, c)$ algorithms being developed with generally decreasing c [30, 31, 34, 321]. These algorithms are mainly motivated by pairing-based cryptography. On the other hand, the latter possibility would seem at first to not be possible, thanks to the argument of the previous paragraph. Let $G_{p,n}$ denote the subgroup of $\mathbb{F}_{p^n}^\times$ of order $\Phi_n(p)$. If there were a hypothetical non-generic algorithm for solving the DLP in $G_{p,n}$ which was more efficient than solving

---

[1] Only if $\Phi_n(p) \le n$, which is never the case for cryptographic applications, may this subgroup embed into a proper subfield of $\mathbb{F}_{p^n}$ [366].

the DLP via the embedding into $\mathbb{F}_{p^n}$, then by solving it there and also in $\mathbb{F}_{p^d}$ for each $d \mid n$, $d < n$, by Eq. (5.1) and the Chinese remainder theorem,[2] one would have solved the DLP in $\mathbb{F}_{p^n}$ more efficiently than was thought possible. It has been argued that with such a security reduction one can be confident in the DLP security of $G_{p,n}$, as it is equivalent to the DLP security of $\mathbb{F}_{p^n}$, which is well studied.

However, this reduction can be viewed in another way: to attack the DLP in $\mathbb{F}_{p^n}$ one can try to invent algorithms for attacking the DLP $G_{p,n}$ directly. Indeed, this is what Granger and Vercauteren did in 2005 [238], as we explain in Section 5.4.3.

5.4.1 Algebraic Tori, Rationality and Compression

In their original paper, Lenstra and Verheul proposed allowing the base field for XTR to itself be an extension field, so henceforth we allow the base field to be $\mathbb{F}_q$, where q is a prime power p^m with $m \geq 1$.

In 2003, Rubin and Silverberg proposed torus-based cryptography, based on the observation that $G_{q,n}$ can be identified with the $\mathbb{F}_q$-rational points on the algebraic torus T_n of dimension $\phi(n)$, which has some cryptographically exploitable properties [503]. As well as showing that one can interpret LUC and XTR in terms of quotients of the algebraic tori T_2 and T_6 by certain actions of the symmetric groups S_2 and S_3 respectively, they observed that, whenever T_n is rational, i.e., there exists a rational map to $\phi(n)$-dimensional affine space, one can compress (almost all of) its elements by a factor of $n/\phi(n)$ relative to the $\mathbb{F}_{q^n}$ representation and use this smaller representation for communications. The torus T_n is known to be rational if n is the product of at most two prime powers, and is conjectured to be rational for all n [334, 608], although no other examples are currently known. Were this conjecture to be proven then one could obtain arbitrarily large compression factors for elements of the cyclotomic subgroup of $\mathbb{F}_{q^n}^{\times}$. The rationality of T_2 gives a simple analogue to LUC, while the rationality of T_6 gives an analogue to XTR, with an advantage that in these analogues one can freely multiply elements, unlike in LUC and XTR.

We now formally define the algebraic torus.

Definition 5.2. Let $k = \mathbb{F}_q$ and $L = \mathbb{F}_{q^n}$. The torus T_n is the intersection of the kernels of the norm maps $N_{L/F}$, for all subfields $k \subset F \subsetneq L$:

$$T_n(k) := \bigcap_{k \subset F \subsetneq L} \mathrm{Ker}[N_{L/F}].$$

---

[2] One may ignore the cryptographically small GCDs of such factors, as they can be computed with generic methods.

The following lemma provides two useful properties of T_n [503].

Lemma 5.3. *(1)* $T_n(\mathbb{F}_q) \cong G_{q,n}$, *and thus* $\#T_n(\mathbb{F}_q) = \Phi_n(q)$.
(2) If $h \in T_n(\mathbb{F}_q)$ *is an element of prime order not dividing n, then h does not lie in a proper subfield of* $\mathbb{F}_{q^n}/\mathbb{F}_q$.

5.4.2 Gaudry's Algorithm

The upcoming attack of Granger and Vercauteren [238] (see Section 5.4.3) may be seen as a version of an algorithm due to Gaudry, which is a general index calculus algorithm that may be applied to any abelian variety once a computationally convenient element representation and group law have been specified [210]. We briefly recall it here. Let $A/\mathbb{F}_q$ be an abelian variety of dimension d on which we would like to solve the DLP. We assume that, except for a negligible proportion of elements, there is an explicit embedding of A into affine space of dimension $d + d'$, i.e., an element $P \in A$ defined over $\mathbb{F}_q$ can be represented as

$$P = (x_1, \ldots, x_d, y_1, \ldots, y_{d'}),$$

where $x_i, y_i \in \mathbb{F}_q$. Because A has dimension d, we assume that for any $x_1, \ldots, x_d \in \overline{\mathbb{F}}_q$ there are only finitely many $y_1, \ldots, y_{d'} \in \overline{\mathbb{F}}_q$ such that the corresponding P is on A. For the factor base, or more appropriately the decomposition base, let

$$\mathcal{F} := \{(x_1, 0, \ldots, 0, y_1, \ldots, y_{d'}) \in A : x_1, y_i \in \mathbb{F}_q\},$$

which one may assume is an absolutely irreducible curve whose closure under the group law is not a strict abelian subvariety of A; for, otherwise, a random linear change of variables can be applied to the x_i-coordinates until these two properties hold. Hence, one may assume that $|\mathcal{F}| \approx q$.

Let $P \in A$ and let $Q \in \langle P \rangle$, with the group operation written additively. In order to find $\log_P Q$ we construct linear combinations $R = aP + bQ$ with a, b uniformly random integers modulo the group order and attempt to express R as a sum of d elements of $\mathcal{F}$, i.e.,

$$R = aP + bQ = P_1 + \cdots + P_d, \qquad (5.2)$$

where $P_i \in \mathcal{F}$, since this will heuristically occur with probability $\approx 1/d!$ as $q \to \infty$. When this occurs, we call Eq. (5.2) a relation. One can then proceed with the usual index calculus method. The crux of this method is how to test whether a random element of A decomposes over $\mathcal{F}$. Because A is an abelian variety and therefore an algebraic group, one can express the right-hand side of Eq. (5.2) as

$$P_1 + \cdots + P_d = (\phi_1(P_1, \ldots, P_d), \ldots, \phi_{d+d'}(P_1, \ldots, P_d)),$$

where $\phi_1, \ldots, \phi_{d+d'}$ are rational functions of the coordinates used. By setting this expression equal to R one obtains a set of equations, which together with the equations arising from membership of A or $\mathcal{F}$ results in a system that will generically be of dimension zero, whose solutions can be found by a Gröbner basis computation, or sometimes by faster methods, depending on A and its element and group law representation.

5.4.3 The Granger–Vercauteren Attack

The algorithm of Granger and Vercauteren [238] uses the affine representation of elements of an algebraic torus, and the group law induced in this representation by field multiplication, i.e., the usual group law. The key insight of the work is that, for a T_n which possesses a rational parameterisation, only $\phi(n)$ elements of a factor base need to be added in order to generate a random element of the group with constant probability. In comparison with using the field representation and defining the decomposition base to be the set of monic linear polynomials, for example, the probability of generating a relation is $1/\phi(n)!$ rather than $1/n!$. Therefore, it is the very compression that made torus-based cryptography attractive, that enables a significant speed up to be made when computing discrete logarithms. We now describe the algorithm for T_2 and T_6 respectively. In the following we assume q is odd.

Algorithm for $T_2(\mathbb{F}_{q^n}) \subset \mathbb{F}_{q^{2n}}^{\times}$ By the previous discussion in Section 5.4 the prime-order subgroup would be in $T_{2n}(\mathbb{F}_q) \subsetneq T_2(\mathbb{F}_{q^n})$, but because we exploit the rationality of T_2 rather than T_{2n}, we work with $T_2(\mathbb{F}_{q^n})$, or more precisely the dimension n variety $(\mathrm{Res}_{\mathbb{F}_{q^n}/\mathbb{F}_q} T_2)(\mathbb{F}_q)$, where Res denotes the Weil restriction of scalars (see [503]).

Let $\mathbb{F}_{q^n} \cong \mathbb{F}_q[t]/(f(t))$ with $f(t) \in \mathbb{F}_q[t]$ an irreducible monic polynomial of degree n. We shall use the polynomial basis $\{1, t, \ldots, t^{n-1}\}$. For a non-square $\delta \in \mathbb{F}_{q^n} \setminus \mathbb{F}_q$, let $\mathbb{F}_{q^{2n}} = \mathbb{F}_{q^n}[\gamma]/(\gamma^2 - \delta)$, with basis $\{1, \gamma\}$. From Definition 5.2 we have

$$T_2(\mathbb{F}_{q^n}) = \{(x, y) \in \mathbb{F}_{q^n} \times \mathbb{F}_{q^n} : x^2 - \delta y^2 = 1\}.$$

Rather than use two elements of $\mathbb{F}_{q^n}$ to represent each point, as the torus T_2 is one-dimensional and rational, one can use the following affine representation:

$$T_2(\mathbb{F}_{q^n}) = \left\{ \frac{z - \gamma}{z + \gamma} : z \in \mathbb{F}_{q^n} \right\} \cup \{O\}, \tag{5.3}$$

where O is the point at infinity. Note that, for $g = g_0 + g_1\gamma \in T_2(\mathbb{F}_{q^n})$ in the $\mathbb{F}_{q^{2n}}$ representation, the corresponding affine representation is $z = -(1 + g_0)/g_1$ if $g_1 \neq 0$, whereas -1 and 1 map to $z = 0$ and $z = O$ respectively. Since $T_2(\mathbb{F}_{q^n})$ has $q^n + 1$ elements, this representation is optimal: note that this map is really from $T_2(\mathbb{F}_{q^n})$ to $\mathbb{P}^1(\mathbb{F}_{q^n})$.

We define the decomposition base as follows:

$$\mathcal{F} = \left\{ \frac{a - \gamma}{a + \gamma} : a \in \mathbb{F}_q \right\} \subset T_2(\mathbb{F}_{q^n}),$$

which contains precisely q elements because the above map is a birational isomorphism from T_2 to $\mathbb{A}^1$. Now let P be a generator and $Q \in \langle P \rangle$. To find relations we test whether for random integers a, b modulo the group order, $R = aP + bQ$ decomposes as a sum of n points in $\mathcal{F}$, i.e.,

$$R = P_1 + \cdots + P_n, \tag{5.4}$$

with $P_1, \ldots, P_n \in \mathcal{F}$. In the affine representation this becomes

$$\frac{r - \gamma}{r + \gamma} = \prod_{i=1}^{n} \left(\frac{a_i - \gamma}{a_i + \gamma} \right),$$

where the $a_i \in \mathbb{F}_q$ are unknowns and $r \in \mathbb{F}_{q^n}$ is the affine representation of R. As the right-hand side is symmetric in the a_i we may expand it in terms of the elementary symmetric polynomials $\sigma_i(a_1, \ldots, a_n)$ of the a_i

$$\frac{r - \gamma}{r + \gamma} = \frac{\sigma_n - \sigma_{n-1}\gamma + \cdots + (-1)^n\gamma^n}{\sigma_n + \sigma_{n-1}\gamma + \cdots + \gamma^n}.$$

Reducing modulo the defining polynomial of γ, we obtain

$$\frac{r - \gamma}{r + \gamma} = \frac{b_0(\sigma_1, \ldots, \sigma_n) - b_1(\sigma_1, \ldots, \sigma_n)\gamma}{b_0(\sigma_1, \ldots, \sigma_n) + b_1(\sigma_1, \ldots, \sigma_n)\gamma},$$

where b_0, b_1 are linear in the σ_i and have coefficients in $\mathbb{F}_{q^n}$. Reducing the right-hand side to the affine representation (Eq. (5.3)) we obtain the equation

$$b_0(\sigma_1, \ldots, \sigma_n) - b_1(\sigma_1, \ldots, \sigma_n)r = 0.$$

Because the σ_i are in $\mathbb{F}_q$, by expressing this equation on the polynomial basis of $\mathbb{F}_{q^n}$ we obtain n linear equations over $\mathbb{F}_q$ in the n unknowns σ_i. If there is a solution $(\sigma_1, \ldots, \sigma_n)^T$ to this linear system, we see whether it corresponds to a solution of Eq. (5.4) by checking whether the following polynomial splits completely over $\mathbb{F}_q$

$$p(x) := x^n - \sigma_1 x^{n-1} + \sigma_2 x^{n-2} - \cdots + (-1)^n \sigma_n.$$

Whenever it does, the roots $a_1, \ldots, a_n$ will be the affine representations of elements of $\mathcal{F}$ which sum to R, i.e., we have found a relation.

In terms of complexity, when $n! \approx q$ the full algorithm runs in expected time $L_{q^n}(1/2, c)$ for some $c > 0$. Experiments in the computer algebra system Magma [98] reported in 2005 [238], demonstrated that it would be faster than Pollard's rho in a (at the time standard) 160-bit subgroup, when q^{2n} was between 400 and 1000 bits, thus indicating its efficacy for some practical parameters.

Algorithm for $T_6(\mathbb{F}_{q^n}) \subset \mathbb{F}_{q^{6n}}^{\times}$ As we saw before, the prime-order subgroup would be in $T_{6n}(\mathbb{F}_q) \subset T_6(\mathbb{F}_{q^n})$, but since we exploit the rationality of T_6 and not T_{6n} we shall work with $T_6(F_{q^n})$, or rather its Weil restriction $(\mathrm{Res}_{\mathbb{F}_{q^n}/\mathbb{F}_q} T_6)(\mathbb{F}_q)$. The central difference between this algorithm and the T_2 algorithm is that for the present case the equations to be solved in the decomposition step are no longer linear.

Let $\mathbb{F}_{q^n} \cong \mathbb{F}_q[t]/(f(t))$, with $f(t)$ an irreducible polynomial of degree n, and use the polynomial basis $\{1, t, t^2, \ldots, t^{n-1}\}$. For the birational map from $T_6(\mathbb{F}_{q^n})$ to $\mathbb{A}^2(\mathbb{F}_{q^n})$ we use the specifications for CEILIDH, the compression mechanism described by Rubin and Silverberg [503]. Assume that $q^n \equiv 2$ or $5 \bmod 9$, and for $(r, q) = 1$ let ζ_r denote a primitive rth root of unity in $\overline{\mathbb{F}}_{q^n}$. Let $x = \zeta_3$ and let $y = \zeta_9 + \zeta_9^{-1}$. Then $x^2 + x + 1 = 0$ and $y^3 - 3y + 1 = 0$. Furthermore let $\mathbb{F}_{q^{3n}} = \mathbb{F}_{q^n}(y)$ and $\mathbb{F}_{q^{6n}} = \mathbb{F}_{q^{3n}}(x)$. The bases we use are $\{1, y, y^2 - 2\}$ for the degree-three extension and $\{1, x\}$ for the degree-two extension. Let $V(f)$ be the zero set of $f(\alpha_1, \alpha_2) = 1 - \alpha_1^2 - \alpha_2^2 + \alpha_1 \alpha_2$ in $\mathbb{A}^2(\mathbb{F}_{q^n})$. The following are inverse birational maps.

- $\psi : \mathbb{A}^2(\mathbb{F}_{q^n}) \setminus V(f) \xrightarrow{\sim} T_6(\mathbb{F}_{q^n}) \setminus \{1, x^2\}$, defined by

$$\psi(\alpha_1, \alpha_2) = \frac{1 + \alpha_1 y + \alpha_2(y^2 - 2) + (1 - \alpha_1^2 - \alpha_2^2 + \alpha_1 \alpha_2)x}{1 + \alpha_1 y + \alpha_2(y^2 - 2) + (1 - \alpha_1^2 - \alpha_2^2 + \alpha_1 \alpha_2)x^2}. \tag{5.5}$$

- $\rho : T_6(\mathbb{F}_{q^n}) \setminus \{1, x^2\} \xrightarrow{\sim} \mathbb{A}^2(\mathbb{F}_{q^n}) \setminus V(f)$, which is defined as follows: for $\beta = \beta_1 + \beta_2 x$, with $\beta_1, \beta_2 \in \mathbb{F}_{q^{3n}}$, let $(1 + \beta_1)/\beta_2 = u_1 + u_2 y + u_3(y^2 - 2)$, then $\rho(\beta) = (u_2/u_1, u_3/u_1)$.

We define the decomposition base as follows:

$$\mathcal{F} = \left\{ \frac{1 + (at)y + (1 - (at)^2)x}{1 + (at)y + (1 - (at)^2)x^2} : a \in \mathbb{F}_q \right\},$$

which clearly contains q elements. Note that we use $\psi(at, 0)$ rather than $\psi(a, 0)$ since the latter would map to the strict subvariety $T_6(\mathbb{F}_q)$. Since $(\mathrm{Res}_{\mathbb{F}_{q^n}/\mathbb{F}_q} T_6)(\mathbb{F}_q)$ is $2n$-dimensional, to find relations we need to solve

$$R = P_1 + \cdots + P_{2n}, \tag{5.6}$$

with $P_1, \ldots, P_{2n} \in \mathcal{F}$. Assuming that R is expressed in affine form, i.e., $R = \psi(r_1, r_2)$, we obtain

$$\frac{1 + r_1 y + r_2(y^2 - 2) + (1 - r_1^2 - r_2^2 + r_1 r_2)x}{1 + r_1 y + r_2(y^2 - 2) + (1 - r_1^2 - r_2^2 + r_1 r_2)x^2} = \prod_{i=1}^{2n} \left(\frac{1 + (a_i t)y + (1 - (a_i t)^2)x}{1 + (a_i t)y + (1 - (a_i t)^2)x^2} \right).$$

Upon expanding the product of the numerators and denominators, the right-hand side becomes

$$\frac{b_0 + b_1 y + b_2(y^2 - 2) + \left(c_0 + c_1 y + c_2(y^2 - 2)\right) x}{b_0 + b_1 y + b_2(y^2 - 2) + \left(c_0 + c_1 y + c_2(y^2 - 2)\right) x^2} \tag{5.7}$$

with b_i, c_i polynomials over $\mathbb{F}_{q^n}$ of degree $4n$ in $a_1, \ldots, a_{2n}$. In general, these polynomials have a large number of terms and are thus slow to compute with. However, as before, by construction these polynomials are symmetric in the $a_1, \ldots, a_{2n}$, so one can rewrite the b_i and c_i in terms of the $2n$ elementary symmetric polynomials $\sigma_j(a_1, \ldots, a_{2n})$ for $j = 1, \ldots, 2n$. This dramatically reduces the degree and size of these polynomials: in particular they become quadratic and, as a consequence, the number of terms is much lower, being bounded by $4n + \binom{2n}{2} + 1$.

To generate a system of quadratic equations, we use the embedding of $T_6(\mathbb{F}_{q^n})$ into $T_2(\mathbb{F}_{q^{3n}})$ and consider the Weil restriction of the following equality:

$$\frac{b_0 + b_1 y + b_2(y^2 - 2)}{c_0 + c_1 y + c_2(y^2 - 2)} = \frac{1 + r_1 y + r_2(y^2 - 2)}{1 - r_1^2 - r_2^2 + r_1 r_2}.$$

This leads to three quadratic equations over $\mathbb{F}_{q^n}$ or, equivalently, to $3n$ quadratic equations over $\mathbb{F}_q$ in the $2n$ unknowns $\sigma_1, \ldots, \sigma_{2n}$. Observe that amongst these equations there must be at least n dependencies arising from the fact that we used the embedding into T_2 rather than T_6.

The properties of such systems, which have the same structure but differ only by the coefficients of R, were investigated using the Magma implementation of the F4 algorithm [185]. It was found that: the ideal generated is zero-dimensional; the Gröbner basis with respect to the lexicographic ordering satisfies the so-called Shape Lemma, i.e., the basis is of the form:

$$\sigma_1 - g_1(\sigma_{2n}), \ \sigma_2 - g_2(\sigma_{2n}), \ \ldots, \ \sigma_{2n-1} - g_{2n-1}(\sigma_{2n}), \ g_{2n}(\sigma_{2n}),$$

where $g_i(\sigma_{2n})$ is a univariate polynomial in σ_{2n} for each i; and in all cases it holds that $\deg(g_{2n}) = 3^n$, rather than the bound of 2^{2n} that one would expect from Bezout's theorem.

Provided that n is not prohibitively large, such systems can be solved in a reasonable time. To test if a random point R decomposes over $\mathcal{F}$, one computes the roots of $g_{2n}(\sigma_{2n})$ in $\mathbb{F}_q$, and then substitutes these in the other g_i to find the values of the other σ_i. For each such solution we then test if the polynomial

$$p(x) := x^{2n} - \sigma_1 x^{2n-1} + \sigma_2 x^{2n-2} - \cdots + (-1)^{2n} \sigma_{2n}$$

splits completely over $\mathbb{F}_q$. Whenever it does, the roots a_i for $i = 1, \ldots, 2n$ lead to a relation of the form in Eq. (5.6).

In terms of complexity, when $n! \approx q$ the full algorithm runs in expected time $L_{q^n}(1/2, c')$ for some $c' > 0$. In terms of experimental results, Table 2 of [238] gave expected running times for attacking $T_6(\mathbb{F}_{q^n})$, for $n = 1, \ldots, 5$. In particular, it showed that in the group $T_{30}(\mathbb{F}_q)$ – which can be embedded into $T_6(\mathbb{F}_{q^5})$ – discrete logarithms are easier than previously expected. Indeed, this group was proposed in [366] and [598] for cryptographic use, and keys of length 960 bits were recommended, i.e., with $q \approx 2^{32}$. The experiments showed that even with a Magma implementation it would be feasible to compute discrete logarithms in $T_{30}(\mathbb{F}_q)$ with $q \approx 2^{20}$, and the attack for $q \approx 2^{32}$ would be about 1000 times faster than Pollard's rho, albeit with a far larger memory constraint. In light of this attack, the security offered by the DLP in finite fields of the form $\mathbb{F}_{q^{30}}$ needed to be reassessed.

5.5 Discrete Logarithms in Finite Fields of Fixed Characteristic

Progress in cryptanalytic algorithms, just as in science more generally, usually evolves by small increments but with occasional revolutionary steps forward [347]. One example of such a step forward could arguably be the rapid development of efficient algorithms for solving the DLP in finite fields of fixed characteristic, that took place from late 2012 to mid 2014, thanks to the present authors and their collaborators. Between these times, the fastest algorithm for solving this problem went from having complexity $L(1/3)$ to being quasi-polynomial [32, 226, 227, 243, 244, 246, 294], rendering such fields entirely unsuitable for discrete logarithm-based cryptography, including pairing-based cryptography over small characteristic supersingular curves. These events constitute a perfect example of Lenstra's (perhaps jocular, but no doubt, in part, quite serious) contention that no problem based on number theory should ever be considered truly secure, even if it has remained impenetrable for several decades.[3]

Since 2014, there have been many surveys of the state of the art in discrete logarithm algorithms for finite fields [245, 299, 303]. Therefore, in the present section we focus only on the key ideas behind the fixed characteristic breakthroughs, to give a flavour of what was behind them, as well as the central results.

5.5.1 Key Insights

If one performs the basic index calculus method as described in Section 5.1.2, in $\mathbb{F}_{q^n}$ for fixed q and $n \to \infty$, then, by using a theorem due to Odlyzko [460] and Lovorn [400], the distribution of smooth polynomials naturally leads to an $L(1/2)$ complexity algorithm. It may therefore have been assumed for some years that this is the best complexity that can be achieved.

A key avenue to improving these index calculus algorithms is to find an approach that generates the relations faster. A first idea is to create relations between elements with smaller norms, in order to increase the smoothness probability and also to reduce the size of the factor base. This is the basis of Coppersmith's celebrated 1984 algorithm [134, 135] and all subsequent $L(1/3)$ algorithms for larger characteristic. These algorithms become heuristic because the considered relations involve equality between elements that are neither independent nor uniformly distributed. A long-standing open problem is to find a way to lift these heuristics.

Alternatively, it turned out that the lack of independence can be used to speed up index calculus for certain fields. This was first described by Joux in 2012 [293], building upon Joux and Lercier's medium prime function field sieve method [296].

This gives the hope of being able to generate field equations between elements that have better than expected smoothness properties. This idea is more subtle than one might assume in retrospect, seeing the breakthroughs it led to. Indeed, before these breakthroughs, this possibility does not even seem to have ever been considered. Most likely, this is because from a complexity analysis perspective it seemed essential that the expected smoothness properties of generated elements hold.

It is an instantiation of this second idea that initiated the aforementioned progress in this area. The technique was discovered independently and at

---

[3]   This perspective is attributed to A. K. Lenstra by the first-listed author, having worked with him for four years and having discussed such matters a few times; any error in attribution is entirely his.

approximately the same time by the present authors, in two different but essentially isomorphic approaches. In particular, it was shown how to produce a family of polynomials of high degree that are smooth by construction, and which thus lead to useful relations, in contrast to uniformly generated polynomials of the same degree, which have only an exponentially small probability of being smooth. The family of polynomials and its exploitation lay the foundation for two independent and theoretically distinct quasi-polynomial algorithms: the first due to Barbulescu, Gaudry, Joux and Thomé (BGJT) in 2013 [32] and the second due to Granger, Kleinjung and Zumbrägel (GKZ) in 2014 [246]. Here, and in the following, the years refer to the initial announcement of related records (see also Table 5.1), typically preceding formal publication.

In terms of historical development, the approach of the BGJT quasi-polynomial algorithm grew naturally from Joux's 2013 paper [294], which is an extension of the previously mentioned method of [293] for medium characteristic. On the other hand, the GKZ quasi-polynomial algorithm grew from observations in [243] and the techniques of [226], which combined independent observations with a specialisation of the the field representation in [296], which was itself motivated by the Granger–Vercauteren algorithm [238].

5.5.2 Polynomial-Time Relation Generation

In order to give a flavour of the ideas behind the breakthrough techniques, we now describe the polynomial-time relation generation methods published by Göloğlu, Granger, McGuire and Zumbrägel (GGMZ) on the one hand [226], and Joux on the other [294], both announced in February 2013.

Both methods start with a family of finite fields $\mathbb{F}_{p^n}$ in which the DLP is to be solved, with p fixed and $n \to \infty$. Each of these is embedded into a corresponding field of the form $\mathbb{F}_Q = \mathbb{F}_{q^{kn}}$, with $k \geq 2$ fixed and $n \approx q$, by setting $q = p^\ell$ with $\ell = \lceil \log_p n \rceil$, increasing the extension degree by a factor of $k\lceil \log_p n \rceil$, which does not significantly impact the complexity of the resulting algorithms.

The GGMZ Method

Let $\mathbb{F}_{p^n} \hookrightarrow \mathbb{F}_{q^{kn}}$ be the target field and let $\mathbb{F}_{q^k}$ be represented arbitrarily. In order to represent $\mathbb{F}_{q^{kn}}$, the GGMZ method uses an extremely unbalanced version of the field representation employed in the Joux–Lercier function field sieve [296]. In particular, let $f = X^q$ and $g = \frac{h_0}{h_1}$ for some $h_0, h_1 \in \mathbb{F}_{q^k}[X]$ of low degree[4] $\leq d_h$, so that there exists a monic irreducible $I \in \mathbb{F}_{q^k}[X]$ of

---

[4] In [226] h_1 was not specified and was thus implicitly 1; h_1 was introduced in [243] in order to increase the number of representable extension degrees.

degree n such that $h_1(X^q)X - h_0(X^q) \equiv 0 \pmod{I}$. For such h_0, h_1, I we define $\mathbb{F}_{q^{kn}} := \mathbb{F}_{q^k}[X]/(I)$.

Let x be a root of I in $\mathbb{F}_{q^{kn}}$ and let $y = f(x) = x^q$. Then by construction we have $x = g(y) = \frac{h_0(y)}{h_1(y)}$, giving an isomorphism between two representations of $\mathbb{F}_{q^{kn}}$, namely $\mathbb{F}_{q^k}(x)$ and $\mathbb{F}_{q^k}(y)$. The factor base is as simple as could be expected, consisting of $h_1(x^q)$ and all linear polynomials on the x-side; the y-side factor base is unnecessary because for all $d \in \mathbb{F}_{q^k}$ one has $(y + d) = (x + d^{1/q})^q$.

For $a, b, c \in \mathbb{F}_{q^k}$, consider elements of $\mathbb{F}_{q^{kn}}$ of the form $xy + ay + bx + c$. Using the above field isomorphisms we have the following identity:

$$x^{q+1} + ax^q + bx + c = \frac{1}{h_1(y)}\left(yh_0(y) + ayh_1(y) + bh_0(y) + ch_1(y)\right). \tag{5.8}$$

A key observation is that the left-hand side of Eq. (5.8) has a very special form, and provably splits completely over $\mathbb{F}_{q^k}$ with probability $\approx 1/q^3$, which is exponentially higher than the probability that a uniformly random polynomial of the same degree splits completely over $\mathbb{F}_{q^k}$, which is $\approx 1/(q + 1)!$. Indeed, for $k \geq 3$, consider the polynomial $X^{q+1} + aX^q + bX + c$. For $ab \neq c$ and $a^q \neq b$, this polynomial may be transformed (up to a scalar) into

$$F_B(\overline{X}) = \overline{X}^{q+1} + B\overline{X} + B, \quad \text{with} \quad B = \frac{(b - a^q)^{q+1}}{(c - ab)^q},$$

via $X = \frac{c-ab}{b-a^q}\overline{X} - a$. Observe that the original polynomial splits completely over $\mathbb{F}_{q^k}$ whenever F_B splits completely over $\mathbb{F}_{q^k}$ and we have a valid transformation from $\overline{X}$ to X. The following theorem provides the precise number of $B \in \mathbb{F}_{q^k}$ for which $F_B(\overline{X})$ splits completely over $\mathbb{F}_{q^k}$.

Theorem 5.4. (Bluher [71]) *The number of elements $B \in \mathbb{F}_{q^k}^\times$ such that the polynomial $F_B(\overline{X}) \in \mathbb{F}_{q^k}[\overline{X}]$ splits completely over $\mathbb{F}_{q^k}$ equals*

$$\frac{q^{k-1} - 1}{q^2 - 1} \quad \textit{if k is odd}, \qquad \frac{q^{k-1} - q}{q^2 - 1} \quad \textit{if k is even}.$$

By using the expression for B in terms of a, b, c one can generate triples (a, b, c) for which the left-hand side of Eq. (5.8) always splits over $\mathbb{F}_{q^k}$. In particular, first compute the set $\mathcal{B}$ of all $B \in \mathbb{F}_{q^k}$ for which F_B splits over $\mathbb{F}_{q^k}$. Then for any $a, b \neq a^q$ and $B \in \mathcal{B}$ there is a uniquely determined c for which the left-hand side splits. By Theorem 5.4, there are $\approx q^{3k-3}$ such triples, giving the aforementioned probability $1/q^3$. For such triples, whenever the right-hand side of Eq. (5.8) splits, one obtains a relation amongst the factor base elements.

This just leaves the case $k = 2$, for which there are no such F_B. However, the set of triples for which the left-hand side splits non-trivially can be shown to be

$$\{(a, a^q, c) \mid a \in \mathbb{F}_{q^2} \text{ and } c \in \mathbb{F}_q, c \neq a^{q+1}\}.$$

So for $k \geq 2$, assuming the right-hand side splits with probability $1/(d_h + 1)!$, there will be sufficiently many relations when $q^{2k-3} > (d_h + 1)!$. Then, for fixed d_h and $q \to \infty$, the cost of computing the logarithms of all of the factor base elements is heuristically $O(q^{2k+1})$ operations in $\mathbb{Z}/(q^{kn} - 1)\mathbb{Z}$ as one can use sparse (weight q) linear algebra techniques; for fixed k this complexity is polynomial in $\log Q = q^{1+o(1)}$, as claimed.

Joux's Method

Joux's method [294] also applies to fields of the form $\mathbb{F}_Q = \mathbb{F}_{q^{kn}}$ (with $k = 2$ being used for the exposition and initial examples), but the crucial degree-n extension is built in a slightly different, but analogous manner. In particular, we have $\mathbb{F}_Q = \mathbb{F}_{q^k}(x) = \mathbb{F}_{q^k}[X]/(I)$, where $I \mid h_1(X)X^q - h_0(X)$ for some $h_0, h_1 \in \mathbb{F}_{q^k}[X]$ of low degree $\leq d_h$. This leads to the field equation $x^q = \frac{h_0(x)}{h_1(x)}$. The factor base consists of $h_1(x)$ and all linear polynomials in x.

Joux's method starts with the identity

$$\prod_{\mu \in \mathbb{F}_q} (X - \mu) = X^q - X.$$

If one substitutes X by $\frac{\alpha X+\beta}{\gamma X+\delta}$ with $\alpha, \beta, \gamma, \delta \in \mathbb{F}_{q^k}$ and $\alpha\delta - \beta\gamma \neq 0$, multiplying by $(\gamma X + \delta)^{q+1}$ gives

$$(\gamma X+\delta) \prod_{\mu \in \mathbb{F}_q} ((\alpha X+\beta) - \mu(\gamma X+\delta)) = (\alpha X+\beta)^q(\gamma X+\delta) - (\alpha X+\beta)(\gamma X+\delta)^q. \quad (5.9)$$

Observe that the right-hand side of Eq. (5.9) has the same monomial degrees as the left-hand side of Eq. (5.8), and automatically splits completely over $\mathbb{F}_{q^k}$ by virtue of the left-hand side of Eq. (5.9). Applying the field equation $x^q = \frac{h_0(x)}{h_1(x)}$ to the right-hand side of Eq. (5.9) produces

$$\frac{1}{h_1(x)}(\alpha^q h_0(x) + \beta^q h_1(x))(\gamma x + \delta) - (\alpha x + \beta)(\gamma^q h_0(x) + \delta^q h_1(x)),$$

and if this degree-$d_h + 1$ polynomial also splits over $\mathbb{F}_{q^k}$ then one has a relation amongst factor base elements.

In order to count the number of distinct splitting polynomials that one can obtain in this manner, first observe that the total number of $(\alpha, \beta, \gamma, \delta)$-transformations is $|\mathrm{PGL}_2(\mathbb{F}_{q^k})| = q^{3k} - q^k$. Second, observe that two transformations will give the same relation (up to multiplication by a scalar in $\mathbb{F}_{q^k}^\times$) if there exists an element of $\mathrm{PGL}_2(\mathbb{F}_q)$, which when multiplied by the first transformation gives the second. Hence the total number of distinct transformations is $\approx q^{3k-3}$, just as we found for the GGMZ method. From a practical perspective, in order to avoid repetitions one should compute a set of co-set representatives for the quotient $\mathrm{PGL}_2(\mathbb{F}_{q^k})/\mathrm{PGL}_2(\mathbb{F}_q)$; by contrast the GGMZ method already achieves this implicitly.

5.5.3 $L(1/4 + o(1))$ and Quasi-Polynomial Algorithms

The two methods just described mean that the first stage of index calculus is (at least heuristically) solvable in polynomial time. So the remaining problem is to compute individual logarithms. However, owing to the extension degree being $n = O(q)$ and the factor base being polynomial only in the size of the field, this is now much harder than before. In particular, if one uses the usual descent method from [296] then the elimination of an element – i.e., expressing it as a product of elements of lower degree, modulo the field polynomial – becomes harder as the degree becomes smaller, with degree-two eliminations being the bottleneck. However, with independent and distinct methods GGMZ [226] and Joux [294] showed how to eliminate degree-two elements efficiently. For reasons of space we refer the reader to the original papers for their expositions (or to the survey article [245]), and note that these methods spawned the building blocks of the individual logarithm stages of the two aforementioned quasi-polynomial algorithms.

Joux [294] also gave a new elimination method that relies on solving multivariate bilinear quadratic systems via Gröbner basis computations, whose cost increases with the degree. Balancing the costs of the Gröbner basis descent and the classical descent (whose cost decreases with the degree) results in a heuristic $L(1/4 + o(1))$ algorithm, which was the first algorithm to break the long-standing $L(1/3)$ barrier. This can be tweaked for fields of the present form to obtain an $L(1/4)$ algorithm [227].

Soon afterwards, in June 2013, Barbulescu, Gaudry, Joux and Thomé announced an algorithm for solving the DLP [32] in the fields $\mathbb{F}_{q^{kn}}$ with $k \geq 2$ fixed and $n \leq q + d$ with d very small, which for $n \approx q$ has heuristic quasi-polynomial time complexity

$$(\log q^{kn})^{O(\log \log q^{kn})} = \exp(O((\log \log q^{kn})^2)). \tag{5.10}$$

Because Eq. (5.10) is smaller than $L(\alpha)$ for any $\alpha > 0$, this constituted a very significant breakthrough for the DLP in finite fields of fixed characteristic. Moreover, when the cardinality of the base field $\mathbb{F}_{q^k}$ can be written as $q^k = L_{q^{kn}}(\alpha)$, the algorithm results in complexity $L(\alpha)$, thus providing an improvement over the original function field sieve algorithms whenever $\alpha < 1/3$. As for the $L(1/4)$ method, this algorithm relies on unproven heuristics. Moreover, it is an asymptotic improvement whose cross-over point with previous techniques is too high to make it usable in record computations.

In February 2014, Granger, Kleinjung and Zumbrägel developed an alternative quasi-polynomial algorithm for fields of essentially the same form. Just as the BGJT elimination step may be viewed as a generalisation of Joux's degree-two elimination method, the GKZ elimination step depends on the degree-two elimination method of GGMZ (albeit combined with another crucial but simple idea). Thanks to the algebraic nature of the elimination method, the only assumption required for the algorithm to be rigorously proven to work is one regarding the existence of a suitable field representation. In particular, the following theorems were proven in [246].

Theorem 5.5. *Given a prime power $q > 61$ that is not a power of 4, an integer $k \geq 18$, co-prime polynomials $h_0, h_1 \in \mathbb{F}_{q^k}[X]$ of degree at most two and an irreducible degree-n factor I of $h_1 X^q - h_0$, the DLP in $\mathbb{F}_{q^{kn}} \cong \mathbb{F}_{q^k}[X]/(I)$ can be solved in expected time*

$$q^{\log_2 n + O(k)}.$$

That the degree of h_0, h_1 is at most two is essential to eliminating smoothness heuristics, since this ensures that the co-factor of the right-hand side of Eq. (5.8) has degree at most one, and is thus automatically 1-smooth. This theorem is reproved by a slightly easier approach that gives better parameters for q and k in [225]. A simple application of Kummer theory shows that such h_1, h_0 exist when $n = q - 1$, which gives the following easy corollary when $m = ik(p^i - 1)$.

Theorem 5.6. *For every prime p there exist infinitely many explicit extension fields $\mathbb{F}_{p^m}$ in which the DLP can be solved in expected quasi-polynomial time*

$$\exp\left((1/\log 2 + o(1))(\log m)^2\right).$$

In practice it is very easy to find polynomials h_0, h_1 for general extension degrees according to Theorem 5.5, and heuristically it would appear to be all but guaranteed. However, proving their existence seems to be a hard problem. The idea of using an alternative field representation arising from torsion points of elliptic curves to obviate this issue is a very natural one. Such field representation were initially introduced by Couveignes and Lercier in [142]. At least three teams of researchers have developed this idea [300, 326, 393] in order to build an analogue of the GKZ algorithm using this alternative field representation. As of the time of writing, the pre-print of Kleinjung and Wesolowski [326] is the only one to claim a full proof. Previously, only an $L(1/2)$ complexity had been proven rigorously for arbitrary extension degrees, so this is a very significant theoretical result. More precisely, [326] states the following.

Theorem 5.7. *Given any prime number p and any positive integer n, the discrete logarithm problem in the group $\mathbb{F}_{p^n}$ can be solved in expected time $(pn)^{2\log_2(n)+O(1)}$.*

5.5.4 Practical Impact

From the perspective of mathematical cryptology, rigorously proving the correctness of new DLP algorithms is of central theoretical interest. However, in terms of real-world cryptographic impact, what matters far more is how practical the algorithms are and whether they can be used to solve previously unsolvable DLP instances. Furthermore, as is well known to practitioners and computational number theorists, carrying out large-scale implementations often leads to new theoretical insights that can, in turn, result in superior algorithms. Hence, the value of practical considerations should not be overlooked.

Shortly after GGMZ and Joux discovered their methods, a period of intense competition began, both in theory [32, 226, 227, 243, 244, 246, 294, 298] as already alluded to, and in practice; see Table 5.1. As one can see these computational records dwarfed those that had been set previously, leading small characteristic pairing-based cryptography to be entirely eschewed by the cryptographic community. Without doubt, this 'academic arms race' accelerated and stretched the development of the new discrete logarithm algorithms, and as such were scientifically extremely beneficial.

As of the time of writing the largest such (publicly known) DLP computation was completed in the field $\mathbb{F}_{2^{30750}}$, by Granger, Kleinjung, Lenstra, Wesolowski and Zumbrägel; this was announced in July 2019 and required approximately 2900 core years [239]. The main purpose of the computation was to test the GKZ quasi-polynomial descent method at scale for the first time, in order to assess its reach when the number of core hours expended is comparable to the number expended during the largest DLP computations in prime fields and integer factorisation efforts. In July 2019, at the time of the announcement [239], the record for the former was in a field of bit-length 768, set in June 2016 [330]; the current record is in a field of bit-length 795, announced in December 2019 [100]. For the latter, back in July 2019 at the time of the announcement [239], the record was the factorisation of a 768-bit RSA challenge modulus, set in December 2009 [327]. Also in December 2019, the factorisation of a 795-bit RSA challenge modulus was announced [100], which was swiftly improved upon in February 2020 by the solving of an 829-bit RSA

Bitlength	Char.	Kummer	Who/when	Running time
127	2	no	Coppersmith, 1984 [134]	$L(1/3, 1.526..1.587)$
401	2	no	Gordon and McCurley, 1992 [233]	$L(1/3, 1.526..1.587)$
521	2	no	Joux and Lercier, 2001	$L(1/3, 1.526)$
607	2	no	Thomé, 2002	$L(1/3, 1.526..1.587)$
613	2	no	Joux and Lercier, 2005	$L(1/3, 1.526)$
556	m	yes	Joux and Lercier, 2006 [296]	$L(1/3, 1.442)$
676	3	no	Hayashi *et al.*, 2010 [265]	$L(1/3, 1.442)$
923	3	no	Hayashi *et al.*, 2012 [266]	$L(1/3, 1.442)$
1 175	m	yes	Joux, 24 Dec 2012	$L(1/3, 1.260)$
619	2	no	CARAMEL, 29 Dec 2012	$L(1/3, 1.526)$
1 425	m	yes	Joux, 6 Jan 2013	$L(1/3, 1.260)$
1 778	2	yes	Joux, 11 Feb 2013	$L(1/4 + o(1))$
1 971	2	yes	GGMZ, 19 Feb 2013	$L(1/3, 0.763)$
4 080	2	yes	Joux, 22 Mar 2013	$L(1/4 + o(1))$
809	2	no	CARAMEL, 6 Apr 2013	$L(1/3, 1.526)$
6 120	2	yes	GGMZ, 11 Apr 2013	$L(1/4)$
3 164	2	yes	GGMZ, May 2013	$L(1/3, 0.763)$
6 168	2	yes	Joux, 21 May 2013	$L(1/4 + o(1))$
1 303	3	no	AMOR, 27 Jan 2014	$L(1/4 + o(1))$
4 404	2	no	GKZ, 30 Jan 2014	$L(1/4 + o(1))$
9 234	2	yes	GKZ, 31 Jan 2014	$L(1/4 + o(1))$
1 551	3	no	AMOR, 26 Feb 2014	$L(1/4 + o(1))$
3 796	3	no	Joux and Pierrot, 15 Sep 2014	$L(0 + o(1))$
1 279	2	no	Kleinjung, 17 Oct 2014	$L(0 + o(1))$
4 841	3	no	ACCMORR, 18 Jul 2016	$L(0 + o(1))$
30 750	2	yes	GKLWZ, 10 July 2019	$L(0 + o(1))$

Table 5.1 *Large-scale discrete logarithm computations in finite fields of small or medium ('m') characteristic (Char.). Details of uncited results can be found in the number theory mailing list [458].*

challenge [102]. For Mersenne numbers, an implementation of Coppersmith's factorisation factory idea resulted in January 2015 in the factorisation of the 17 remaining unfactored moduli of the form $2^n - 1$ with $1007 \le n \le 1199$ [329].

In terms of remaining hard open problems in the area of finite-field discrete logarithms, there are two central – and natural – ones. The first challenging problem is to find a classical polynomial time algorithm for fixed-characteristic DLPs, either heuristic or rigorous. The second, probably far more challenging, problem is to develop quasi-polynomial classical algorithms for medium and large characteristic fields. As there is far less structure for prime fields in particular, it seems that fundamentally new ideas will be required.

5.6 Conclusion

As is well known, the DLP and the integer factorisation problem can be solved in polynomial time using a sufficiently large quantum computer [539]. At present, such computers are not available, despite a widespread worry or excitement that they might come soon. To be ready when this occurs, a large part of the cryptographic community is currently working on post-quantum secure alternatives. However, the flexibility of discrete logarithms for constructing cryptographic protocols is currently unsurpassed. As a consequence, it remains essential to study the security of discrete logarithms against classical computers. New sporadic breakthroughs could possibily occur and would likely also affect factoring and the RSA cryptosystem.

Acknowledgements This work has been supported in part by the European Union's H2020 Programme under grant agreement number ERC-669891.

6

RSA, DH and DSA in the Wild

Nadia Heninger

6.1 Introduction

The previous chapters discussed breaking practical cryptographic systems by solving the mathematical problems directly. This chapter outlines techniques for breaking cryptography by taking advantage of implementation mistakes made in practice, with a focus on those that exploit the mathematical structure of the most widely used public-key primitives.

In this chapter, we will mostly focus on public-key cryptography as it is used on the Internet, because as attack researchers we have the greatest visibility into Internet-connected hosts speaking cryptographic protocols.

While this chapter was being written, prior to the standardisation or wide-scale deployment of post-quantum cryptography, the set of algorithms used for public-key cryptography on the Internet was surprisingly limited.

In the context of communications protocols such as the transport layer security (TLS), secure shell (SSH), and Internet protocol security (IPsec), key exchange is accomplished by using finite-field Diffie–Hellman (DH), elliptic-curve Diffie–Hellman (ECDH), or Rivest–Shamir–Adleman (RSA) encryption. Historically, RSA encryption was the most popular key-exchange method used for TLS, while SSH and IPsec only supported Diffie–Hellman. It is only in the past few years that ECDH key-exchange has begun to be more popular for all three protocols. The digital signature algorithms (DSA) most widely used in practice are RSA, elliptic-curve DSA (ECDSA), and prime field DSA. RSA is by far the most popular digital signature algorithm in use, and is only just now beginning to be supplanted by ECDSA for network protocols.

6.2 RSA

Historically, RSA was by far the most common public-key encryption method, as well as the most popular digital signature scheme, and it remains extremely commonly used in the wild. Ninety per cent of the HTTPS certificates seen by the ICSI Certificate Notary use RSA for digital signatures [281]. Both TLS 1.1 [160] (standardised in 2006) and 1.2 [161] (standardised in 2008) require TLS-compliant implementations to support RSA for key-exchange. RSA encryption and signatures are also still widely used for PGP. Both host and client authentication via RSA digital signatures were recommended in the original SSH specifications [629], and remain in common use today [430]. RSA also remains ubiquitous for other authentication scenarios, including smart cards and code signing.

6.2.1 RSA Key Generation

RSA Key Generation, in Theory Textbook RSA key generation works as follows: first, one generates two primes p and q of equal size, and verifies that $p - 1$ and $q - 1$ are relatively prime to the desired public exponent e. Then one computes the modulus $N = pq$ and private exponent $d = e^{-1} \bmod (p-1)(q-1)$. The public key is then the pair (e, N) and the private key is the pair (d, N) [501].

Factorisation of the modulus N remains the most straightforward method of attack against RSA, although this is not known to be equivalent to breaking RSA [8, 82].

RSA Key Generation, in Practice Both textbook RSA and the description of RSA in many standards leave a number of choices to implementers. In addition, there are several unusual properties of most RSA implementations that are surprising from a purely theoretical point of view.

6.2.2 Prime Generation

The simplest method for an implementation to generate a random prime is to seed a pseudorandom-number generator (PRNG), read a bit string out of the PRNG of the desired length of the prime, interpret the bit string as an integer, and then test the integer for primality by using, typically, a combination of trial division and some number of iterations of a probabilistic primality test such as Miller–Rabin [431, 490]. If the number is determined to be composite, a new sequence of bits is read from the pseudorandom-number generator and the primality tests are repeated until a prime is found. Alternatively, an

implementation may start at a pseudorandomly generated integer and increment or sieve through successive integers within some range until a prime is
found. The prime-generation process varies across implementations in different cryptographic libraries [16].

6.2.3 Prime Structure

Many implementations enforce additional structural properties of the primes
that they generate for RSA. Common properties include being 'safe' primes,
that is, that $(p - 1)/2$ is also a prime, or that $(p - 1)$ and/or $(p + 1)$ have known
prime factors of some minimum size. For example, the National Institute of
Standards and Technology (NIST) recommends that these 'auxiliary' prime
factors should be at least 140 bits for the 1024-bit primes used to construct a
2048-bit RSA key [283]. This is intended to protect against Pollard's $p - 1$ and
$p + 1$ factoring algorithms [475]. For the 1024-bit and larger key sizes common
in practice, random primes are unlikely to have the required structure to admit
an efficient attack via these algorithms, so enforcing a particular form for the
primes used for RSA is less important.

Implementation Fingerprints The different choices that implementations
make in generating primes can result in quite different distributions of properties among the primes generated by these distinct implementations. Some of
these are detectable only from the prime factorisation, others are evident from
the public key.

Mironov [435] observed that OpenSSL (Open Secure Sockets Layer), by
default, generates primes p that have the property that $p \neq 1 \bmod 3, p \neq 1 \bmod$
$5, \ldots, p \neq 1 \bmod 17\,863$; that is, that $p - 1$ is not divisible by any of the first
2048 primes p_i. The probability that a random 512-bit prime has this property
is $\prod_{i=2}^{2048}(p_i - 2)/(p_i - 1) \approx 7.5$ per cent, so the probability that a random 1024-
bit RSA modulus has this property is 0.05625 per cent, and thus a factored
RSA key can be identified as having been generated by OpenSSL or not with
good probability.

Svenda *et al.* [578] documented numerous implementation artifacts that
could be used to identify the implementation used to generate a key or a collection of keys. Many libraries generated primes with distinctive distributions of
prime factors of $p-1$ and $p+1$, either eliminating small primes, enforcing large
prime divisors of a given size, or some combination of the two; some libraries
clamped the most significant bits of the primes p and q to fixed values such as
1111 and 1001, resulting in a predictable distribution of most significant bits
of the public modulus.

The ability to fingerprint implementations does not seem to be an immediate vulnerability for RSA encryption or signatures, although for some applications of RSA, where, for example, a user might use a public key as a pseudonymous identifier, the loss of privacy due to implementation fingerprinting may constitute a vulnerability. However, this analysis later led to the discovery of ROCA ('Return of Coppersmith's Attack'), which exploited the fact that the specific prime-generation process used by Infineon smart cards resulted in so much structure to the prime that the resulting RSA keys were efficiently factorable using a variant of Coppersmith's lattice methods [451]. See Chapter 4 for a detailed explanation about Coppersmith's method.

6.2.4 Prime Sizes

RSA moduli are almost universally generated from two equal-sized primes. However, exceptions occasionally arise. There are a small number of RSA moduli that have been found to be divisible by very small primes: Lenstra *et al.* [386] report finding 171 RSA moduli used for HTTPS and PGP with prime factors less than 2^{24} in 2012, and Heninger *et al.* [269] report finding 12 SSH host keys that were divisible by very small primes. Many of these may be due to copy–paste or memory errors in an otherwise well-formed modulus; several of the moduli with small factors were one hex character different from another valid modulus they observed in use, or contained unlikely sequences of bytes.

Anecdotally, at least one RSA implementation accidentally generated prime factors of unequal sizes, because the library had accidentally fixed the size of one prime to 256 bits, and then generated an n-bit RSA modulus by generating a second prime of length $n - 256$ [342].

6.2.5 Short Modulus Lengths

In 1999, 512-bit RSA was first factored [118] and, by 2015, 512-bit factorisation was achievable within a few hours using relatively modest computing resources [593]. The use of 1024-bit RSA moduli was allowed by NIST recommendations until 2010, deprecated until 2013, and disallowed after 2013 [39]. In 2009, 1024-bit RSA was already believed to be feasible in principle to a powerful adversary using only general-purpose computing, although such a calculation was thought to be infeasible for an academic community effort for the near future, and as of this writing no factorisation of a generic 1024-bit RSA modulus has been reported yet in the public literature [88].

However, 512- and 1024-bit RSA keys remained in common use well after these dates for a number of reasons including hard-coded limits on key sizes and long-term keys that were difficult to deprecate. Multiple long-term TLS certificate authority root certificates with 1024-bit RSA keys and multi-decade validity periods remained trusted by major browsers as late as 2015 [624], and the process of certificate authorities phasing out end-user certificates with 1024-bit RSA keys was still in process years afterward. DNSSEC keys are practically limited to 1024-bit or shorter RSA signing keys because many older router and network middlebox implementations cannot handle User Datagram Protocol (UDP) packets larger than 1500 bytes [36]. Because of these limitations, and the fact that DNSSEC predates widespread support for ECDSA, which has much shorter keys and signatures, tens of thousands of 512-bit RSA public keys were still in use for DNSSEC in 2015 [593]. A few thousand 512-bit RSA keys were still in use in 2015 for both HTTPS and mail protocols, and millions of HTTPS and mail servers still supported insecure 512-bit 'export'-grade RSA cipher suites [13, 175, 593]. As of 2020, support for 512-bit RSA has dropped to less than 1 per cent of popular HTTPS sites [486].

There are multiple structural causes contributing to the long lifespans of short keys. The choice of public-key length is typically left to the party generating the key. Since RSA keys were historically considered to be computationally expensive to generate, RSA key pairs tend to be infrequently generated and valid for long periods. In order to maintain interoperability, most implementations have traditionally been permissive in the lengths of RSA moduli they will accept.

Pathologically short keys are also occasionally found in practice, presumably as a result of implementers who do not understand the security requirements for factorisation-based cryptosystems. A 128-bit RSA modulus used for the DKIM protocol used to authenticate email senders was in use in 2015 [593]. In 2016, an implantable cardiac device made by St. Jude Medical was found to be secured using 24-bit RSA [397]. (The US FDA later issued a recall.)

6.2.6 Public Exponent Choice

In theory, the public exponent e could have any length, and RSA is not known to be insecure in general with most possible choices of e, either large or small. In practice, however, implementations nearly universally use short exponents, and in fact typically restrict themselves to a handful of extremely common values.

Common Exponents The public exponent e does not contain any secret information, and does not need to be unique across keys. By far the most commonly used RSA public exponent is the value $65\,537 = 2^{16} + 1$, which has the virtue of being relatively small and has low Hamming weight, so that encryption and signature verification are both fast.

Short Public Exponents Although very small e such as $e = 3$ are not known to be broken in general, the existence of several low-exponent RSA attacks such as Coppersmith's low-exponent decryption attacks [138] and Bleichenbacher's low-exponent signature forgery attacks [188] makes many practitioners nervous about such values. For example, NIST requires $e > 2^{16}$ [38], which is believed to be large enough to render even hypothesised improvements to these attacks infeasible.

There are multiple barriers to using larger RSA public exponents in practice. The Windows CryptoAPI used in Internet Explorer until as late as Windows 7 encodes RSA public exponents into a single 32-bit word, and cannot process a public key with a larger exponent. NIST requires that RSA public exponents be at most 256 bits [38].

The Case $e = 1$ Occasionally implementers do choose terrible values for e: in 2013, the Python SaltStack project fixed a bug that had set $e = 1$ [586], and Lenstra *et al.* [386] found eight PGP RSA keys using exponent $e = 1$ in 2012. In this case, an encrypted 'ciphertext' would simply be the padded message itself, thus visible to any attacker, and a 'signature' would be the padded hash of the message, trivially forgeable to any attacker.

6.2.7 Repeated RSA Moduli

If two parties share the same public RSA modulus N, then both parties know the corresponding private keys, and thus may decrypt each others' traffic and forge digital signatures for each other. Thus, in theory, one would expect RSA public moduli to be unique in the wild. In practice, however, it turns out that repeated RSA public keys are quite common across the Internet.

In 2012, Lenstra *et al.* [386] found that 4 per cent of the distinct certificates used for HTTPS shared an RSA modulus N with another certificate. They also found a very small number (28 out of 400 000) RSA public keys that were shared among PGP users. Heninger *et al.* [269] performed a similar independent analysis in 2012, and found a rate of 5 per cent of public key sharing in distinct HTTPS certificates. Among 12.8 million distinct hosts on the Internet who successfully completed a HTTPS TLS handshake in a single scan, there were 5.9 million distinct certificates, of which 5.6 million contained distinct public keys. Similarly, for the 10.2 million hosts who successfully completed an SSH handshake, there were only 6.6 million distinct RSA public host keys. In other words, 60 per cent of HTTPS IPv4 hosts and 65 per cent of SSH IPv4 hosts shared their RSA public keys with some other host on the Internet.

There are numerous reasons why keys are shared across distinct certificates, certificates are shared across seemingly unrelated hosts, and host keys are shared across thousands of hosts, not all of which are vulnerabilities. Many large hosting providers use the same backend infrastructure for large ranges of IP addresses or seemingly distinct websites. However, there are also many common reasons for shared public keys that do constitute vulnerabilities. Many home routers, firewalls, and 'Internet of things' devices come with pre-configured manufacturer default public and private keys and certificates, which may be shared across the entire device or product line for that manufacturer. These private keys are thus accessible to anyone who extracts them from one of these devices. Knowledge of the private key would allow an attacker to decrypt ciphertexts encrypted to the public key, or forge signatures that will validate with the public key. Databases of such keys have been published on the Internet [267].

Random Number Generation Vulnerabilities Different entities may share identical public moduli because of random-number generation (RNG) vulnerabilities. If two different entities seed the same PRNG algorithm with the same value, then they will each obtain the same sequence of output bits from the algorithm. If two entities use the same PRNG seeds in the course of generating primes for RSA key generation, then they will both obtain the same resulting primes as output, and thus generate the same public modulus.

In 2008, Luciano Bello discovered that the Debian version of the OpenSSL library had accidentally removed all sources of entropy for the random-number generator except for the process ID [628]. This meant that only 16 384 possible RSA moduli could be generated for a given CPU architecture (32-bit or 64-bit, and big or little endian). Reportedly, he discovered the vulnerability after observing the same public keys being generated in the wild. This bug affected all cryptographic key generation between 2006 and 2008 on affected systems, and vulnerable keys were still being found in the wild years after the bug was fixed [174].

The analyses of Lenstra *et al.* [386] and Heninger *et al.* [269] showed that repeated RSA public keys due to random-number generation vulnerabilities occur surprisingly often in practice. Heninger *et al.* [269] traced many of the vulnerable keys back to a flaw in the Linux random-number generator that affected many network devices. First, cryptographic keys for network services like HTTPS and SSH were often generated the first time a machine boots.

Second, the Linux operating system PRNG enforced delays between seeding intervals in order to prevent attacks where a local attacker could brute force individual inputs and thus predict future outputs. Third, network devices lacked the entropy sources such as keyboard timings, mouse movements, or hard-disk timings that the operating system was using to seed the random-number generator. This resulted in a vulnerability where small devices may not yet have seeded the PRNG with any inputs from the environment when the key generation process was run.

In principle, an attacker with the ability to study a given implementation using such poor randomness could reverse-engineer the possible seeds used to generate the key pairs, and thus compute the private key corresponding to a vulnerable public key. This has been done for vulnerable Debian OpenSSL keys [628], but we are not aware of attempts to do this for implementations affected by the Linux boot-time kernel vulnerability.

The presence of a vulnerable public key is also a signal to an attacker that other non-public values that were generated by the same random-number generator are likely to be vulnerable to enumeration as well. The collisions between public keys signal that the random-number generator design or implementation is not incorporating enough entropy to have forward or backward secrecy. This means that an attacker may be able to use a public value to verify that they have successfully recovered the state of the random-number generator, and then wind the state of the random-number generator forward or backward to recover other sensitive values. This is similar in spirit to attacks on cryptographic protocols targeting flawed random-number generation designs that exploit nonces, cookies, or other public values to recover secret values generated later [119, 120, 129, 130].

The Linux kernel vulnerability was patched in 2012, and in 2014 Linux introduced a new system call with a safer interface for generating pseudorandom numbers [179].

6.2.8 RSA Moduli with Common Factors

A more serious version of the repeated-key vulnerability arises if two different parties have different public RSA moduli N_1 and N_2 that share exactly one prime factor p in common, but have distinct second prime factors q_1 and q_2. In that case, any external attacker can compute the private keys for each party outright by computing $p = \gcd(N_1, N_2)$.

Lenstra *et al.* [386] and Heninger *et al.* [269] both independently searched for the existence of such keys in PGP, HTTPS, and SSH RSA public keys in 2012, and found that two PGP users ([386]), 0.2 per cent of HTTPS certificates

([386]) or 0.5 per cent of HTTPS hosts ([269]), and 0.03 per cent of SSH hosts ([269]) had RSA public keys that were completely factored by sharing one prime factor with another RSA public key in their datasets.

This vulnerability was traced in many cases to a variant of the PRNG implementation problems above [269]. Most practical PRNG implementations mix new entropy inputs into their state during the course of normal operation. For example, the OpenSSL PRNG mixes the current time in seconds into the state of its PRNG after every call to extract an output to generate a bignum integer. If two different entities begin the RSA key-generation process using the same initial PRNG seed values, but sometime during key generation mix in different entropy inputs, the stream of PRNG outputs, and therefore the values of the primes generated, will diverge from that point onward. If the PRNG states are identical during the generation of the first prime p but diverge during the generation of the second prime q, then this results in exactly the GCD vulnerability.

A 2016 followup study by Hastings *et al.* [263] gives details on the rates of vulnerable keys over time, broken down by product vendor. They observed no evidence of end users patching devices to remove vulnerable keys after vendors released vulnerability disclosures and patches, and found multiple implementations that had newly introduced GCD vulnerabilities since 2012, most likely the result of using old versions of the Linux kernel. They were able to fingerprint 95 per cent of the vulnerable HTTPS certificates as having been generated by OpenSSL using the OpenSSL prime fingerprint discussed previously in Section 6.2.3. Thus the popularity of OpenSSL, together with its vulnerable pattern of behaviour, appears to have contributed to the different vulnerability rates between keys used for HTTPS and SSH.

Anecdotally, this vulnerability has also arisen in an implementation that periodically generated new RSA keys in an idle process in the background, but sometimes only one new prime was swapped out of memory before the key was exported for use [342].

6.2.9 RSA Primes with Shared or Predictable Bits

Heninger *et al.* [269] and Bernstein *et al.* [52] document a further variant of these shared-key vulnerabilities: they observed RSA keys in the wild whose prime factors shared most significant bits in common. If enough bits are shared in common, these RSA keys can be efficiently factored using lattice basis reduction, using techniques from Coppersmith [138] or Howgrave-Graham [279] (see Chapter 4 for more details).

For the vulnerable HTTPS and SSH RSA keys of this form documented by Heninger *et al.* [269], these primes may be due to an implementation that uses a PRNG whose output length is less than the length of the prime factor to be generated, so that multiple PRNG outputs are concatenated to generate a single prime, and the states of two different entities' PRNGs diverged during the generation of this prime. Bernstein *et al.* [52] found such primes generated by smart cards, where the prime factors appeared to be generated by a faulty physical random-number generator process that would sometimes generate predictable repeating sequences of bits in the resulting primes.

Heninger *et al.* [269] report several SSH public host keys with prime factors that were all zeros except that the first two bytes and last three bytes were set. This may have resulted from an implementation that generated primes by setting the most significant bits of a buffer as many popular implementations do [578], reading intermediate bits from a broken random-number generator that returned all zeros, and then incrementing until a prime was found. Keys of this form would be vulnerable to factoring via brute-force enumeration as well as variants of lattice attacks exploiting known bits [52, 279].

6.2.10 RSA Encryption and Signing

Encryption and Signing, in Theory In theory, one encrypts a message m by calculating the ciphertext $c = m^e \bmod N$. The plaintext can be recovered by computing $m = c^d \bmod N$. For digital signatures, the signature is $s = m^d \bmod N$ and one can verify the signature by verifying that $m = s^e \bmod N$.

Encryption and Signing, in Practice In practice, RSA encryption and digital signatures must use a padding scheme to avoid a wide variety of malleability and signature forgery attacks. The PKCS#1v1.5 padding scheme [308] remains almost universally in use for both encryption and digital signature padding despite the fact that PKCS#1v1.5 is not CCA secure and later versions of the PKCS#1 standard [443] included padding schemes for RSA encryption and signatures that were designed to be provably secure (OAEP [44] and PSS [43]). Although the publication of practical padding oracle attacks against PKCS#1v1.5 [64] and the development of the provably secure OAEP padding scheme both pre-dated the standardisation of TLS 1.0 in 1999, TLS 1.0–1.2 continued to use PKCS#1v1.5 in order to preserve backwards compatibility, and attempted to mitigate the threat of padding oracles with protocol-level countermeasures [162].

Hybrid Encryption RSA public-key encryption is almost universally used for key encapsulation for hybrid encryption in practice, where the public-key encryption operation is used only to transmit symmetric key material, and the actual encrypted content is encrypted using a symmetric encryption scheme. One prominent counterexample is the original Apple iMessage protocol, which encrypted the AES session key as well as the first 101 bytes of the symmetrically encrypted ciphertext in the RSA-OAEP-encrypted message. Unfortunately, the protocol was insecure against chosen ciphertext attacks because it did not authenticate the symmetric encryption properly [209].

6.2.11 Key Re-Use across Protocols

Cryptographic best practice dictates that cryptographic keys should be used for a single purpose, but this principle is not always followed. In TLS versions 1.2 [161] and below, a server typically has one certificate containing its public key, and this key is used both for digital signatures and encryption. The server generates digital signatures to sign Diffie–Hellman key-exchange parameters when Diffie–Hellman key-exchange is used, and uses the same RSA key to decrypt secrets when RSA is used as a key exchange mechanism.

RSA keys are also universally re-used by web servers across different versions of SSL/TLS, and it is quite common for web servers to support many old protocol versions for backwards compatibility. This led to the DROWN vulnerability [25] discussed in Section 6.2.13, where an attacker could exploit a protocol flaw in SSLv2 to compromise an otherwise secure RSA key-exchange message from a server using the same key with TLS 1.2.

There are also sporadic cases of keys shared across entirely different protocols: Heninger *et al.* [269] document two RSA public keys that were used for both HTTPS and SSH in 2012.

6.2.12 Encryption Padding

In PKCS#1v1.5 [308], data D is padded before encryption as

$$\text{EB} = 00 \ || \ 02 \ || \ \text{PS} \ || \ 00 \ || \ D$$

where 00 and 02 represent byte strings, PS is a pseudorandomly generated padding string that contains no null bytes, and the D is the data to be encrypted, typically a value like a TLS premaster secret from which symmetric encryption and authentication keys will be derived. Textbook RSA encryption is applied to the padded value EB. For decryption, textbook RSA decryption is applied to recover the padded plaintext, and the decrypter must then check that the padding is valid before stripping off the padding and returning the unpadded data.

6.2.13 RSA Encryption Padding Oracle Attacks

If a decryption implementation returns an error when the plaintext padding is incorrect, then it may be exploitable as an oracle for a chosen ciphertext attack. Bleichenbacher developed a padding oracle attack in 1998 against the PKCS#1v1.5 encryption padding scheme as used in SSL [64]. Bleichenbacher's attack exploits implementations where the decrypter (usually a server receiving ciphertexts from a client) provides an error message if decryption fails due to incorrect plaintext padding. The attacker begins with a target ciphertext c that they wish to decrypt to recover padded message m, and submits mauled ciphertexts $a^e c \bmod N$ to the decryption oracle for carefully chosen values of a. The error messages from the padding oracle reveal whether the most significant bytes of the mauled plaintext am are 00 || 02, which allows the attacker to iteratively narrow down possible values for m. The originally published attack required millions of queries to recover a plaintext.

Despite the existence of this attack, the designers of TLS 1.0 through 1.2 decided against using a CCA-secure RSA padding scheme, and instead continued to specify PKCS#1v1.5 padding for RSA encryption for backwards compatibility reasons. To attempt to mitigate padding oracle attacks, the TLS standards required that when implementations encountered RSA padding errors on decryption, they should avoid signaling the error to the attacker by simply generating a placeholder plaintext and continuing with the handshake using this value. In this case, the handshake should naturally fail when the client sends its authentication of the handshake. However, numerous studies have found that implementation support for this countermeasure is often incomplete, and many implementations of this countermeasure result in side channels that lead to practically exploitable padding oracle attacks in the wild [73, 424].

The 2015 DROWN attack [25] exploited a confluence of vulnerabilities in RSA usage in the wild: (1) RSA keys were universally used by servers across different versions of the TLS protocol including old versions of SSL, (2) the SSLv2 protocol supported weakened key strengths at the protocol level in order to conform to the US government regulations on the export of cryptography, (3) many servers never disabled support for SSLv2 for backwards compatibility reasons and (4) both a protocol flaw in SSLv2 and implementation flaws in the OpenSSL library could serve as particularly strong padding oracles that could allow an attacker to decrypt TLS 1.2 ciphertexts or forge digital signatures after tens of thousands of connections. In 2015, 33 per cent of HTTPS servers were vulnerable to this attack [25].

OAEP is also vulnerable to padding oracle attacks if implemented improperly. In 2001, Manger gave a chosen ciphertext attack against OAEP that can recover plaintext after as few as 1000 queries if the implementation allows an attacker to distinguish between integer-byte conversion errors and integrity check errors, a situation that was not ruled out by PKCS#1v2.0 [402].

6.2.14 Signature Padding

The PKCS#1v1.5 [308] padding for a digital signature on message m is

$$\texttt{EB = 00 || 01 || FF ... FF || 00 || ASN.1 || H(m)}$$

where `ASN.1` is a sequence of bytes that encodes an OID (object identifier string) that describes the hash function and signature algorithm, encoded using ASN.1, and H is a hash function like SHA-256. The textbook RSA signing procedure is applied to the padded value `EB`. Some variants of PKCS#1v1.5 signature padding may be secure in some models [287].

PKCS#1v1.5 also specifies a padding type that uses `00` bytes for padding, but we have never seen it used.

The IPsec IKE (Internet Key-Exchange) RFC [259] specifies that RSA signatures used for IPsec authentication should be encoded as private key decryptions rather than PKCS#1v1.5 signatures, but the implementations we examined appear to use signature padding.

Bleichenbacher's Low-Exponent PKCS#1v1.5 Signature Forgery Bleichenbacher observed in a 2006 Crypto rump session [188] that, for small public exponents, the PKCS#1v1.5 padding function can be vulnerable to a signature forgery attack against implementations that do not check the full length of the `FF ... FF` padding string. Let us specialise to the case $e = 3$. Then to generate a forged signature, the attacker simply needs to find a string

$$c = \texttt{00 || 01 || FF || FF || FF || 00 || ASN.1 || H(m) || G}$$

where $c < N$, `ASN.1` and H are as above, and the vaue G is chosen so that when the value c is interpreted as the hexadecimal representation of an integer, it is a perfect cube over the integers. Then the attacker can 'forge' a signature s that will validate against these implementations by returning the value $s = c^{1/3}$ over the integers. Lazy implementations that simply begin matching the padding format from the most significant bits of $s^e \bmod N$ without verifying that the length of the padding string `FF ... FF` is correct for the key length will validate this signature as valid. A simple method that will often succeed in generating a perfect cube is to start with the integer corresponding to

$$b = \texttt{00 || 01 || FF FF FF || 00 || ASN.1 || H(m) || FF ... FF}$$

where the number of trailing bits is chosen so that the the integer corresponding to this value is less than the modulus N. Then the forged signature is $s = \lfloor b^{1/3} \rfloor$, where the cube root is computed over $\mathbb{R}$ and rounded down to the nearest integer. The value c above is s^3.

Numerous RSA signature implementations, including OpenSSL, were vulnerable to this attack in 2006; in 2014 this vulnerability was found in the Mozilla NSS library [153].

6.2.15 Cross-Protocol Attacks

Wagner and Schneier [610] found a theoretical protocol vulnerability with the SSL 3.0 specification that could allow an attacker to substitute signed ephemeral RSA and Diffie–Hellman public keys for each other, because the server's digital signature on the keys did not include the key-exchange algorithm. In their attack, a message containing a Diffie–Hellman prime modulus p and generator g would be interpreted as an RSA modulus p with exponent g. If the client encrypts the key-exchange message m to this public key by computing $c = m^g \bmod p$, an attacker can easily recover m by computing $c^{1/g} \bmod p$, since p is prime. Although it was not ruled out by the specification, implementations at the time apparently did not allow this attack in practice.

A 2015 vulnerability in JSON Web Token libraries allowed attackers to forge authentication tokens because the libraries did not tie the algorithm type to the verification key, and used both public and private-key verification algorithms. In the attack, libraries could be confused into using an RSA public key as an HMAC secret key [414].

6.2.16 Key Theft and Cryptanalysis

When RSA is used as a key-exchange or key-encapsulation mechanism, the same long-term public RSA key is typically used to encrypt many messages or sessions. In the context of a protocol like TLS, long-term RSA public keys are validated by using digital signatures from certificate authorities, and certificate validity periods are typically months to years in duration. If an adversary is ever able to steal or mount a cryptanalytic attack against the private key corresponding to one of these long-term RSA public keys, then the adversary would be able to passively decrypt any sessions or messages that had ever used that public key for key-exchange.

National Security Agency (NSA) slides leaked by Edward Snowden in 2013 mentioned using known RSA private keys to passively decrypt SSL/TLS network traffic [526] as well as targeted hacking operations to learn pre-shared key values to enable passive decryption of VPN connections [609].

Because of the risk of passive decryption attacks, as well as persistent cryptographic vulnerabilities resulting from PKCS#1v1.5 padding, TLS 1.3 removed RSA key-exchange entirely from the protocol, allowing only elliptic-curve or prime-field Diffie–Hellman key-exchange [220].

There have also been multiple documented instances of attackers compromising certificate authority digital signing infrastructure to issue fraudulent certificates, including the Comodo [132] and DigiNotar [5] hacks in 2011 that enabled man-in-the-middle attacks against Google in Iran. Browser vendors responded to these compromises by adopting public-key pinning, which ties the hash of a public key to its associated domains [445], and certificate transparency [353], in which all valid issued certificates are published in a tamper-proof log to provide defense in depth against the risk of signing key compromise.

6.2.17 Countermeasures

The most straightforward countermeasure against the attacks discussed in this section is to avoid using RSA entirely. For key-exchange or key encapsulation, elliptic-curve cryptography offers smaller key sizes and more efficient operation. Eliminating RSA can be difficult for legacy protocols, unfortunately: there is evidence that the adoption of TLS 1.3 has been delayed because of the decision to eliminate RSA key-exchange. If RSA must be used, then Shoup's RSA-KEM scheme [544] sidesteps padding implementation issues entirely: one uses RSA to encrypt a random string of the same length as the key, and derives a symmetric key by applying a key derivation function to this random message. RSA signatures may still be more efficient to verify than elliptic-curve signatures despite their increased size; in this case, a padding scheme like RSA-PSS [45] was designed to be provably secure.

6.3 Diffie–Hellman

Diffie–Hellman key-exchange is a required step in many network protocols, including SSH [629], the IPsec IKE handshake [259, 317], and TLS 1.3 [220]. It has also been an optional key-exchange method in SSL/TLS since SSLv3, and a cipher suite including Diffie–Hellman key-exchange was required by the TLS 1.0 specification [162]. In this section, we focus on Diffie–Hellman over prime fields, which was historically the only option, and is still supported by the protocols we list above. In the past five years, elliptic-curve Diffie–Hellman (ECDH) has replaced RSA key-exchange and prime-field Diffie–Hellman as the most popular key-exchange method; we cover ECDH in Section 6.4.

6.3.1 Diffie–Hellman Key-Exchange

Diffie–Hellman, in Theory Textbook prime-field Diffie–Hellman key-exchange works as follows [163]: first, the two parties agree somehow on a prime p and a generator g of a multiplicative group mod p. To carry out the key-exchange, Alice generates a secret integer exponent a, and sends the value $y_a = g^a \bmod p$ to Bob. Bob responds with the value $y_b = g^b \bmod p$. Alice computes the value $y_b^a = g^{ab} \bmod p$, and Bob computes the value $y_a^b = g^{ba} \bmod p = g^{ab} \bmod p$, so they have a shared value.

The most straightforward means of attack for an attacker is to compute the discrete logarithm of one of the key-exchange messages, although this is not known to be equivalent to computing the Diffie–Hellman shared secret [404].

Diffie–Hellman, in Practice The textbook description of Diffie–Hellman leaves a wide number of choices to implementers, including the type of group, how groups are agreed on, and exponent generation. Different standards and recommendations differ on all of these choices, and implementations differ further from standards.

In the context of protocols, symmetric-key material is typically computed by applying a key derivation function to the Diffie–Hellman shared secret together with other messages from the client–server handshake, and digital signatures are used to authenticate the key-exchange against man-in-the-middle attacks. The details of authentication vary across protocols, resulting in a different set of vulnerabilities for different protocols: in SSH [629] and TLS 1.3 [495], the server digitally signs the entire handshake; in TLS versions 1.2 [161] and below, the server digitally signs the Diffie–Hellman key-exchange values and handshake nonces, and the handshake is authenticated by using the derived symmetric keys. IPsec offers numerous authentication options negotiated as part of the cipher suite [259, 317].

ElGamal Encryption ElGamal public-key encryption is not commonly supported among most of the network protocols we discuss in this chapter, although it was historically a popular option for PGP. Lenstra *et al.* [386] report that 47 per cent of the PGP public keys in a public repository in 2012 were ElGamal public keys. Because the structure of ElGamal is very close to Diffie–

Hellman, and thus the implementation issues are related, we include it in this section; we briefly remark on its security in Section 6.3.8.

Textbook ElGamal encryption works as follows [181]. An ElGamal public key contains several parameters specifying the group to be used: a group generator g, a modulus p, and the order q of the subgroup generated by g. Alice's private key is a secret exponent a, and the public key is a value $h_a = g^a \bmod p$. To encrypt a message m to Alice's public key, Bob chooses a secret exponent b and computes the values $h_b = g^b \bmod p$ and $h_{ab} = h_a^b \bmod p$. The ciphertext is the pair $(h_b, m \cdot h_{ab} \bmod p)$. To decrypt, Alice computes the value $h_{ab} = h_b^a \bmod p$ and multiplies the second element of the ciphertext by h_{ab}^{-1} to recover m.

6.3.2 Group Agreement

Before carrying out the key-exchange, the two parties must agree on the group parameters p and g. Different protocols do this differently. In TLS versions 1.2 and earlier [161], the server generates the group parameters and sends them to the client together with the server's key-exchange message. The server signs this key-exchange message and the random handshake nonces using its long-term public key to prevent man-in-the-middle attacks. In TLS version 1.3 [495] servers no longer generate their own Diffie–Hellman group parameters. Instead, the client and server negotiate a group choice from a pre-specified list of Diffie–Hellman groups. The prime-field groups were custom-generated for TLS 1.3 [220]. The IKE key-exhange for the IPsec protocol [259, 317] specifies a pre-defined list of Diffie–Hellman groups, and the client and server negotiate the choice of group while negotiating cipher suites. SSH [629] includes a few pre-generated groups in the specification, but also allows 'negotiated' groups [196], in which the client specifies their desired prime length, and the server responds with one of a custom list of server-chosen group parameters of the desired size.

These differences in how groups are specified and agreed upon mean that the impact of the Diffie–Hellman vulnerabilities discussed below can look very different for these three network protocols, and the threat model is different for clients and servers even though textbook Diffie–Hellman appears to be equally contributory from both parties. The standardised groups chosen for TLS 1.3, IKE, and SSH were generated to avoid some of the vulnerabilities discussed below. There is no vetting process, and no feasible way for clients to evaluate the quality of custom server-generated Diffie–Hellman parameters for SSH or TLS v1.2 and below. The distribution of groups used for these different protocols also looks very different: for servers supporting TLS v1.2 and earlier, there are a handful of common values and a long tail of custom values, while most IKE servers prefer the same group parameters [7].

6.3.3 Prime Lengths

The number field sieve algorithms for factoring and discrete logarithm have the same asymptotic running times, but the discrete logarithm problem is believed to be slightly harder in practice [379], and discrete logarithm records have historically lagged several years behind factoring records of the same length. While 512-bit RSA factorisation was first carried out in 1999, the first discrete logarithm computation exceeding this, at 530 bits, dates to 2007 [325]. There is a similar gap between the 768-bit factoring and discrete log records: 2009 for factoring [327], and 2016 for discrete log [330]. Nevertheless, the official key-size recommendations for prime-field Diffie–Hellman have historically been quite similar to those for RSA [379]. The use of 1024-bit Diffie–Hellman moduli was allowed by NIST until 2010, deprecated until 2013, and disallowed after 2013 [39].

Despite these recommendations, 512-bit and 1024-bit Diffie–Hellman remained in common use well after 2010. In 2015, 8 per cent of popular HTTPS web sites still supported 512-bit 'export'-grade Diffie–Hellman cipher suites [7], even though server maintainers had already been encouraged earlier that year to disable TLS export cipher suites in the wake of the publicity around the FREAK downgrade attack against export-grade RSA [57].

Historically, 1024-bit primes for Diffie–Hellman have been very common. In 2015, 91 per cent of IPsec servers supported 1024-bit primes and 66 per cent of them preferred 1024-bit primes over other options in common client configurations; 99 per cent of SSH servers supported 1024-bit primes and 26 per cent of them preferred 1024-bit primes over other options in common client configurations; and 84 per cent of HTTPS servers supporting prime-field Diffie–Hellman used a 1024-bit prime for key-exchange [7]. In 2017, 95 per cent of HTTPS Diffie–Hellman connections seen by Google Chrome telemetry used 1024-bit primes [125].

Software maintainers had difficulty increasing Diffie–Hellman key sizes due to interoperability issues stemming from hard-coded size limits in libraries and hard-coded 1024-bit primes in libraries and specifications. Java JDK versions prior to version 8 did not support primes larger than 1024 bits for Diffie–Hellman key-exchange and DSA signatures [463]. Servers using Java without further upgrades could not generate or use larger Diffie–Hellman keys, and older Java-based clients could not handle a larger modulus presented by a server. The SSHv2 transport layer specification [629] includes a hard-coded

1024-bit prime group for Diffie–Hellman key-exchange, the Oakley Group 2 discussed in Section 6.3.4, which all implementations were required to support. This group was removed from default support by OpenSSH in 2015, but allowed as a 'legacy' option for backward compatibility [462]. Many major web browsers raised the minimum Diffie–Hellman modulus length for HTTPS to 768 or 1024 bits in 2015 [496], but by 2017, it appeared to be easier for browsers to remove support entirely for prime-field Diffie–Hellman for old versions of TLS in favour of elliptic-curve Diffie–Hellman than to increase minimum key strengths to 2048 bits [125, 590]. This issue has been avoided in TLS 1.3 [220]: the protocol supports only fixed groups, and the shortest Diffie–Hellman prime included in the standard is 2048 bits.

6.3.4 Standardised and Hard-Coded Primes

A number of protocols and implementations have pre-generated primes with various properties for use in Diffie–Hellman. Some of the most widely used primes have been carefully generated to ensure various desirable properties; the provenance of others is less well documented although the group structures are verifiable.

Some of the most commonly used primes for Diffie–Hellman on the Internet originated with RFC 2412, the Oakley key determination protocol [464], which specified three 'mod p' groups and two elliptic-curve groups. These primes have the property that the high and low 64 bits are all clamped to 1, with the explanation that this helps remaindering algorithms [464]. The middle bits are the binary expansion of π, intended to be a 'nothing up my sleeve' number to allay concerns of trapdoors. The middle bits are incremented until the prime is a 'safe' prime, so $p = 2q + 1$ for prime q, and 2 generates the subgroup of order q. These primes were built into the IKEv1 [259], IKEv2 [317] and SSH [629] key-exchange protocols as named groups that should be supported by implementations and that can be negotiated for Diffie–Hellman key exchange, and the 1024-bit and 2048-bit primes of this form were historically some of the most commonly used values for Diffie–Hellman key exchange for these protocols.

Adrian *et al.* [7] estimated that, in 2015, 66 per cent of IPsec servers and 26 per cent of SSH servers used the 1024-bit Oakley prime by default for Diffie–Hellman, and 18 per cent of the Alexa Top 1 Million HTTPS web sites used a 1024-bit prime that was hard-coded into the Apache web server software by default.

The TLS 1.3 protocol includes several named Diffie–Hellman primes that have the same structure as the Oakley groups, except that they use the binary expansion of e instead of π [220].

6.3.5 Fixed Primes and Pre-Computation

The fact that the number field sieve algorithms for factoring and discrete log have the same asymptotic complexity has led to the heuristic estimate that equal key sizes for RSA and prime-field Diffie–Hellman offer approximately the same amount of bit security.

As discussed previously in Section 6.3.3, 1024-bit Diffie–Hellman primes remained in common use for many years after 1024-bit number field sieve computations were believed to be tractable for powerful adversaries. Many implementers justified the continued use of 1024-bit Diffie–Hellman via what they believed to be a calculated risk: given the high estimated cost of 1024-bit number field sieve computations, it was thought that even a powerful adversary would likely be able to carry out at most only a handful of such computations per year. Whereas RSA keys typically have long validity periods and thus a single factoring computation would allow the adversary to decrypt many RSA messages encrypted to the broken private key, implementations typically generate new Diffie–Hellman secrets per session, and thus in principle an attacker would need to carry out a new discrete log computation for each session. Prior to 2015, practitioners believed that adversaries would be unlikely to expend the high cost of a full 1024-bit number field sieve computation to break a single Diffie–Hellman key-exchange, and thus chose to accept the risk of cryptanalytic attacks in order to take advantage of the added efficiency of small keys and avoid the interoperability issues from increasing key sizes.

However, this popular understanding of computational power was incomplete. The most computationally expensive stages of the number field sieve discrete logarithm calculation depend only on the prime modulus, and the final individual logarithm phase of the algorithm that actually computes the log of the target is asymptotically faster [29, 131]. In practice, this asymptotic difference means that the individual log computation is significantly faster than the prime-dependent pre-computation.

This means that a well-resourced adversary could perform a single expensive pre-computation for a given 1024-bit prime modulus, after which the private keys for many individual sessions using that prime would be relatively efficient in practice to compute. This attack is rendered more effective in practice because the adversary could target a small number of hard-coded and standardised 1024-bit primes that account for a relatively large fraction of Diffie–Hellman moduli used in the wild [7]. Adrian *et al.* estimate that the 66 per cent of IPsec servers and 26 per cent of SSH servers that defaulted to the 1024-bit Oakley prime would be vulnerable to passive decryption by an adversary who carried out a single discrete log pre-computation for that prime, and that carrying out ten discrete log pre-computations would allow passive decryption to 24 per cent of the most popular HTTPS sites in 2015 [7].

6.3.6 Short Exponents

It is quite common for implementations to generate 'short' Diffie–Hellman secret exponents by default. In principle for an otherwise secure Diffie–Hellman group, the strongest attack for short exponents is the Pollard lambda algorithm [476], which takes $O\left(2^{n/2}\right)$ time against an n-bit exponent a. Thus implementations that wish to achieve a 128-bit security level often generate 256-bit exponents. The TLS 1.3 specification [220] as well as the SSH group exchange specification [196] both suggest the use of shorter exponents for performance reasons, as long as they are at least twice the length of the derived secret.

Valenta *et al.* [594] examined nine different TLS libraries implementing Diffie–Hellman in 2016 and found that eight of the nine used short exponents. Of these libraries, the Mozilla NSS and libTomCrypt libraries used hardcoded short exponent lengths, the Java OpenJDK uses the max of $n/2$ and 384 for an n-bit prime, OpenSSL and GnuTLS used the bit length of the subgroup order (if specified), with a max of 256 in the latter case, and three libraries used a quadratic curve or work factor calculation to set the bit-length of the exponent to match the expected cryptanalytic complexity of the number field sieve algorithm for the length of the prime.

Short exponent lengths are not a vulnerability on their own, but the use of short exponents in combination with common implementation flaws can make some of the key-recovery attacks described in Sections 6.3.8 and 6.3.10 more severe.

Some implementations appear to generate pathologically short exponents. In 2017, Joshua Fried found that 3 per cent of the 4.3 million hosts that responded to an IKEv1 IPsec handshake and 1.3 per cent of the 2.5 million hosts that responded to an IKEv2 IPsec handshake used exponents that were shorter than 16 bits. These were found by pre-computing 2^{17} public key-exchange values for positive and negative 16-bit exponents for the most common Diffie–Hellman groups, and comparing these to the key-exchange messages transmitted by servers [194].

6.3.7 Exponent Re-Use

Many implementations re-use Diffie–Hellman exponents by default across multiple connections. OpenSSL re-used Diffie–Hellman exponents by default until January 2016, unless a specific SSL_OP_SINGLE_DH_USE flag was set. Springall *et al.* found in 2016 that 7 per cent of the Alexa Top 1 Million HTTPS servers who supported prime-field Diffie–Hellman re-used Diffie–Hellman key-exchange values [557].

Re-use of Diffie–Hellman key-exchange values for multiple connections is not in principle a vulnerability, and should be no less secure than using an RSA public key for encryption across multiple connections but, in practice, key-exchange message re-use can make some of the attacks described in Section 6.3.10 more severe.

6.3.8 Group Structure

For prime-field Diffie–Hellman, there is a variety of choices of primes and group structures that implementations can choose. The security of Diffie–Hellman relies crucially on the structure of the group.

For a prime modulus p, the order of the group generated by g will divide $p - 1$. The Pollard rho [475] and 'baby step giant step' algorithms have running times that depend on the group order: for a group of order q, these algorithms run in time $O\left(\sqrt{q}\right)$.

If the order of the group generated by g has a prime factor q_i, then an adversary can take advantage of the existence of a subgroup of order q_i, and use one of these algorithms to compute the discrete logarithm of a target modulo q_i. If the order of the group generated by g has many such subgroups, the attacker can use the Pohlig–Hellman algorithm [472] to repeat this for many subgroup orders q_i and use the Chinese remainder theorem and Hensel lifting to compute the secret exponent a modulo the product of the known q_i.

To protect against these attacks, implementations should choose g such that g generates a subgroup of large prime-order q, where q should have bit-length at least twice the desired security parameter of the encryption.

Safe Primes A common recommendation is for implementations to use 'safe' primes, where $p = 2q + 1$ for q a prime, and then to use a generator g of the subgroup of order q modulo p. This protects against attacks based on subgroup structure by maximising the order of the subgroup an attacker would need to attack.

Until 2019, OpenSSL would by default generate Diffie–Hellman group parameters where p was a 'safe' prime, but where g generated the 'full' group of order $2q$ modulo p [180]. This meant that an adversary could always compute one bit of information about the exponent in any key-exchange message, by computing the discrete log in the subgroup of order 2. Of around 70 000 distinct group parameters g, p in use by HTTPS servers in 2015, around 64 000 of the prime moduli p were 'safe', and only 1250 of those used a generator g that generated a group of prime-order q [7]. In other words, in practice the Decisional Diffie–Hellman assumption [76] is often false. In particular, textbook ElGamal encryption, where a message is chosen as a non-zero integer modulo p, is not semantically secure as a consequence.

DSA-Style Groups An alternative common structure for primes used for Diffie–Hellman key exchange is to use a prime p generated so that $p - 1$ has a prime factor q of a fixed size much smaller than p, and g is chosen in order to generate the subgroup of order q modulo p. Commonly encountered parameter lengths include choosing the subgroup order q to be 160 bits for a 1024-bit modulus p, or choosing the subgroup order q to be 224 or 256 bits for a 2048-bit prime p [283]. Groups of this type were originally used for DSA, and then recommended for many years for use in Diffie–Hellman by NIST SP800-56A [37].

RFC 5114 specified several pre-generated groups of this form for use in SMIME, SSH, TLS and IPsec to conform to the NIST standard [389]. In 2017, Valenta *et al.* [594] found that 11 per cent of the 10.8 million HTTPS servers that supported prime-field Diffie–Hellman were using the 1024-bit Group 22 specified in RFC 5114. In 2018, NIST updated their recommendation to allow only 'safe' primes for Diffie–Hellman. 'DSA'-style groups are now only permitted for backwards compatibility [37].

In the context of the DSA digital signature algorithm, the presence of these subgroups permits shorter digital signatures, and allows implementations to use hash functions of common lengths. The length of the subgroup is chosen so that the bit complexity of different families of attacks matches the desired overall security of the system [368]: for the 1024-bit parameters, the number field sieve is believed to take about 2^{80} time, which matches the expected complexity of carrying out a Pollard rho attack against the 160-bit subgroup q, or the Pollard lambda attack against the 160-bit secret exponent.

The reasoning behind the recommendation to use smaller subgroups appears to have been a desire to use short secret exponents to make the modular exponentiations required for key-exchange more efficient, combined with concerns about vulnerability to hypothetical attacks if the exponents were much shorter than the subgroup order. Such attacks exist for DSA signatures [277], but this is not known to be a vulnerability in the Diffie–Hellman setting. However, the use of smaller subgroups necessitates additional validity checks to prevent small subgroup attacks, described in Section 6.3.10.

Unstructured Primes and Composite-Order Groups For an arbitrary randomly generated prime p, the prime factorisation of $(p - 1)/2$ is expected to look like a random integer. That is, it is likely to have many small factors, some medium-sized factors, and a few large factors. The expected length of the largest subgroup order for a 1024-bit prime is over 600 bits [492], meaning that using the full Pohlig–Hellman algorithm [472] to compute arbitrary discrete logarithms modulo such a prime is likely to be infeasible due to the square root running time of computing discrete logs in such a large subgroup.

However, if the target of the attack also uses short exponents, van Oorschot and Wiener's attack [601] exploits the fact that an attacker could combine Pollard's lambda algorithm [476] with the Pohlig–Hellman algorithm over a strategically chosen subgroup whose order has relatively small prime factors to uniquely recover a shorter exponent. An adversary would be able to recover a 160-bit exponent with 2^{40} computation for 32 per cent of 1024-bit primes.

The prescribed subgroup structures described above are intended to prevent such attacks, but some relatively rare implementations do not appear to take any steps to generate primes with a cryptographically secure structure.

In 2015, among around 70 000 distinct primes p used for Diffie–Hellman key-exchange for 3.4 million HTTPS servers supporting Diffie–Hellman key exchange on the Internet, there were 750 groups for which $(p - 1)/2$ was not prime, and an opportunistic effort to factor $(p - 1)/2$ using the ECM algorithm [388] revealed prime factors of the order of g. This allowed Adrian *et al.* [7] to apply van Oorschot and Wiener's attack to compute the full exponent for 159 key exchanges (many of which were using 128-bit exponents) and partial information for 460 key exchanges.

Composite Moduli Diffie–Hellman key exchange is almost universally described as being carried out modulo a prime p. However, non-prime moduli have been found in the wild.

If a non-prime Diffie–Hellman modulus can be efficiently factored, then, in general, computing the discrete logarithm, and thus the private key for a Diffie–Hellman key exchange, is only approximately as difficult as the problem of factoring the modulus and computing the discrete logarithm of the target key-exchange message for each prime factor or subgroup of the modulus. The Chinese remainder theorem can then be used to reconstruct the discrete logarithm modulo the least common multiple of the totient function of each of the prime factors.

Not all non-prime Diffie–Hellman moduli are necessarily insecure: a hard-to-factor composite RSA modulus, for example, would likely still be secure to use for Diffie–Hellman, since an adversary would not be able to factor the modulus to learn the group structure. A composite Diffie–Hellman modulus with one small factor where computing the discrete logarithm is easy and one large factor where computing the discrete logarithm is still difficult could allow an adversary to compute partial information about the secret exponent. And a highly composite Diffie–Hellman modulus where every prime factor is relatively small or admits an easy discrete logarithm would allow the adversary to efficiently compute the full secret exponent.

In 2016, the Socat tool was found to be using a hard-coded non-prime modulus p of unknown origin for Diffie–Hellman key exchange [498]. Lenstra *et al.* [386] found 82 ElGamal public keys in the PGP key repository in 2012 using non-prime p values. Many of these shared bit patterns with other group parameters used by PGP users, suggesting they may have been invalid, corrupted keys.

Valenta *et al.* [594] found 717 SMTP servers in 2016 in the wild using a 512-bit composite Diffie–Hellman modulus whose hexadecimal representation differed in one byte to a default Diffie–Hellman prime included in SSLeay, the predecessor to OpenSSL.

6.3.9 Group Parameter Confusion

Adrian *et al.* [7] found 5700 HTTPS hosts that were using DSA-style group parameters that had been hard-coded in Java's `sun.security.provider` package for Diffie–Hellman, except that they were using the group order q in place of the generator g. Adrian *et al.* hypothesised that this stemmed from a usability problem. The ASN.1 representation of Diffie–Hellman key-exchange parameters in PKCS#3 is the sequence (p, g), while DSA parameters are specified as (p, q, g). For the parameters with 512-bit p, the group generated by q would leak 290 bits of information about the secret exponent using 2^{40} computation. Java uses 384-bit exponents for this prime length. Computing discrete logarithms for this group would thus be more efficient using the number field sieve than using Pollard lambda on the remaining bits of the exponent, but this is a near-miss vulnerability.

6.3.10 Small Subgroup Attacks

Diffie–Hellman is vulnerable to a variety of attacks in which one party sends a maliciously crafted key-exchange value that reduces the security of the shared secret by confining it to a small set of values. These attacks could allow a man-in-the-middle attacker to coerce a Diffie–Hellman shared secret to an insecure value, or let an attacker learn information about a victim's secret exponent by carrying out a protocol handshake with them.

Zero-Confinement For prime-field Diffie–Hellman, the value 0 is not a member of the multiplicative group modulo p, and therefore should never be sent as a key-exchange value. However, if Alice sends the value 0 as her key-exchange value to Bob, then Bob will derive the value 0 as his shared secret. For protocols like TLS versions 1.2 and below, where the integrity of the handshake is ensured using a key derived from the shared secret, then this could allow a man-in-the-middle to compromise the security of the key exchange. To protect against this type of attack, implementations must reject the value 0 as a key-exchange value.

Valenta *et al.* [594] scanned the Internet in 2016 and found that around 3 per cent of SSH servers and 0.06 per cent of HTTPS servers were willing to accept a Diffie–Hellman key-exchange value of 0. They note that, until 2015, a vulnerability in the Libreswan and Openswan IPsec VPN implementations caused the daemon to restart when it received 0 as a key-exchange value [147].

Subgroups of Order 1 and 2 Since $(p - 1)$ is even for any prime $p > 2$, there will be a multiplicative subgroup of order 2 modulo p, generated by the value -1, as well as the trivial group of order 1 generated by the value 1. For Diffie–Hellman, if g generates the full group of integers modulo p, then the group of order 2 generated by -1 will be a proper subgroup; if g is a 'safe' prime or a DSA-style prime as described above and g is chosen to generate a subgroup of prime-order q, then the value -1 will not be contained in the group generated by g. The value 1 is contained in the group in either case.

Thus, similar to the case of 0, Alice could send the element 1 as a key exchange value and ensure that the resulting shared secret derived by Bob is the value 1, or send -1 and ensure that Bob's resulting secret is either 1 or -1. In the latter case, if Alice subsequently learns a value derived from Bob's view of the shared secret (for example, a ciphertext or MAC, depending on the protocol) then Alice can learn one bit of information about Bob's secret exponent, whether or not the Diffie–Hellman group Bob intended to use had large prime order.

To prevent these attacks, implementations must reject the values 1 and -1 as key-exchange values.

Valenta *et al.* [594] report that, in 2016, 3 per cent of HTTPS servers and 25 per cent of SSH servers accepted the value 1, and 5 per cent of HTTPS servers and 34 per cent of SSH servers accepted the value -1 as a Diffie–Hellman key-exchange value.

Subgroups of Larger Order For 'safe' primes p, the validation checks eliminating the values 0, 1 and -1 are the only checks necessary to eliminate small subgroup confinement attacks. For DSA-style primes where g generates a group of order q where q is much less than p, then there are subgroups modulo p for each of the factors of $(p-1)/q$. The recommended prime-generation procedures do not require the co-factor $(p-1)/(2q)$ to be prime, so in practice many primes in use have co-factors that are random integers with many small prime factors. For example, the 1024-bit prime p specified in RFC 5114 [389] was chosen to have a 160-bit prime-order subgroup, but it also has a subgroup of order 7.

Let g_7 be a generator of the group of order 7 modulo p. If Alice sends g_7 as her key-exchange value, then the resulting shared secret derived by Bob will be confined to this subgroup of order 7. If Alice subsequently learns a value derived from Bob's view of the shared secret, then Alice can compute Bob's secret exponent b mod 7 by brute forcing over the subgroup size.

Lim and Lee [395] developed a full secret key-recovery attack based on these principles. In the Lim–Lee attack, the victim Bob re-uses the secret exponent for multiple connections, and the attacker Alice wishes to recover this secret. Alice finds many small subgroups of order q_i modulo Bob's choice of Diffie–Hellman prime p, sends generators g_{q_i} of each subgroup order in sequence, and receives a value derived from Bob's view of the secret shared key in return. Alice then uses this information to recover Bob's secret exponent b mod q_i for each q_i. Depending on the protocol, this may take $O\left(\sum_i \sqrt{(q_i)}\right)$ time if Bob were to directly send back his view of the key share $g_{q_i}^b$, or $O\left(\sum_i q_i\right)$ time if Bob sends back a ciphertext or MAC whose secret key is derived from Bob's key share $g_{q_i}^b$.

To prevent these small subgroup attacks when using groups of this form, implementations must validate that received key-exchange values y are in the correct subgroup of order q by checking that $y^q \equiv 1 \bmod p$. Unfortunately, although TLS versions 1.2 and below and SSH allow servers to generate their own Diffie–Hellman groups, the data structures used for transmitting those groups to the client do not include a field to specify the subgroup order, so it is not possible for clients to perform these validation checks. In principle, it should be feasible for servers to perform these validation checks on key-exchange values received by clients, but in practice, Valenta *et al.* found in 2016 that almost no servers in the wild actually performed these checks [594].

In 2016, OpenSSL's implementation of the RFC 5114 primes was vulnerable to a full Lim–Lee key-recovery attack, because it failed to validate that Diffie–Hellman key-exchange values were contained in the correct subgroup, and servers re-used exponents by default. For the 2048-bit prime with a 224-bit subgroup specified in RFC 5114, a full key-recovery attack would require 2^{33} online work and 2^{47} offline work [594].

These attacks could also have been mitigated by requiring the co-factor of these groups to have no small factors, so that primes would have the form $p = 2qh + 1$, where the co-factor h is prime or has no factors smaller than the subgroup order q [395]. This was not suggested by the relevant recommendations [37].

6.3.11 Cross-Protocol Attacks

In TLS versions 1.2 and earlier, the server signs its Diffie–Hellman key-exchange message to prevent man-in-the-middle attacks. However, the signed message does not include the specific cipher suite negotiated by the two parties, which enables multiple types of cross-protocol attacks.

A simple case is the Logjam attack of Adrian *et al.* [7]. In this attack, a man-in-the-middle attacker impersonates the client, negotiates an 'export-grade' Diffie–Hellman cipher suite with the server, and receives the server's Diffie–Hellman key-exchange message using a 512-bit prime, signed with the server's certificate private key. This message does not include any indication that it should be bound to an 'export-grade' Diffie–Hellman cipher suite. The attacker can then impersonate the server and forward this weak key-exchange message to the victim client, who believes that they have negotiated a normal-strength Diffie–Hellman key-exchange with the server. From the client's perspective, this message is indistinguishable from a server who always uses a 512-bit prime for Diffie–Hellman. In order to be successful in completing the man-in-the-middle attack undetected, the attacker must compute the 512-bit discrete log of the client or server's key-exchange message in order to compute the symmetric authentication keys, forge the client and server MAC contained in the 'Finished' messages sent at the end of the handshake, and decrypt the symmetrically encrypted data. This was feasible or close to feasible, in practice, for 512-bit primes because of the benefits of pre-computation, discussed previously in Section 6.3.5.

Mavrogiannopoulos *et al.* [405] observed that TLS versions 1.2 and prior are vulnerable to a more sophisticated cross-cipher-suite attack in which an attacker convinces the victim to interpret a signed elliptic-curve Diffie–Hellman key-exchange message as a valid prime-field Diffie–Hellman key-exchange message. This attack exploits three protocol features: first, for prime-field

Diffie–Hellman, the server includes its choice of prime and group generator as explicit parameters along with the ephemeral Diffie–Hellman key-exchange value; second, for elliptic-curve Diffie–Hellman, the key-exchange message included an option to specify an explicit elliptic-curve via the curve parameters; and third, these messages do not include an indication of the server's choice of cipher suite.

Putting these properties together for an attack, the man-in-the-middle impersonates the client and initiates many connections to the server requesting an ECDH cipher suite until it receives an ECDH key-exchange message that can be parsed as a prime-field Diffie–Hellman key-exchange message by using a weak modulus. At this point, the man-in-the-middle attacker forwards this weak message to the victim client, who believes that they have negotiated a prime-field Diffie–Hellman key-exchange. The attacker must compute the discrete logarithm online in order to forge the 'Finished' messages and complete the handshake. The key to the attack is that the random ECDH key-exchange values are parsed as a 'random'-looking modulus and group generator for prime-field Diffie–Hellman, which could be exploited as described for the composite-order groups in Section 6.3.8 above. Mavrogiannopoulos *et al.* estimate that the attacker is likely to succeed after 2^{40} connection attempts with the server.

6.3.12 Nearby Primes

Many of the most commonly used prime moduli used for Diffie–Hellman share many bits in common with each other, because they have the same fixed most and least significant bits, and use the digits of π or e for the middle bits.

In addition to the fixed, named groups of this form included in the SSH specification, SSH also allows the client and server to negotiate group parameters by specifying approved bit-lengths for keys [196]. When this group negotiation is carried out, SSH servers choose group parameters from a pre-generated file (e.g., /etc/ssh/moduli) of primes and group generators. Default files of these parameters are distributed along with the operating system.

These files contain dozens of primes of each specified key length that all differ in only the least-significant handful of bits. They appear to have been generated by starting at a given starting point, and incrementing to output 'safe' primes.

It is not known whether an adversary would be able amortise cryptanalytic attacks over many nearby primes of this form via the number field sieve or other discrete logarithm algorithms.

6.3.13 Special Number Field Sieve

The number field sieve (NFS) is particularly efficient for integers that have a special form. This algorithm is known as the special number field sieve (SNFS). The improved running time applies for integers p where the attacker knows a pair of polynomials f, g, of relatively low degree and small coefficients, such that they share a common root modulo p. A simple case would be a p of the form $m^6 + c$ for small c; this results in the polynomial pair $f(x) = x^6 + c$ and $g(x) = x - m$.

Although neither a general 1024-bit factorisation nor discrete logarithm have been known to have been carried out in public as of this writing (2020), a 1039-bit SNFS factorisation was completed in 2007 by Aoki *et al.* [22], and a 1024-bit SNFS discrete logarithm was carried out in 2016 by Fried *et al.* in a few calendar months [195].

The latter note that they found several Diffie–Hellman primes in the wild with forms that were clearly amenable to SNFS computations. These included 150 HTTPS hosts using the 512-bit prime $2^{512} - 38\,117$, and 170 hosts using the 1024-bit prime $2^{1024} - 1\,093\,337$ in March 2015. They also report that the LibTomCrypt library included several hard-coded Diffie–Hellman groups ranging in sizes from 768 to 4096 bits, with a readily apparent special form amenable to the SNFS. The justification for using such primes appears to have been to make modular exponentiation more efficient [394].

SNFS Trapdoors In 1991, in the context of debates surrounding the standardisation of the Digital Signature Algorithm, Lenstra and Haber [370] and Gordon [232] raised the possibility that an adversary could construct a malicious prime with a hidden trapdoor structure that rendered it amenable to the SNFS algorithm, but where the trapdoor structure would be infeasible for an outside observer to discover. This could be accomplished by constructing a pair of SNFS polynomials first, with resultant an appropriately structured prime p, and publishing only p, while keeping the polynomial pair secret. Computing discrete logarithms would then be easy for this group for the attacker, but infeasible for anyone who did not possess the secret. Although this type of trapdoor would be difficult to hide computationally for the 512-bit primes in use in the early 1990s, Fried *et al.* [195] argue that such a trapdoor would be possible to hide for 1024-bit primes.

In response to these concerns about trapdoor primes, the Digital Signature Standard suggests methods to generate primes for DSA in a 'verifiably random' way, with a published seed value [283]. Alternatively, the use of 'nothing up my sleeve' values such as the digits of π or e for the Oakley [464] and TLS 1.3 [220] primes is intended to certify that these primes are not trap-doored, because it would not be possible for an attacker to have embedded the necessary hidden structures.

The use of 'nothing up my sleeve' numbers is easily verifiable for those primes, but for 'verifiably random' primes generated according to FIPS guidelines, publishing the seeds is optional, and almost none of the primes used for either DSA or Diffie–Hellman in the wild is published with the seeds used to generate them. These include several examples of widely used standardised and hard-coded primes whose generation is undocumented. An example would be the primes included in RFC 5114 [389], which were taken from the FIPS 186-2 [282] test vectors, but were published without the seeds used to generate them. These primes were used by 2 per cent of HTTPS hosts and 13 per cent of IPsec hosts in 2016 [195].

6.3.14 Countermeasures

The most straightforward countermeasure against the attacks we describe against prime-field Diffie–Hellman is to avoid its use entirely, in favour of elliptic-curve Diffie–Hellman. Given the current state of the art, elliptic-curve Diffie–Hellman appears to be more efficient, permit shorter key lengths, and allow less freedom for implementation errors than prime-field Diffie–Hellman. Most major web browsers have dropped support for ephemeral prime-field Diffie–Hellman cipher suites for TLS 1.2 and below [125, 590], and elliptic-curve Diffie–Hellman now represents the majority of HTTPS handshakes carried out in the wild.

If prime-field Diffie–Hellman must be supported, then the implementation choices made by TLS 1.3 are a good blueprint for avoiding known attacks: groups should use a fixed, pre-generated 'safe' prime deterministically produced from a 'nothing up my sleeve' number, the minimum acceptable prime length is 2048 bits, and any Diffie–Hellman-based protocol handshake must be authenticated via a digital signature over the full handshake.

6.4 Elliptic-Curve Diffie–Hellman

Elliptic-curve cryptography offers smaller key sizes and thus more efficient operation compared with factoring and finite-field-based public-key cryptography, because sub-exponential-time index calculus-type algorithms for the elliptic-curve discrete log problem are not known to exist. While the idea of using elliptic curves for cryptography dates to the 1980s [340, 433], adoption of elliptic-curve cryptography has been relatively slow, and elliptic-curve cryptography did not begin to see widespread use in network protocols until after 2010. As of this writing, however, elliptic-curve Diffie–Hellman is used in more than 75 per cent of the HTTPS key-exchanges observed by the ICSI Certificate Notary [281].

There were numerous contributing factors to the delays in wide adoption of elliptic-curve cryptography, including concerns about patents, suspicions of the NSA's role in the development of standardised curves, competition from RSA, and the belief among practitioners that elliptic curves were poorly understood compared with the more 'approachable' mathematics of modular exponentiation for RSA and prime-field Diffie–Hellman [339].

The NSA has been actively involved in the development and standardisation efforts for elliptic-curve cryptography, and the original version of the NSA's Suite B algorithm recommendations in 2005 for classified US government communications included only elliptic-curve algorithms for key agreement and digital signatures. Thus it came as quite a surprise to the community when the NSA released an announcement in 2015 that owing to a 'transition to quantum resistant algorithms in the not too distant future', those 'that have not yet made the transition to Suite B elliptic-curve algorithms' were recommended to not make 'a significant expenditure to do so at this point but instead to prepare for the upcoming quantum resistant algorithm transition' [449]. The updated recommendations included 3072-bit RSA and prime-field Diffie–Hellman; the algorithm suite has been renamed but the recommended algorithms and key sizes remained the same. Koblitz and Menezes evaluate the suspicions of an algorithmic break or backdoor that this announcement sparked [341].

ECDH, in Theory An elliptic-curve group is defined by a set of domain parameters: the order q of the field $\mathbb{F}_q$, the coefficients of the curve equation $y^2 = x^3 + ax + b \bmod q$, and a generator G of a subgroup of prime-order n on the curve. The co-factor h is equal to the number of curve points divided by n.

Alice and Bob carry out an elliptic-curve Diffie–Hellman key-exchange in theory as follows: Alice generates a secret exponent k_a and sends the public value $Q_a = k_a G$. Bob generates a secret exponent b and sends the public value $Q_b = k_b G$. Alice then computes the shared secret as $k_a Q_b$, and Bob computes the shared secret as $k_b Q_a$.

Elliptic-curve public keys can be represented in uncompressed form by providing both the x and y coordinates of the public point, or in compressed form by providing the x coordinate and a bit to specify the y value.

6.4.1 Standardised Curves

In contrast to the situation with prime-field Diffie–Hellman, end users almost never generate their own elliptic curves, and instead rely on collections of standardised curve parameters. Several collections of curve parameters have been published. The SEC 2 [525] recommendation published by Certicom includes 'verifiably random' and Koblitz curve parameters for 192-, 224-, 256-, 384- and 521-bit prime field sizes, and 163-, 233-, 239-, 283-, 409- and 571-bit binary field sizes. NIST included the 224-, 256-, 384- and 521-bit random curves in their elliptic-curve recommendations; the 256-bit prime-field 'verifiably random' NIST P-256 curve is by far the most commonly used one in practice [594]. The 'verifiably random' curve-generation procedure has been criticized for simply hashing opaquely specified values [53]. Although the likelihood of undetectably backdoored standardised curves is believed to be small [341], distrust of standardised elliptic curves has slowed the adoption of elliptic-curve cryptography more generally.

Another prominent family of standardised curves are the Brainpool curves authored by the German ECC Brainpool consortium [398], which give 'verifiably pseudorandom' curves over prime fields of 160-, 192-, 224-, 256-, 320-, 384- and 512-bit lengths. These curves are supported by many cryptographic libraries, although are less popular than the NIST curves [594].

Curve25519 [49] is a 256-bit curve that was developed by Bernstein to avoid numerous usability and security issues that affect the SEC 2/NIST and Brainpool curves, and has been growing in popularity in network protocols.

TLS versions 1.2 and below included a mechanism by which individual servers could configure custom elliptic-curve parameters, but this does not seem to be used at all for HTTPS servers reachable on the public Internet [594].

6.4.2 Invalid Curve Attacks

In principle, a small subgroup attack analogous to the small subgroup attacks described in Section 6.3.10 for prime-field Diffie–Hellman is also possible against elliptic curves. To mitigate these attacks, the elliptic curves that have been standardised for cryptography have typically been chosen to have small co-factors. NIST recommends a maximum co-factor for various curve sizes.

However, there is a much more severe variant of this attack that is due to Antipa *et al.* [21]: an attacker could send an elliptic-curve point of small order q_i that lies on a different curve entirely. If the victim does not verify that the received key-exchange value lies on the correct curve, the attacker can use the victim's response to compute the victim's secret modulo q_i. If the victim is using static Diffie–Hellman secrets, then the attacker can repeat this for many chosen curves and curve points of small prime-order to recover the full secret.

To mitigate this attack, implementations must validate that the points it receives lie on the correct curve, or use a scalar multiplication algorithm that computes only on the x-coordinate together with a curve that is secure against curve twist attacks.

Elliptic-curve implementations have suffered from repeated vulnerabilities due to failure to validate that elliptic-curve points are on the correct curve. Jager *et al.* [286] found that three out of eight popular TLS libraries did not validate elliptic-curve points in 2015. Valenta *et al.* [594] scanned TLS, SSH and IKE addresses in 2016 using a point of small order on an invalid curve and estimated 0.8 per cent of HTTPS hosts and 10 per cent of IKEv2 hosts did not validate ECDH key-exchange messages.

6.4.3 Countermeasures

Elliptic-curve Diffie–Hellman has been growing in popularity, and currently appears to offer the best security and performance for key exchange in modern network protocols. Compared with the long history of implementation messes involving RSA and prime-field Diffie–Hellman, ECDH seems relatively unscathed. The one dark spot is the relatively expensive validation checks required to protect against the invalid curve and twist attacks that have plagued some implementations of the NIST curves. As a countermeasure, Curve25519 was designed to require only a minimal set of validation checks.

Elliptic-curve Diffie–Hellman (ECDH) implementations may paradoxically be 'protected' from implementation mistakes by a form of security through obscurity. Because the mathematics of elliptic curves is so much more complex than RSA or prime-field Diffie–Hellman, implementers seem empirically less likely to attempt to design their own curves, or deviate from standard recommendations for secure implementations. Another protective factor may have been the relatively late dates of ECDH adoption. By the time elliptic-curves began to be used on any scale in the wild, well after 2010, the cryptographic community had already discovered the analogues of many of the basic cryptographic vulnerabilities that early RSA and Diffie–Hellman implementations suffered from.

Nadia Heninger

6.5 (EC)DSA

The DSA algorithm was originally standardised in the early 1990s. Prime-field DSA never found widespread use for SSL/TLS (Lenstra *et al.* [386] report finding only 141 DSA keys out of more than 6 million distinct HTTPS certificates in 2012; nearly all of the rest were RSA). However, the SSH specification requires implementations to support DSA as a public-key format [629], and it was almost universally supported by SSH servers until it began to be replaced by ECDSA. DSA public keys were also widely used for PGP: Lenstra *et al.* [386] report that 46 per cent of 5.4 million PGP public keys they scraped in the wild in 2012 were DSA public keys. The relative popularity of DSA compared with RSA in these different protocols is likely due to the fact that RSA was protected by a patent that did not expire until the year 2000.

In the past handful of years, ECDSA, the elliptic-curve digital signature algorithm, has rapidly grown in popularity. According to the ICSI Certificate Notary, 10 per cent of the HTTPS certificates seen in the wild as of this writing use ECDSA as their signature algorithm [281]. Most of the major cryptocurrencies use ECDSA signatures to authenticate transactions.

DSA Key Generation, in Theory The DSA public parameters include a prime p chosen so that $(p - 1)$ is divisible by a smaller prime q of specified length, and g a generator of the subgroup of order q modulo p. To generate a DSA public key, the signer generates a random secret exponent x, computes the value $y = g^x \bmod p$, and publishes the values (p, q, g, y) [283].

DSA Signatures and Verification, in Theory As originally published, the DSA is randomised. To sign the hash of a message $H(m)$, the signer chooses an integer k, which is often called a signature nonce, although it serves as an ephemeral private key. The signer computes the values $r = g^k \bmod p \bmod q$ and $s = k^{-1}(H(m) + xr) \bmod q$, and publishes the values (r, s) [283]. To mitigate vulnerabilities due to random-number generation failures in generating k, many implementations now use 'deterministic' nonce generation, where the nonce value k is generated pseudorandomly and deterministically by applying a pseudorandom function, typically based on HMAC, to the message m and the secret key x [483].

To verify a signature (r, s) with a public key (p, q, g, y), the verifier computes the values $w = s^{-1} \bmod q$, $u_1 = H(m)w \bmod q$, and $u_2 = rw \bmod q$. The verifier then verifies that $r = g^{u_1} y^{u_2} \bmod p \bmod q$ [283].

6.5.1 Distinct Primes

Perhaps in response to the trapdoor prime controversy surrounding the standardisation of DSA described in Section 6.3.13, custom primes for DSA have been historically more common than custom primes for Diffie–Hellman. Lenstra *et al.* [386] note that, of the more than 2.5 million DSA public keys in the PGP key database in 2012, only 1900 (0.07 per cent) used the same prime as another public key in the database. This is in contrast to the choices for ElGamal PGP public keys, where 66 per cent of the more than 2.5 million ElGamal public keys used a prime that was used by another key in the database, and there were only 93 such primes that were shared.

6.5.2 Repeated Public Keys

Repeated public keys are also common across hosts. Heninger *et al.* [269] documented extensive repetition of DSA public keys across SSH hosts in 2012, with common keys being served by thousands of IP addresses. Many of these were large hosting providers with presumably shared infrastructure, but they also found evidence of hard-coded keys baked into network device firmware, as well as evidence of random-number generation issues similar to those affecting RSA. In these cases, the owner of a vulnerable key will know the corresponding secret key for any host sharing the same public key, and could thus generate valid signatures for these vulnerable hosts. For keys that have been hard-coded into firmware images or repeated due to poor randomness, the secret keys can be compromised by an attacker who reverse-engineers the implementation.

6.5.3 ECDSA

ECDSA Key Generation, in Theory The public-domain parameters for an ECDSA public key are the same as for elliptic-curve Diffie–Hellman: for the purposes of this section, the relevant curve parameters are a specification of a finite field $\mathbb{F}$ and an elliptic curve E, together with a generator G of a subgroup of prime-order q on the curve. The private signing key is an integer x, and the public signature key is the point $Y = xG$ [283].

ECDSA Key Generation, in Practice The NIST P-256 curve is by far the most commonly used curve for ECDSA by SSH and HTTPS servers in the wild. The most common cryptocurrencies including Bitcoin, Ethereum and Ripple all use the curve secp256k1, which is a 256-bit Koblitz curve described in SEC 2 [525].

The Ed25519 signature scheme [51] is a variant of EdDSA, the DSA signature scheme adapted to Edwards curves, specialised to Curve25519. It includes countermeasures against many common implementation vulnerabilities: in particular, signature nonces are defined as being generated deterministically, and it is implemented without secret-dependent branches that might introduce side channels.

ECDSA Signatures and Verification, in Theory To sign a message hash $H(m)$, the signer chooses an integer k, computes the point $(x_r, y_r) = kG$, and outputs $r = x_r$ and $s = k^{-1}(H(m) + xr) \bmod q$. The signature is the pair (r, s) [283]. In 'deterministic' ECDSA, the nonce k is generated pseudorandomly and deterministically by applying an HMAC-based pseudorandom function to the message m and the secret key x [483].

To verify a message hash using a public key Y, the verifier computes $(x'_r, y'_r) = hs^{-1}G + rs^{-1}Y$ and verifies that $x'_r \equiv r \bmod q$.

Signature Normalisation ECDSA signatures have the property that the signatures (r, s) and $(r, -s)$ will validate with the same public key. In order to ensure that signatures are not malleable, Bitcoin, Ethereum and Ripple all use 'signature normalisation', which uses the smaller of s and $-s$ for a signature. This is a mitigation against attacks in the cryptocurrency context in which an attacker duplicates a transaction under a different transaction identifier (computed as the deterministic hash over all of the transaction data including the signature) by modifying the signature [333].

6.5.4 Curve-Replacement Attacks

In 2020, the NSA announced a critical security vulnerability in Microsoft's Windows 10 certificate validation code [448]. The code flaw failed to validate that the elliptic-curve group parameters in the signature match the curve parameters in a trusted certificate. This allowed an adversary to forge a signature that would validate for any given key under this code by giving the public-key value Y in place of the generator G of the curve, and generating a signature using the secret key $x = 1$.

6.5.5 Small Secret Keys

Because the security of (EC)DSA relies on the (elliptic-curve) discrete logarithm of the public key y or Y being hard to compute, the secret key x should be difficult to guess. However, numerous implementation vulnerabilities appear to have led to predictable secret exponents being used for ECDSA public keys in the context of Bitcoin. The Large Bitcoin Collider is a project that is using a linear brute-force search algorithm to search for Bitcoin keys with a small public exponent, and has found several secret keys used in the wild after searching a 55-bit space [497].

6.5.6 Predictable Secret Keys

A Bitcoin 'brainwallet' is a tool that derives an ECDSA public–private key pair from a passphrase provided by a user. A typical key derivation might apply a cryptographic hash function H to a passphrase to obtain the ECDSA secret key x. This allows the user to use this public–private key pair without having to store an opaque blob of cryptographic parameters in a secure fashion. However, attackers have carried out dictionary attacks and successfully recovered numerous Bitcoin secret keys, together with associated funds [115].

6.5.7 Predictable Nonces

The security of (EC)DSA also relies crucially on the signature nonce (actually a one-time secret key) for every signature remaining secret and hard to predict. If an attacker can guess or predict the nonce value k used to generate a signature (r, s), the attacker can compute the signer's long-term secret key

$$x = r^{-1}(ks - H(m)) \bmod q. \qquad (6.1)$$

Breitner and Heninger [107] found numerous Bitcoin signatures using 64-bit or smaller signature nonces, including several whose values appeared to be hand-generated.

Several million signatures in the Bitcoin blockchain use the signature nonce value $(q-1)/2$, where q is the order of the secp256k1 curve. The x-coordinate of $(q - 1)/2 \cdot G$ is 166 bits long, where one would expect a random point to have 256 bits, resulting in a much shorter signature than expected. This value is apparently used intentionally to collect small amounts of funds from addresses that will be then abandoned, and appears to be used because reducing the length of the digital signature reduces the transaction costs [68]. Many Bitcoin private keys are intended to be used only once, so the users of this nonce value appear to be using it intentionally because they do not care about compromising the secret key.

6.5.8 Repeated Nonces

If a victim ever signs two distinct message hashes $H(m_1)$ and $H(m_2)$ using the same (EC)DSA private key and signature nonce k to generate signatures (r_1, s_1) and (r_2, s_2), then it is trivial to recover the long-term private signing key from the messages. If the signatures have been generated using the same k value, then this is easy to recognise because $r_1 = r_2$. Then the value k can be recovered as

$$k = (H(m_1) - H(m_2))(s_1 - s_2)^{-1} \bmod q.$$

Once k has been recovered, then the secret key x can be recovered as in Eq. (6.1).

Repeated DSA and ECDSA signature nonces have been found numerous times in the wild.

In 2011, the ECDSA signature implementation used by the Sony PS3 was found to always use the same nonce k, revealing the code-signing key.

In 2012, the private keys of 1.6 per cent, around 100 000, of SSH hosts with DSA host keys were compromised via repeated signature nonces [269]. Most of these vulnerable hosts appeared to be compromised due to poorly seeded deterministic pseudorandom-number generators. First, numerous hosts appear to have generated identical public keys using a pseudorandom-number generator seeded with the same value. Next, if the same poorly seeded pseudorandom-number generator is used for signature generation, then multiple distinct hosts will generate the same deterministic sequence of signature nonces. There were also multiple SSH implementations that were observed always using the same signature nonce for all signatures. In 2018, Breitner and Heninger [107] found 80 SSH host public keys were compromised by repeated signature nonces from ECDSA or DSA signatures, suggesting that the random-number generation issues found in 2012 may have been mitigated. They found a small number of repeated ECDSA signature nonces from HTTPS servers that compromised 7 distinct private keys used on 97 distinct IP addresses.

Repeated signature nonces in the Bitcoin blockchain have been well documented over the course of many years [95, 108, 116, 140], and empirical evidence suggests that attackers are regularly scanning the Bitcoin blockchain and immediately stealing Bitcoins from addresses whose private keys are revealed through repeated signature nonces. These vulnerabilities have been traced to at least two high-profile random-number generation vulnerabilities: a 2013 bug in the Android SecureRandom random-number generator [335, 426] and a 2015 incident in which the Blockchain.info Android application had been attempting to seed from `random.org`, but was instead seeding from a 403 Redirect page to the HTTPS URL [585].

To prevent this vulnerability, (EC)DSA implementations are recommended to avoid generating nonces using a random-number generator, and instead to derive signature nonces deterministically using a cryptographically secure-key derivation function applied to the message and the secret key [483].

6.5.9 Nonces with Shared Prefixes and Suffixes

If an implementation generates signature nonces where some bits of the nonce are known, predictable, or shared across multiple signatures from the same private key, then there are multiple algorithms that can recover the private key from these signatures. Let us specialise to the case where the signature nonce k has some most significant bits that are 0, so there is a bound $B < q$ such that $k < B$. For a signature (r_i, s_i) on message hash h_i, the signature satisfies the relation

$$k_i = s_i^{-1} r_i x - s_i^{-1} h_i \bmod q,$$

where the k_i is known to be small, the secret key x is unknown, and the other values are known.

Howgrave-Graham and Smart [277] showed that one can solve for the secret key using a lattice attack due to Boneh and Venkatesan [81]. Generate the lattice basis

$$M = \begin{bmatrix} q & & & & & \\ & q & & & & \\ & & \ddots & & & \\ & & & q & & \\ s_1^{-1} r_1 & s_2^{-1} r_2 & \cdots & s_m^{-1} r_m & B/q & \\ s_1^{-1} h_1 & s_2^{-1} h_2 & \cdots & s_m^{-1} h_m & & B \end{bmatrix}.$$

The vector $v_x = (k_1, k_2, \ldots, k_m, Bx/q, B)$ is a short vector generated by the rows of M, and when $|v_x| < \det M^{1/\dim M}$, we hope to recover v_x using a lattice reduction algorithm like LLL [380] or BKZ [512, 520].

In 2018, Breitner and Heninger found 302 Bitcoin keys that were compromised because they had generated signatures with nonces that were short, or shared a prefix or a suffix with many other nonces [107]. These were hypothesized to be due to implementation flaws that used the wrong length for nonces, or memory-bounds errors that caused the buffer used to store the nonce to overlap with other information in memory.

They also found three private keys used by SSH hosts whose signature nonces all shared the suffix f27871c6, which is one of the constant words used in the calculation of a SHA-2 hash, suggesting that the implementation generating these nonces was attempting to use a buggy implementation of SHA-2. In addition, there were keys used for both Bitcoin and SSH that used 160-bit nonces, suggesting that a 160-bit hash function like SHA-1 may have been used to generate the nonces.

6.5.10 Bleichenbacher Biased Nonces

Bleichenbacher developed an algorithm that uses Fourier analysis to recover the secret key in cases of even smaller bias than the lattice attack described above works for [65]. The main idea is to define a function for signature samples (r_i, s_i) on message hash h_i

$$f(s_i^{-1} r_i) = e^{\frac{-2\pi i s_i^{-1} h_i}{q}}.$$

It can be shown that this function will have a large Fourier coefficient $\hat{f}(x)$ at the secret key x, and will be close to 0 everywhere else.

Bleichenbacher's algorithm then uses a technique for finding significant Fourier coefficients even when one cannot compute the full Fourier transform of the function to find the secret key x.

Bleichenbacher's algorithm was inspired by the observation that many DSA and ECDSA implementations will simply generate an n-bit random integer as the nonce for an n-bit group order, and do not apply rejection sampling if the nonce is larger than the group order. This means that the distribution of the nonce k over many signatures will be non-uniform. That is, if the group order $q = 2^n - t$ and an implementation naively generates an n-bit nonce k, then the values between 0 and t are twice as likely to occur than all other values.

The countermeasure against this attack is for implementations to use rejection sampling to sample a uniform distribution of nonces modulo q.

Most common elliptic-curves have group orders that are very close to powers of two, so a practical attack is likely infeasible for these curves, but prime-field DSA group-generation procedures generally do not put such constraints on the subgroup order.

6.5.11 Countermeasures

The fact that a long-term (EC)DSA private key is compromised if that key is ever used to generate signatures by using a faulty or even very slightly biased random-number generator is a severe usability problem for the signature scheme, and has led to numerous problems in the real world. To protect against this vulnerability, implementations must always use 'deterministic' (EC)DSA to generate nonces. Fortunately, this countermeasure is becoming more popular, and more recent signature schemes such as Ed25519 build this nonce-generation procedure directly into the signature-generation scheme from the start.

6.6 Conclusion

The real-world implementation landscape for public-key cryptography contains numerous surprises, strange choices, catastrophic vulnerabilities, oversights and mathematical puzzles. Many of the measurement studies we toured evaluating real-world security have led to improved standards, more secure libraries, and a slowly improving implementation landscape for cryptography in network protocols.

The fact that it has taken decades to iron out the implementation flaws for our oldest and most well-understood public-key cryptographic primitives such as RSA and Diffie–Hellman raises some questions about this process in the future: with new public-key standards on the horizon, are we doomed to repeat the past several decades of implementation chaos and catastrophic vulnerabilities?

Fortunately, the future seems a bit brighter than the complicated history of public-key cryptographic deployments might suggest. Although the mathematical structure of our future public-key cryptography standards will look rather different from the factoring and cyclic-group based cryptography we use now, many of the general classes of vulnerabilities are likely to remain the same: random-number generation issues, subtle biases and rounding errors, pre-computation trade-offs, backdoored constants and substitution attacks, parameter negotiation, omitted validation checks and error-based side channels. With an improved understanding of protocol integration, a focus on real-world threat models, developer usability, and provably secure implementations, we have a chance to get these new schemes mostly right before they are deployed into the real world.

Acknowledgements

I am grateful to Shaanan Cohney, Matthew Green, Paul Kocher, Daniel Moghimi, Keegan Ryan and the anonymous reviewers for helpful suggestions, anecdotes and feedback.

7

A Survey of Chosen-Prefix Collision Attacks

Marc Stevens

7.1 Cryptographic Hash Functions

A cryptographic hash function $H : \{0, 1\}^* \rightarrow \{0, 1\}^n$ is a function that computes a fixed-length hash value of n bits for any arbitrary long input message M. For cryptographic purposes, H should satisfy at least the following security properties.

- *Pre-image resistance*: given a hash value h, it should be infeasible to find a message x such that $H(x) = h$.
- *Second pre-image resistance*: given a message x, it should be infeasible to find a message $y \neq x$ such that $H(x) = H(y)$.
- *Collision resistance*: it should be infeasible to find collisions, i.e., two distinct messages x and y that hash to the same value $H(x) = H(y)$.

Cryptographic hash functions are the Swiss army knives within cryptography. They are used in many applications including digital-signature schemes, message-authentication codes, password hashing, cryptocurrencies and content-addressable storage. The security or even the proper functioning of these applications relies on the security property that is the main focus of this chapter: collision resistance. For instance, all major digital-signature schemes rely on the hash-then-sign paradigm. This implies that, for any colliding pair $x \neq y$ with $H(x) = H(y)$, any signature for x is also an unwanted valid signature for y, and vice versa. When finding meaningful collision pairs (x, y) is practical, this can have grave implications, as will become clear below.

Collision Resistance The best generic attack to find collisions for a cryptographic hash function is the birthday search. For an output hash of n bits, on average $\sqrt{\pi/2} \cdot 2^{n/2}$ hash evaluations are needed to find a collision among them.

This can be achieved in practice using little memory and is easily parallelisable [602]. We call a cryptographic hash function collision resistant if it is not possible to find collisions faster than this generic attack. In contrast to essentially all security properties of other cryptographic functionalities, collision resistance of a cryptographic hash function defies formalisation as a mathematical security property. Such formalisations dictate the non-existence of an efficient algorithm that solves the security problem with non-negligbile probability. The underlying issue, dubbed the foundations of hashing dilemma [502], is the following. For any given cryptographic hash function H there exists a collision $x \neq y$ with $H(x) = H(y)$ by the pigeon-hole principle. This implies there actually exists a trivial collision-finding algorithm $A_{x,y}$ that succeeds with probability 1 by simply printing a collision pair x, y. Collision resistance thus remains an informal security property that there does not exist a known algorithm to find collisions for H faster than the generic attack.

Standards In practice, the SHA-2 and SHA-3 families of hash functions, each consisting of 4 hash functions with output sizes 224, 256, 384 and 512 bits, are current National Institute of Standards and Technology (NIST) standards that are recommended for cryptographic purposes. Their predecessors, namely the MD5 and SHA-1 hash functions, have been used for decades as the software industry's de facto standards.

With the exception of the recent SHA-3 family, MD5, SHA-1 and SHA-2 are all based on the Merkle–Damgård framework [149, 423]. This framework builds a cryptographic hash function H for arbitrary-size inputs from a compression function $f : \{0, 1\}^n \times \{0, 1\}^m \to \{0, 1\}^n$ with fixed-size inputs. Any input message M is unambiguously padded to a multiple of m bits and then split into a sequence of m-bit blocks $padding(M) = m_1 \| \cdots \| m_k$. The hash function initialises a chaining value CV_0 with a fixed public value IV and iteratively updates this chaining value using $f: CV_i = f(CV_{i-1}, m_i)$, for $i = 1, \ldots, k$. The final chaining value CV_k is used as the hash output $H(M) := CV_k$ of M.

The Merkle–Damgård construction admits a security reduction of the collision resistance of H to the collision resistance of f, i.e., it implies that it is at least as hard to find collisions for the hash function H as it is hard to find collisions for its compression function f. Hence, an early practical collision attack on MD5's compression function by den Boer and Bosselaers [154], should have been an important warning against MD5's widespread adoption till 2004.

Collision Attacks The first collision attack for MD5 was presented in 2004 by Wang *et al.* [616], together with an actual example collision consisting of two different 128-byte random looking files. Their collision attack can be used

Year	Identical-prefix collision cost	Chosen-prefix collision cost
< 2004	2^{64} generic	2^{64} generic
2004	$\mathbf{2^{40}}$ **[616]**	–
2005	2^{37} [331]	–
2006	2^{32} [332, 564]	$\mathbf{2^{49}}$ **[570]**
2007	2^{25} [565]	–
2008	2^{21} [626]	–
2009	2^{16} [573]	2^{39} [573]
2020	2^{16} [573]	2^{39} [573]

Table 7.1 *Historical overview of MD5 collision attack complexities.*[a]

[a] References in bold type = first collision example.

to craft so-called identical-prefix collisions (IPC) of the form $H(P\|C_1\|S) = H(P\|C_2\|S)$, where P is a shared prefix with byte length a multiple of 64, $C_1 \neq C_2$ are two 128-byte collision bit strings (dependent of P), and S is a shared suffix.

As identical-prefix collisions have differences only in 128-byte random-looking strings C_1 and C_2, these initially did not seem very harmful for real-world applications. Yet several abuse scenarios for identical-prefix collisions were quickly demonstrated, e.g., to mislead integrity software [310, 429] in 2004, X.509 certificates with distinct public keys [369] in 2005, visually distinct Word, TIFF, and black-and-white PDF files [214] in 2006. Although these were working proofs of concept of meaningful collisions, due to their limitations, they did not form very convincing real-world threats.

In 2007, a stronger and more expensive attack called chosen-prefix collision (CPC) attack [570] was introduced that can produce collisions of the form $H(P_1\|C_1\|S) = H(P_2\|C_2\|S)$, where the prefixes P_1 and P_2 can be arbitrarily and independently chosen. The main topic of this chapter is the development of practical chosen-prefix collision attacks and their applications. A historical overview of the identical-prefix and chosen-prefix collision attack costs for MD5 and SHA-1 can be found in Tables 7.1 and 7.2.

Near-Collision Attacks Both identical-prefix collisions and chosen-prefix collisions are built from near-collision attacks on the compression function. Differential cryptanalysis is used to construct such attacks against the compression function, where differences between two related evaluations $f(CV, B)$ and $f(CV', B')$ are analysed. For any variable X in the evaluation of $f(CV, B)$,

Year	Identical-prefix collision cost		Chosen-prefix collision cost	
< 2005	2^{80}	generic	2^{80}	generic
2005	2^{69}	[618]	–	
	$(u : 2^{63}$	[617])	–	
2007	$(u : 2^{61}$	[415])	–	
2009	$(w : 2^{52}$	[412])	–	
2013	2^{61}	[569]	2^{77}	[569]
2017	**G : $2^{63.1}$**	**[575]**	–	
2019	–		$G : 2^{67}$	[390]
2020	2^{61} / $G : 2^{61.2}$	[569] / [391]	**G : $2^{63.4}$**	**[391]**

Table 7.2 *Historical overview of SHA-1 collision attack complexities.*[a]

[a] Abbreviations: u = unpublished; w = withdrawn; G = using GPU. References in bold = first collision example.

we denote by X' the respective variable in the evaluation of $f(CV', B')$ and by $\delta X = X' - X$ the difference between these two variables.

The compression function is essentially a block cipher $E_K(M)$ in Davies–Meyer feed-forward mode $f(CV, B) = CV + E_B(CV)$. More specifically, a near-collision attack employs a differential characteristic against the block cipher, i.e., a valid trail of differences for all internal values of all rounds of the compression function. The first part of the differential characteristic over about 1/4 of the rounds can be dense and is required to start with the given input chaining value difference. But it is the last 3/4 of the rounds that is the critical part that mostly determines the attack cost, which needs to be optimised: sparse and high probability. Owing to Davies–Meyer, the final difference D of the differential characteristic is added to the input chaining-value difference: $\delta CV_1 = \delta CV_0 + D$.

For instance, an identical-prefix collision requires a partial differential characteristic over the last 3/4 of the rounds with high probability. This partial characteristic with final difference D will be used for both near-collision attacks as follows. The first near-collision attack begins with chaining-value difference $\delta CV_0 = 0$. An appropriate full differential characteristic is used, which can be pre-computed, and dedicated collision search techniques are used to find a solution: a pair of blocks that fulfills the differential characteristic. After this first attack a chaining-value difference $\delta CV_1 = \delta CV_0 + D = D$ is obtained.

The second attack continues with $\delta CV_1 = D$ and now instead 'subtracts' the difference D. It re-uses the exact same critical part of the differential characteristic of the first block, except all differences are negated. The first part of the second attack's differential characteristic needs to be constructed for the input chaining-value difference $\delta CV_1 = D$. After the second near-collision attack an internal-state collision $\delta CV_2 = \delta CV_1 - D = 0$ is achieved. Owing to the Merkle–Damgård structure any internal-state collision directly results in a hash collision, even after appending an arbitrary shared suffix S.

7.2 Chosen-Prefix Collisions

Chosen-prefix collision attacks were introduced by Stevens, Lenstra and de Weger [570] in 2007. Given two distinct prefixes P and P' of equal bit-length $|P| = |P'|$ a multiple of the block-size of 512 bits, a chosen-prefix collision attack constructs two suffixes $C \neq C'$ such that $H(P\|C) = H(P'\|C')$ in general in two phases.

(1) *Birthday search phase* to find bit strings B and B' such that appending these to P and P' results in a chaining value difference in a target set $\mathcal{D}$:

$$\delta CV_0 = \hat{H}(P\|B) - \hat{H}(P'\|B') \in \mathcal{D},$$

where $\mathcal{D}$ is determined by the next phase.[1]

(2) *Near-collision attacks phase* to find a sequence of near-collision attack block pairs $(M_0, M_0'), \ldots, (M_k, M_k')$ that each add specific differences to the latest chaining-value difference in order to reduce the chaining-value difference δCV_0 to 0 in the end:

$$\hat{H}(P\|B\|M_0\|\cdots\|M_k) = \hat{H}(P'\|B'\|M_0'\|\cdots\|M_k').$$

The desired suffixes are thus $C = B\|M_0\|\cdots\|M_k$ and $C' = B'\|M_0'\|\cdots\|M_k'$. These two phases are elaborated below.

Birthday Search Phase Van Oorschot and Wiener [602] remains the state-of-the-art work on how a birthday search can be executed in practice. This is done by iterating a properly chosen deterministic function $f : V \to V$ on a certain space $V = \{0, 1\}^k$, assuming that the trail of points of V thus visited form a pseudorandom walk. To take advantage of parallelism, many different pseudorandom walks are generated, of which only the startpoints, lengths and endpoints are kept. All walks end at a 'distinguished point'; these form a subset of points with an easily recognisable pattern that is chosen depending on $|V|$, available storage and other characteristics. Because any two walks that intersect due to a collision will have the same endpoint, they can be easily detected.

---

[1] By $\hat{H}$ we mean the hash function H without any padding for all messages of bit-length a multiple of 512 bits, thus resulting in the chaining value directly after processing the entire input.

The collision point can then be recomputed given the startpoints and lengths of the two colliding trails.

If there are additional constraints that could not be encoded into the choice of f and V then the birthday search has to continue until a collision point has been found that satisfies these additional constraints. Assume a collision point has a probability of p of satisfying these additional constraints, then on average one has to find $1/p$ collision points. In this case the expected total cost C_{total} is:

$$C_{total} = C_{tr} + C_{coll}, \quad C_{tr} = \sqrt{\frac{\pi \cdot |V|}{2 \cdot p}}, \quad C_{coll} = \frac{2.5 \cdot C_{tr}}{p \cdot M},$$

where C_{tr} is the cost before a collision occurs and C_{coll} is the cost of recomputing the collision point(s) and M is the average amount of stored walks. It follows that, to ensure $C_{coll} \ll C_{tr}$, i.e., the overhead in recomputing costs remains a small fraction of the total cost, one needs to ensure $M \gg 2.5/p$ by choosing a suitable distinguished points pattern.

In the case of a chosen-prefix collision attack, given the set $\mathcal{D}$ of target chaining-value differences we can expect a complexity $C_{tr} \approx \sqrt{\pi \cdot 2^n / |\mathcal{D}|}$ by appropriately choosing f and V as follows. We need to partition V into two equally sized disjoint subsets V_1 and V_2 in order to map half of the inputs to a computation related to prefix P and the other half to P', e.g., by partitioning on the first bit of each $v \in V = \{0, 1\}^k$. The function f is thus actually split into two functions $f_1 : V_1 \to V$ and $f_2 : V_2 \to V$, and each typically is composed of three functions: $f_i = g_i \circ MD5Compress \circ h_i$.

- $h_i : V \to \{0, 1\}^n \times \{0, 1\}^{512}$ maps every point to a compression function input pair (CV, B) of chaining value CV and message block B. For example, $h_i(v) = (CV_i, 0^{512-k}\|v)$, where $CV_1 = \hat{H}(P)$ and $CV_2 = \hat{H}(P')$ (see also footnote 1).[2]
- $g_i : \{0, 1\}^n \to V$ maps a chaining value x to an element of V by taking a fixed subsequence of k bits of x indexed by I denoted as $v = (x_j)_{j \in I}$, which is the same for both g_1 and g_2. To allow more efficient encodings, one can consider g_2 that first adds a fixed offset d to x: $v = ((x + d)_j)_{j \in I}$. Given bit-selection I and offset d one can efficiently approximate the corresponding probability

---

[2] Instead of prepending zero bits, one can also use any (more complex) injective map onto the block space $\{0, 1\}^{512}$. It is even possible to generalise to encode the input x into the entire prefix P_x or P'_x, but this is not advisable as this will require many *MD5Compress* calls per f-call and significantly increase the birthday-search cost.

$p_{I,d}$ of a collision point being useful:[3]

$$p_{I,d} := 1/2 \cdot \Pr[y - x \in \mathcal{D} \mid (y_j)_{j \in I} = ((x + d)_j)_{j \in I}].$$

The goal is to choose (I, d) with high probability $p_{I,d}$, ensuring low memory requirements, from those that achieve near-optimal birthday-search cost

$$\sqrt{\pi \cdot 2^{|I|}/(2 \cdot p_{I,d})} \approx \sqrt{\pi \cdot 2^n/|\mathcal{D}|}.$$

Example I and d as chosen for actual chosen-prefix collisions attacks are shown in Section 7.4.

Near-Collision Attacks Phase Once the above birthday search has succeeded in finding messages $P\|B$ and $P'\|B'$ with a target chaining-value difference $\delta CV_0 = \hat{H}(P'\|B') - \hat{H}(P\|B) \in \mathcal{D}$, the next goal is to find a sequence of near-collision blocks that reduces the chaining-value difference to 0. These attacks are designed in the following steps.

(1) First, one needs one or more high-probability core differential characteristic over all but the first $1/4$ of the rounds. The success probability of this part mainly determines the near-collision attack complexity. For MD5, the main family of core differential characteristics considered is based on message block difference $\delta m_{11} = \pm 2^b$ and $\delta m_i = 0$ for $i \neq 11$, see Table 7.8. However, MD5 chosen-prefix collision attacks with other core differential characteristics have been demonstrated [187, 573].

(2) Second, one determines which set of output differences to use for each core characteristic. Each of the chosen core differential characteristics $\mathcal{P}$ results in a single output difference $D_{\mathcal{P}}$ (on top of which the Davies–Meyer feed-forward adds the input chaining-value difference). To increase significantly the set $\mathcal{D}$ beyond these $D_{\mathcal{P}}$, one chooses a lower bound on the differential characteristic probability p_{min}. Next, one searches for variants of the core differential characteristics $\mathcal{P}'$ where only the very last few steps are changed (resulting in different output difference $D_{\mathcal{P}'}$) with success probability $p \geq p_{min}$.

(3) Third, a suitable strategy has to be chosen. During the actual attack one has to perform a series of near-collision attacks and one needs an overall strategy how to choose, for each near-collision attack, which one or more variant(s) $\mathcal{P}'$ of a single core $\mathcal{P}$ to use. The most common strategy is a systematic one, where essentially each bit position of a given δCV_0 with a possible

---

[3] Besides requiring $f(y) - f(x) \in \mathcal{D}$, we additionally need these inputs belonging to different prefixes, i.e., $x \in V_1 \Leftrightarrow y \in V_2$, which happens with probability $p = 1/2$.

non-zero difference is assigned to a specific core $\mathcal{P}$. In this case, determining the set of variant characteristics $\mathcal{P}'$ whose output differences sum up to $-\delta CV_0$ is an almost trivial matter, and one can execute these in arbitrary order (cf. [187, 570, 573]). When there exist many distinct combinations of output differences that sum up to $-\delta CV_0$, a more complex strategy may exploit this to achieve a lower overall attack complexity as described in [390]. Based on the resulting set of output differences $\mathcal{D}$ that may be cancelled in this manner, one can now determine suitable values for I, d for the birthday-search step function as described earlier.

After the birthday search has resulted in a suitable chaining-value difference δCV_0, one applies the chosen strategy and constructs and executes a sequence of near-collisions. For each near-collision attack one chooses a differential characteristic $\mathcal{P}$ based on the strategy, and executes the near-collision attack in the following steps.

(1) Construct full differential characteristic $\mathcal{P}_{full}$ over all steps. The initial state is determined by the current input chaining-value pair, while most later steps are fixed by $\mathcal{P}_{full}$. What remains is to find a partial differential characteristic that connects these two. The first successful connection by Wang *et al.* was done entirely by hand. Since then two main algorithmic approaches have been developed: guess-and-determine, proposed by De Cannière and Rechberger [152], and the meet-in-the-middle approach [566, 567, 570]. In the case of SHA-1, even if the partial differential characteristic found is locally valid over those steps, there may be global incompatibilities due to dependencies between the expanded message-difference equations over all steps. This has been successfully addressed by using a SAT solver to tweak the full-differential characteristic to find a complete valid differential characteristic [575]. Finally, the full differential characteristic between two compression function evaluations can be transformed into an equivalent set of conditions on a single compression function evaluation.

(2) Search for a suitable combination of collision search speed-up techniques. The basic idea behind these techniques in collision search is to change a currently considered message block value using available remaining freedom that does not interfere with the conditions of the full differential characteristic up to some step (preferably as far as possible), possibly making use of additional conditions to make it work with high probability. The most simple is called neutral bits [62], which only flips a single message bit. Advanced message modification [616] / tunnels [332] / boomerangs [297] aim to cover more steps by changing message and state bits together, essentially applying a local very sparse differential characteristic that is independent of $\mathcal{P}_{full}$.

(3) Search for a solution. Finally, the actual near-collision search is most commonly implemented as a depth-first tree search. Here, each level of the tree is related to a single step or a (combination of) message modifications and, for each level, the degrees of freedom create new branches, that may be immediately pruned by the conditions. The first phase over the first, say 16, levels consists of finding a solution for the first steps that uses the entire input message block exactly once, which is typically quite easy and cheap. But if it is especially dense in number of conditions it may be that no solutions are possible, in which case another variant differential characteristic needs to be tried. The second phase over the next few levels consists of applying speed-up techniques that modify the current solution up to a certain step in a very controlled manner resulting in another solution up to that step. In the last phase, no degrees of freedom are left and it remains to verify all remaining conditions which are fulfilled probabilitistically, hence all conditions in this phase directly contribute to the collision search's complexity.

To achieve this, one needs a family of near-collision attacks, of which the first $1/4$ of the rounds are adjusted to the problem. This can be a family built around a single core differential characteristic, where the final few rounds are adjusted to allow for a large set of final differences at a cost in increased attack complexity. Or, even better, a family built around several core differential characteristics, each with many variant final few rounds and final differences.

7.3 Chosen-Prefix Collision Abuse Scenarios

When exploiting collisions in real-world applications, two major obstacles must be overcome.

- The problem of constructing meaningful collisions. Given current methods, collisions require appendages consisting of unpredictable and mostly uncontrollable bit strings. These must be hidden in the usually heavily formatted application data structure without raising suspicion.
- The problem of constructing realistic attack scenarios. As we do not have effective attacks against MD5's (second) pre-image resistance but only collision attacks, we cannot target existing MD5 hash values. In particular, the colliding data structures must be generated simultaneously, along with their shared hash, by the adversary.

In this section, several chosen-prefix collision applications are surveyed where these problems are addressed with varying degrees of success.

- Section 7.3.1: attacks on X.509 certificates;
- Section 7.3.2: rogue X.509 Certification Authority certificate;
- Section 7.3.3: super malware FLAME's malicious Windows Update;
- Section 7.3.4: breaking the PGP web of trust with colliding keys;
- Section 7.3.5: man-in-the-middle attacks on TLS, SSH;
- Section 7.3.6: Nostradamus attack on hash-based commitments;
- Section 7.3.7: colliding documents;
- Section 7.3.8: PDF files and a practical Nostradamus attack;
- Section 7.3.9: colliding executables and software integrity checking;
- Section 7.3.10: digital forensics;
- Section 7.3.11: peer-to-peer software;
- Section 7.3.12: content addressed storage;
- Section 7.3.13: polyglots – multi-format collisions;
- Section 7.3.14: hashquines – embedding the MD5 hash in documents.

7.3.1 Digital Certificates

In [369] it was shown how identical-prefix collisions can be used to construct colliding X.509 certificates with different RSA moduli but identical Distinguished Names. Here the RSA moduli absorbed the random-looking near-collision blocks, thus inconspicuously and elegantly solving the meaningfulness problem. Identical Distinguished Names do not enable very realistic threat scenarios. Different Distinguished Names can be achieved by using chosen-prefix collisions, as was have shown in [570]. The certificates of these forms do not contain any spurious bits, so superficial inspection at bit level of either of the certificates does not reveal the existence of a sibling certificate that collides with it signature-wise. In Sections 7.3.2, 7.3.3 and 7.3.4 are additional interesting colliding certificate constructions that are more intricate and achieve different certificate usage properties.

7.3.2 Creating a Rogue Certification Authority Certificate

One of the most impactful demonstrations of the threat of collision attacks in the real world, was the construction of a rogue Certification Authority (CA) [573]. It thus directly undermined the core of public key infrastructure (PKI): to provide a relying party with trust, beyond reasonable cryptographic doubt, that the person identified by the Distinguished Name field has exclusive control over the private key corresponding to the public key in the certificate. With the private key of the rogue CA, and the rogue CA certificate having a valid signature of a commercial CA that was trusted by all major browsers (at that time), 'trusted' certificates could be created at will. Any website secured using TLS can be impersonated by using a rogue certificate issued by a rogue CA (nowadays such impersonation can be thwarted by using so-called certificate pinning). The forgery works irrespective of which CA issued the website's true certificate and of any property of that certificate, such as the hash function it is based upon – SHA-256 is not any better in this context than MD4. Combined with redirection attacks where `http` requests are redirected to rogue web servers, this leads to virtually undetectable phishing attacks.

In fact, any application involving a CA that provides certificates with sufficiently predictable serial number and validity period, and using a non-collision-resistant hash function, may be vulnerable, see for instance Section 7.3.3. This type of attack relies on the ability to predict the content of the certificate fields inserted by the CA upon certification: if the prediction is correct with non-negligible probability, a rogue certificate can be generated with the same non-negligible probability. Irrespective of the weaknesses of the cryptographic hash function used for digital-signature generation, our type of attack becomes effectively impossible if the CA adds a sufficient amount of fresh randomness to the certificate fields before the actual collision bit strings hidden in the public-key fields.

To compare with prior X.509 certificate collisions, the first colliding X.509 certificate construction was based on an identical-prefix collision, and resulted in two certificates with different public keys, but identical Distinguished Name fields (cf. Section 7.3.1). Then as a first application of chosen-prefix collisions it was shown how the Distinguished Name fields could be chosen differently as well [570]. The rogue CA goes one step further by also allowing different 'basic constraints' fields, where one of the certificates is an ordinary website certificate, but the other one is a CA certificate. Unlike the previous colliding certificate constructions where the CA was under the researchers' control, a commercial CA provided the digital signature for the (legitimate) website certificate.

In short, the following weaknesses of the commercial CA that carried out the legitimate certification request were exploited.

- Its usage of the cryptographic hash function MD5 to generate digital signatures for new certificates.
- Its fully automated way to process online certification requests that fails to recognise anomalous behaviour of requesting parties.
- Its usage of sequential serial numbers and its usage of validity periods that are determined entirely by the date and time in seconds at which the certification request is processed.

- Its failure to enforce, by means of the 'basic constraints' field in its own certificate, a limit on the length of the chain of certificates that it can sign. This, when properly set, could potentially invalidate any certificate signed by the rogue CA, if clients properly enforce this chain-length constraint.

Results were disclosed to the relevant parties to ensure this vulnerability was closed prior to publication. This was done privately and anonymously using the Electronic Frontier Foundation as an intermediary in order to reduce the not-insignificant risk of cumbersome legal procedures to stifle or delay publication. After disclosure, CAs moved quickly and adequately. MD5 was quickly phased out from all public CAs, and later formally forbidden in the CA/Browser Forum's Baseline Requirements. Furthermore, to prevent similar attacks with other hash functions in use, such as SHA-1 (till 2019), a formal requirement was added that certificate serial numbers must be unpredictable and contain at least 64 bits of entropy. Besides the impact on CAs, this work clearly and strongly demonstrated the significant real-world threat of chosen-prefix collisions in general and created a stronger push for the deprecation of MD5 in other applications.

Certificate Construction We summarise the construction of the colliding certificates in the sequence of steps below, and then describe each step in more detail.

(1) Construction of templates for the two to-be-signed parts, as outlined in Table 7.3. Note that we distinguish between a 'legitimate' to-be-signed part on the left-hand side, and a 'rogue' to-be-signed part on the other side.
(2) Prediction of serial number and validity period for the legitimate part, thereby completing the chosen prefixes of both to-be-signed parts.
(3) Computation of the two different collision-causing appendages.
(4) Computation of a single collision-maintaining appendage that will be appended to both sides, thereby completing both to-be-signed parts.
(5) Request certification of the legitimate to-be-signed part, if unsuccessful return to Step (2).

Step 1. Templates for the to-be-signed parts Table 7.3 shows the templates for the to-be-signed parts of the legitimate and rogue certificates. On the legitimate side, the chosen prefix contains space for serial number and validity period, along with the exact Distinguished Name of the commercial CA where the certification request will be submitted. This is followed by a subject Distinguished Name that contains a legitimate website domain name (i.broke.the.internet.and.all.i.got.was.this.

	Legitimate website certificate (A)	Rogue CA certificate (B)
prefix	serial number A	serial number B
	signing CA name	signing CA name
	validity period A	validity period B
	long domain name A	rogue CA name B
	"	RSA public key B
	"	X.509v3 ext: CA=true
	start RSA public key A	start X.509 comment
collision	collision data A	collision data B
common suffix	tail of RSA public key	<copy from A>
	X.509v3 ext: CA=false	<copy from A>

Table 7.3 *The to-be-signed parts of the colliding certificates.*

t-shirt.phreedom.org) consisting of as many characters as allowed by the commercial CA (in this case 64), and concluded by the first 208 bits of an RSA modulus, the latter all chosen at random after the leading '1'-bit.

The corresponding bits on the rogue side contain an arbitrarily chosen serial number, the same commercial CA's Distinguished Name, an arbitrarily chosen validity period (actually chosen as indicating 'August 2004', to avoid abuse of the rogue certificate), a short rogue CA name, a 1024-bit RSA public key generated using standard software, and the beginning of the X.509v3 extension fields. One of these fields is the 'basic constraints' field, a bit that was set to indicate that the rogue certificate will be a CA certificate (in Table 7.3 this bit is denoted by 'CA=TRUE').

At this point the entire chosen prefix is known on the rogue side, but on the legitimate side predictions for the serial number and validity period still need to be decided; this is done in Step 2.

The various field sizes were selected so that on both sides the chosen prefixes are now 96 bits short of the same MD5 block boundary. On both sides these 96 bit positions are reserved for the birthday bits of the chosen-prefix collision attack. On the legitimate side these 96 bits are part of the RSA modulus, on the rogue side they are part of an extension field of the type 'Netscape Comment' described as 'X.509 comment' in Table 7.3.

After the chosen-prefix collision has been constructed in Step 3, the certificates need to use a shared suffix. This part is determined by the legitimate certificate and simply copied to the rogue certificate, since on the rogue side this suffix is part of the 'Netscape Comment' field and is ignored by all clients. On the legitimate side after the three near-collision blocks of 512 bits each, another 208 bits are used to complete a 2048-bit RSA modulus (determined in Step 4). This is followed by the RSA public exponent (the common value of 65 537) and the X.509v3 extensions including the bit indicating that the legitimate certificate will be an end-user certificate (in Table 7.3 denoted by 'CA=FALSE').

Note that the legitimate certificate looks inconspicuous to the CA, except maybe the long and weird requested domain name. In contrast, the rogue certificate looks very suspicious under manual inspection: 'Netscape Comment' fields are uncommon for CA certificates, and its weird structured content that belongs to an end-user certificate is a big red flag. However, users rarely inspect certificates closely, and software clients simply ignore these suspicious elements.

Step 2. Prediction of serial number and validity period Based on repeated certification requests submitted to the targeted commercial CA, it turned out that the validity period can very reliably be predicted as the period of precisely one year plus one day, starting exactly six seconds after a request is submitted. So, to control that field is quite easy: all one needs to do is select a validity period of the right length, and submit the legitimate certification request precisely six seconds before it starts.

Predicting the serial number is harder but not impossible. In the first place, it was found that the targeted commercial CA uses sequential serial numbers. Being able to predict the next serial number, however, is not enough: the construction of the collision can be expected to take at least a day, before which the serial number and validity period have to be fixed, and only after which the to-be-signed part of the certificate will be entirely known. As a consequence, there will have been a substantial and uncertain increment in the serial number by the time the collision construction is finished. So, another essential ingredient of the construction was the fact that the CA's weekend workload is quite stable: it was observed during several weekends that the increment in serial number over a weekend does not vary a lot. This allowed us to predict fairly reliably Monday morning's serial numbers on the Friday afternoon before. See in Step 6 how the precise selected serial number and validity period was targeted.

Step 3. Computation of the collision At this point both chosen prefixes have been fully determined so the chosen-prefix collision can be computed: first the birthday bits per side, followed by the calculation of three pairs of 512-bit near-collision blocks. The constraint of just three pairs of near-collision blocks causes the birthday search to be significantly more costly compared to the near-collision block construction, whereas normally these can be balanced against each other by allowing more near-collision block pairs. However, the available resources in the form of Arjen Lenstra's cluster of 215 PlayStation 3 (PS3) game consoles at EPFL was very suitable for the job. The entire calculation takes on average about a day on this cluster using suitable chosen-prefix collision parameters as described below.

When running Linux on a PS3, applications have access to 6 Synergistic Processing Units (SPUs), a general purpose CPU, and about 150MB of RAM per PS3. For the most costly phase, the birthday-search phase, the 6×215 SPUs are computationally equivalent to approximately 8600 regular 32-bit cores, due to each SPU's 4×32-bit wide SIMD architecture. The other parts of the chosen-prefix collision construction (differential characteristic construction and near-collision block search) are much harder to implement efficiently on SPUs, but for these we were able to use the 215 PS3 CPUs effectively. With these resources, the choice for a much larger differential characteristic family (cf. Table 7.8 using $w = 5$) than normal still turned out to be acceptable despite the 1000-fold increase in the cost of the actual near-collision block construction.

We optimised the overall birthday-search complexity for the plausible case that the birthday search takes $\sqrt{2}$ times longer than expected. Limited to 150MB per PS3, for a total of about 30GB, the choice was made to force $k = 8$ additional colliding bits in the birthday search (cf. Section 7.4.2).

Given the available timeline of a weekend, and that the calculation can be expected to take just a day, a number of chosen-prefixes were processed sequentially, each corresponding to different serial numbers and validity periods (targetting both Monday and Tuesday mornings). Because the PS3 SPUs are solely used for the birthday phase, and the PS3 CPUs solely for the near-collision attack phase, several attempts could be pipelined. So, a near-collision block calculation on the CPUs would always run simultaneously with a birthday search on the SPUs for the 'next' attempt.

Step 4. Finishing the to-be-signed parts At this point, the legitimate and rogue sides collide under MD5, so that from here only identical bits may be appended to both sides.

With $208 + 24 + 72 + 3 \times 512 = 1840$ bits set, the $2048 - 1840 = 208$ bits that remain need to be set for the 2048-bit RSA modulus on the legitimate side. Because in the next step the RSA private exponent corresponding to the RSA public exponent is needed, the full factorisation of the RSA modulus needs to be known, and the factors must be compatible with the choice of the

RSA public exponent, which can be achieved as follows. Let N be the 2048-bit integer consisting of the 1840 already determined bits of the RSA modulus-to-be, followed by 208 one bits. Select a 224-bit integer p at random until $N = a \cdot p + b$ with $a \in \mathbb{N}$ and $b < 2^{208}$, and keep doing this until both p and $q = \lfloor N/p \rfloor$ are prime and the RSA public exponent is co-prime to $(p-1)(q-1)$. Once such primes p and q have been found, the number pq is the legitimate side's RSA modulus, the leading 1840 bits of which are already present in the legitimate side's to-be-signed part, and the 208 least significant bits of which are inserted in both to-be-signed parts.

To analyse the required effort somewhat more in general, 2^{k-208} integers of k bits (with $k > 208$) need to be selected on average for pq to have the desired 1840 leading bits. Because an ℓ-bit integer is prime with probability approximately $1/\ln(2^{\ell})$, a total of $k(2048 - k)2^{k-208}(\ln 2)^2$ attempts may be expected before a suitable RSA modulus is found. The co-primality requirement is a lower-order effect that is disregarded. Note that for $k(k - 2048)(\ln 2)^2$ of the attempts the k-bit number p has to be tested for primality, and that for $(2048 - k)\ln 2$ of those q needs to be tested as well (on average, obviously). For $k = 224$ this could be done in a few minutes on a standard PC.

This completes the to-be-signed parts on both sides. Now it remains to be hoped that the legitimate part that actually will be signed corresponds, bit for bit, with the legitimate to-be-signed part that was constructed in this manner.

Step 5. Request certification of the legitimate to-be-signed part Using the relevant information from the legitimate side's template, i.e., the subject Distinguished Name and the public key, a PKCS#10 Certificate Signing Request is prepared. The CA requires proof of possession of the private key corresponding to the public key in the request. This is done by signing the request using the private key – this is the sole reason that we need the RSA private exponent.

The targeted legitimate to-be-signed part contains a very specific validity period that leaves no choice for the moment at which the certification request needs to be submitted to the CA. Just hoping that, at that time, the serial number would have precisely the predicted value is unlikely to work, so a somewhat more elaborate approach is used. About half an hour before the targeted submission moment, the same request is submitted, and the serial number in the resulting certificate is inspected. If it is already too high, the entire attempt is abandoned. Otherwise, the request is repeatedly submitted, with a frequency depending on the gap that may still exist between the serial number received and the targeted one, and taking into account possible certification requests by others. In this way the serial number is slowly nudged toward the right value at the right time.

Various types of accidents were experienced, such as another CA customer 'stealing' the targeted serial number just a few moments before the attempt to get it, thereby wasting that weekend's calculations. But, after the fourth weekend it worked as planned to get an actually signed part that exactly matched a predicted legitimate to-be-signed part.

Given the perfect match between the actually signed part and the hoped-for one, and the MD5 collision between the latter and the rogue side's to-be-signed part, the MD5-based digital signature present in the legitimate certificate as provided by the commercial CA is equally valid for the rogue side. To finish the rogue CA certificate it suffices to copy the digital signature to the right spot in the rogue CA certificate.

7.3.3 The FLAME Super Malware

In response to the rogue CA construction, various authorities explicitly disallowed MD5 in digital signatures. For instance, the CA/Browser Forum adopted Baseline Requirements for CAs in 2011,[4] and Microsoft updated its Root CA Program in 2009.[5] Nevertheless, surprisingly, MD5 was not completely removed for digital signatures within Microsoft, which came to light in 2012 with the discovery [146, 316] of the FLAME super malware for espionage in the Middle East, supposedly in a joint USA–Israel effort [589]. This was a highly advanced malware that could collect a vast amount of data including files, keyboard inputs, screen contents, microphone, webcam and network traffic, sometimes even triggered by use of specific applications of interest [146, 316]. In contrast to normal malware, FLAME infections occured with surgical precision to carefully selected targets, which helped evading detection possibly since 2007 [146].

Of cryptanalytic interest is one of its means of infection [556]: FLAME used WPAD (Web Proxy Auto-Discovery Protocol) to register itself as a proxy for the domain `update.windows.com` to launch man-in-the-middle attacks for Windows Update (WU) on other computers on the local network. By forcing a fall-back from the secure HTTPS protocol to the unauthenticated HTTP protocol, FLAME was able to push its own validly signed Windows Update patches to other Windows machines. It was a huge surprise to find that FLAME was actually in possession of its own illegitimately signed Windows Update patch to infect other machines, and after inspection it became clear it was constructed with a collision attack [427]. It was discovered that MD5 signatures were still being used for licensing purposes in their Terminal Server Licensing Service,

---

[4] https://cabforum.org/wp-content/uploads/Baseline_Requirements_V1.pdf
[5] https://technet.microsoft.com/en-us/library/cc751157.aspx

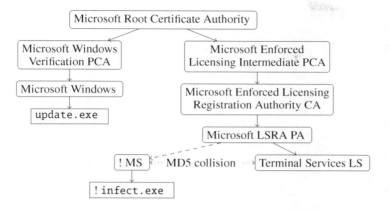

Figure 7.1 Windows Update and FLAME certificate tree.

which automatically generated MD5-based signed certificates that lead up to the Microsoft Root CA. Using a collision attack enabled the attackers to convert a signed license certificate into a validly signed code signing certificate that was accepted for Windows Update by all versions of Windows [427]. Below we go more into detail on two very interesting aspects: the colliding certificates and the used chosen-prefix collision attack.

The Colliding Certificates Normally, Windows Update executables are signed by a code-signing certificate in a dedicated Windows Update certificate chain under the Microsoft Root certificate. However, the FLAME code-signing certificate is connected to a completely different part of the PKI tree leading up to the same Root certificate, as illustrated in Figure 7.1.

As described by Trail-of-Bits [556], there is an automated process whereby a customer-side running License Server (LS) generates a private key and a X.509 certificate signing request with: (1) customer information, (2) machine id and (3) corresponding public key. A Microsoft activation server then returns a certificate with an MD5-based signature from 'Microsoft LSRA PA' (see the right-most branch in Figure 7.1). This signed certificate has the following properties.

- No identifying information, except CommonName (CN) set to 'Terminal Services LS'.
- Fixed validity period until 19 Febuary 2012.
- No Extended Key Usage restrictions, implying that restrictions are inherited from the CA certificate, which includes code signing.

	Lost legitimate LS certificate (A)	Rogue WU signing certificate (B)
prefix	? serial number A ? validity period A CN='Terminal Services LS' start ? 4096-bit RSA key B "	serial number B validity period B CN='MS' 2048-bit RSA key A start issuerUniqueID field to end
collision	collision data A	collision data B
common suffix	tail of RSA key critical Hydra extension	<copy from A> <copy from A>

Table 7.4 *The to-be-signed parts of FLAME colliding certificates.*

- Special critical Microsoft Hydra X.509 extension, marked 'critical' thus will be rejected if not understood by the Crypto library.

The private key and certificate can be recovered by the customer, and as it supports code-signing and chains up to the Microsoft Root CA, it was found out that any such LS certificate is sufficient to sign Windows Update executables for Windows XP. For Windows Vista and Windows 7, however, the critical Hydra extension causes such signed executables to be rejected. It was for this reason that a chosen-prefix collision attack was employed such that a rogue WU signing certificate was obtained with CommonName set to 'MS' and that could sign Windows Update executables for all Windows versions at that time.

From the FLAME malware the full rogue WU signing certificate has been analysed. Although the legitimate, colliding LS certificate has been lost, this analysis sufficed to estimate the layout of the colliding certificates, as depicted in Table 7.4.

The Chosen-Prefix Collision Attack Once the FLAME certificate structure was figured out and it was clear a chosen-prefix collision attack was used, it was immediately thought the publicly available chosen-prefix collision implementation [566] was used. With a novel technique called counter-cryptanalysis [568] we succeeded in recovering all near-collision block data of the lost legitimate certificate from the FLAME rogue certificate and thereby the differential characteristics used for each near-collision attack. Surprisingly, it became immediately clear that these did not match the known chosen-prefix collision attacks: the message block differences used were $\delta m_4 = \delta m_{14} = 2^{31}$ and $\delta m_{11} = \pm 2^{15}$, which so far were used only by Wang *et al.*'s first MD5 (identical prefix) collision attack [616] and improvements thereof. This made

	Key A	Key B	bytes
prefix	header A	header B	4
	timestamp A	timestamp B	4
	key A size: 8192 bits	key B size: 6144 bits	3
collision	collision data A	collision data B	757
common suffix	\<copy from B\>	remaining RSA key	16
	\<copy from B\>	PGP image header	24
	\<copy from B\>	tiny JPEG image	181
	remaining RSA key	\<copy from A\>	51
	PGP user id	\<copy from A\>	30
	\<copy from B\>	signature trailer	18

Table 7.5 *PGP key collision [391].*

it obvious that neither the HashClash implementation [566] nor even the best-known chosen-prefix collision attack [573] were used, but a yet-unknown new variant chosen-prefix collision attack developed by world-class cryptanalists. A closer look on the attack details is described in Section 7.4.4.

7.3.4 Colliding PGP Keys with Different Identities

Recently, Leurent *et al.* [391] demonstrated a pair of PGP keys with different identities that collide under SHA-1. This implies that any certification of one key can be transferred to the other key, directly undermining the web of trust of PGP keys. This was quickly fixed by deprecating SHA-1 in GnuPG in the modern branch, while a legacy branch remains unaltered and insecure. The main structure of these colliding files is depicted in Table 7.5, one of the notable features is that the collision data precedes any user data. The produced collision is thus re-usable for many attacks.

   Their attack re-used the GPU implementation of the first identical-prefix collision for SHA-1 [575] with some refinements and adjustments as the basis for all near-collision attacks, but the main improvement lies in the near-collision attack strategy presented earlier by the same authors [390]. Their strategy exploits two facts.

(1) Each near-collision attack can essentially produce many output differences with almost the same probability.
(2) For each chaining-value difference there are many distinct choices of which differences each near-collision attack eliminates.

   They propose to represent all possible choices in a weighted graph, which they use to compute, for each near-collision attack, which set of possible differences to target, given the remaining chaining-value difference so far. By targetting many output differences, the cost of near-collision attacks drops significantly. Here the last block, which has to eliminate the remaining chaining-value difference and has only one target output difference, is the most costly. In total this first SHA-1 chosen-prefix collision has an estimated cost of $2^{63.4}$ SHA-1 compression functions, which could be executed very economically for the bargain price of \$75 000 by renting a cheap GPU farm (originally built for blockchain mining).

7.3.5 Transcript Collision Attacks on TLS, IKE and SSH

Although consensus was reached among certification authorities and software vendors to stop issuing and accepting MD5-based signatures and certificates, MD5 remained in use in many popular security applications. In 2016, Bhargavan and Leurent [58] demonstrated meet-in-the-middle attacks on the key exchange protocols underlying various version of transport layer security (TLS), Internet protocol security (IPsec) and secure shell (SSH). In particular, they show how to impersonate TLS-1.2 clients, TLS-1.3 servers and IKEv2 (Internet key exchange) initiators, as well as how to downgrade TLS-1.1 and SSHv2 connections to use weak ciphers. In these meet-in-the-middle attacks the two honest parties will observe different protocol execution transcripts. However, online chosen-prefix collision attacks can be exploited to ensure that transcript hashes used for mutual authentication collide, which in turn allows authentication messages to be forwarded successfully.

Most of these attacks are quite involved and depend on the low-level details of those protocols; however, we show a simple example man-in-the-middle attack in Figure 7.2 for the Sign-and-MAC protocol [345]. After the first message m_1 by A, the attacker has to find an online collision $h = H(m_1|m_2') = H(m_1'|m_2)$, where it also needs to predict correctly the first response m_2 by B after receiving m_1'. At that point the attacker can successfully set up a shared session key with A ($g^{xy'}$) and with B ($g^{x'y}$). In the final step, the transcripts are signed by both parties in each connection for mutual authentication. As the to-sign transcript hashes collide, the signatures are valid for both sessions. The attacker can simply forward those signatures between A and B to pretend A and B have an authenticated private channel.

In contrast to most other chosen-prefix collision attacks described here, these transcript collision attacks have to be carried out online. This places heavy constraints on the wallclock time available to compute the collision attack for a successful outcome. Bhargavan and Leurent report special

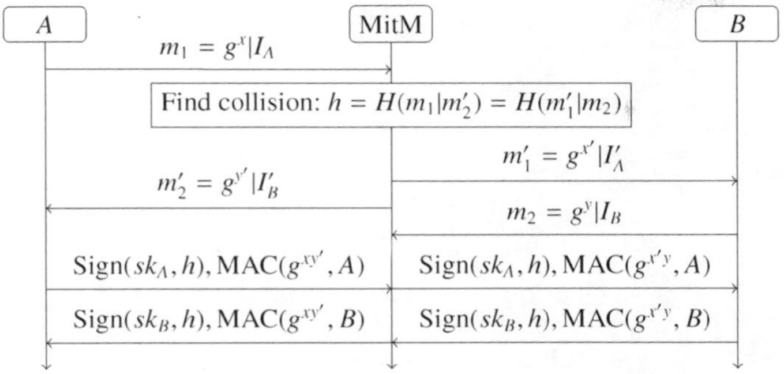

Figure 7.2 Man-in-the-middle attack on the Sign-and-MAC protocol.

modifications to the HashClash software [566] to avoid large memory-disk transfers required for the interaction between the different programs by merging those programs, and by using all possible pre-computations. These modifications reportedly enabled them to reduce the wallclock time of computing chosen-prefix collisions on a 48-core machine from at least three hours to only one hour. Collision finding scales rather well with the amount of computational power available, although it remains unclear how feasible it is to bring the wallclock time down to, say, within a minute. For most online attacks even a minute delay, let alone an one hour delay, in the key exchange is very troublesome, and unthinkable with direct human interaction, as is the case for Internet browsers. But it appears that some TLS client software is willing to wait indefinitely on the key exchange, as long as messages such as warning alerts are being sent regularly; thus such TLS clients that remain unsupervised may be susceptible.

7.3.6 Nostradamus Attack on Hash-based Commitments

Kelsey and Kohno [319] presented a Nostradamus attack to commit to a hash value first, after which a document containing any message of one's choice can be constructed that matches the committed hash value with lower cost than the generic pre-image attack cost. The method applies to any Merkle–Damgård hash function, such as MD5, that given a chaining value and a suffix produces another chaining value. Depicted in Figure 7.3, and omitting details involving message lengths and padding, the idea is to build a tree of 2^d chaining values that all lead to a single root chaining value h_{root} and then to commit to a hash value h_{commit} computed by hashing some chosen suffix starting from h_{root}. The tree is a complete binary tree and is calculated from its leaves up to the root,

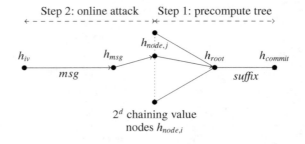

Figure 7.3 Nostradamus attack.

so the root h_{root} will be the last value calculated. This calculation is done in such a way that each node of the tree is associated with a chaining value along with a suffix that together hash to the chaining value associated with the node's parent. Thus, two siblings have chaining values and suffixes that collide under the hash function. The chaining values at the leaves may be arbitrarily chosen but are, preferably, all different.

After committing to h_{commit}, given a prefix *msg* of one's choice, one performs a brute-force search for a suffix x such that hashing $msg\|x$ results in the chaining value of one of the leaves of the tree. Appending the message suffixes that one encounters on the path from that leaf to the root, then results in a final message with the desired prefix and committed hash value.

For MD5, however, it remains far from feasible to carry out the entire construction in practice, as it requires massive birthday searches to find the necessary 2^{d+1} collisions. However, there are two different variants that have been demontrated in practice.

First, for very small d there is a variant that is feasible, as then the construction of the tree can be done efficiently by using chosen-prefix collisions to construct sibling node suffixes based on their chaining values. An example of this variant is discussed in Section 7.3.8 involving 12 PDF documents.

Another actually practical approach is discussed in Section 7.3.14. This variant uses a large 2^d multicollision created from d sequential collision attacks, in combination with file format exploit techniques to allow the resulting files to display 2^d different shown messages when rendered. The examples in Section 7.3.14 use this approach to craft *hashquines*: documents that show their own MD5 hash.

7.3.7 Colliding Documents

In [151] it was shown how to construct a pair of PostScript files that collide under MD5, but that display different messages when viewed or printed. These constructions use identical-prefix collisions and thus the only difference between the colliding files is in the generated collision bit strings. See also [214] for similar constructions for other document formats. These constructions have to rely on the presence of both messages in each of the colliding files and on macro-functionalities of the document format used to show either one of the two messages.

This can be improved with chosen-prefix collisions as one message per colliding document suffices and macro-functionalities are no longer required. For example, by using a document format that allows insertion of colour images (such as Microsoft Word or Adobe PDF), and by inserting one message per document, two documents can be made to collide by appending carefully crafted color images after the messages. A short one pixel wide line will do – for instance, hidden inside a layout element, a company logo, or a nicely coloured barcode – and preferably scaled down to hardly visible size (or completely hidden from view, as is possible in PDF). An extension of this construction is presented in the paragraphs below and set forth in detail in Section 7.3.8.

7.3.8 PDF Files and the Nostradamus Attack

Stevens, Lenstra and de Weger [574] describe a detailed construction for PDF files that also demonstrates a variant scenario of the Nostradamus attack [319]. In the original Nostradamus attack one first commits to a certain hash value, and afterwards constructs, for any message, a document that not only contains that message but that also matches the committed hash value. This attack is at this point in time not feasible for MD5 in its full generality; however, it can easily be done if a limited-size message space has been defined upfront.

Suppose there are messages $m_1, m_2, \ldots, m_r$, then using $r - 1$ chosen-prefix collisions one can construct r suffixes $s_1, s_2, \ldots, s_r$ such that the r documents $d_i = m_i \| s_i$ all have the same hash. After committing to the common hash, any of the r documents $d_1, d_2, \ldots, d_r$ can be shown, possibly, to achieve some malicious goal. The other documents will remain hidden and their prefix parts, i.e., the m_i-parts, cannot be derived from the single published document or from the common hash value. However, given the structure of PDF documents, it is not entirely straightforward to insert different chosen-prefix collision blocks, while keeping the parts following those blocks identical in order to maintain the collision as described below.

Part	Contents
object header	`42 0 obj`
image header	`<< /ColorSpace /DeviceRGB /Subtype /Image`
image size	`/Length 9216 /Width 64 /Height 48 /BitsPerComponent 8`
image contents	`>> stream...endstream`
object footer	`endobj`

Table 7.6 *An example numbered image object in the PDF format.*

In [574], we succeed in constructing 12 different PDF documents with a common MD5 hash value, where each document predicts a different outcome of the 2008 US presidential election. The common MD5 hash value of the 12 colliding PDF documents is

$$3D5\ 15\ DEAD\ 7AA16560ABA3E9DF05CBC80.$$

See [571] for the actual PDF documents, one of which correctly predicted the outcome one year before the elections took place.

A PDF document is built up from the following four consecutive parts: a fixed header, a part consisting of an arbitrary number of numbered 'objects', an object lookup table and, finally, a trailer. The trailer specifies the number of objects, which object is the unique root object (containing the root of the document content tree) and which object is the info object (containing the document's meta information such as authors, title, etc.), and contains a filepointer to the start of the object lookup table.

Given a file containing a PDF document, additional objects can be inserted, as long as they are added to the object lookup table and the corresponding changes are made to the number of objects and the filepointer in the trailer. If these are not referenced by the document content tree anywhere then they will be ignored by any PDF processor. The format is mostly text based, with the exception of binary images where collision data can be safely hidden. The template used for inserted image objects is given in Table 7.6, which offers sufficient room to hide 11 chosen-prefix collisions with 12 message blocks each (1 birthday-search block and 11 near-collision blocks). Any binary image is put between single line-feed characters (ASCII code 10) and the result is encapsulated by the keywords `stream` and `endstream`. The keyword `/Length` must specify the byte length of the image equal to $3 \times width \times height$, since for uncompressed images each pixel requires three bytes ('RGB'). The object number (42 in the example object header) must be set to the next available object number.

When constructing colliding PDF files they must be equal after the collision-causing data. The object lookup tables and trailers for all files must therefore be the same. This was achieved as follows.

- Because all documents must have the same number of objects, dummy objects are inserted where necessary.
- Because all root objects must have the same object number, they can be copied if necessary to objects with the next available object number.
- The info objects are treated in the same way as the root objects.
- To make sure that all object lookup tables and filepointers are identical, the objects can be sorted by object number and if necessary padded with spaces after their `obj` keyword to make sure that all objects with the same object number have the same file position and byte length in all files.
- Finally, the object lookup tables and trailers need to be adapted to reflect the new situation – as a result they should be identical for all files.

Although this procedure works for basic PDF files, such as those produced by pdflatex, it should be noted that the PDF document format allows additional features that may cause obstructions.

Given r LaTeX files with the desired subtle differences (such as names of r different candidates), r different PDF files are produced by using a version of LaTeX that is suitable for our purposes (as discussed above). In all these files a binary image object with a fixed object number is then inserted, and the approach sketched above is followed to make the lookup tables and trailers for all files identical. Because this binary image object is present but not used in the PDF document, it remains hidden from view in a PDF reader. To ensure that the files are identical after the hidden image contents, their corresponding objects were made the last objects in the files. This then leads to r chosen prefixes consisting of the leading parts of the PDF files up to and including the keyword `stream` and the first line-feed character. After determining $r - 1$ chosen-prefix collisions resulting in r collision-causing appendages, the appendages are put in the proper binary image parts, after which all files are completed with a line-feed character, the keywords `endstream` and `endobj`, and the identical lookup tables and trailers.

Note that the `Length`, etc., fields have to be set before collision finding, and that the value of `Length` will grow logarithmically with r and linearly in the number of near-collision blocks one is aiming for.

7.3.9 Colliding Executables and Software Integrity Checking

In [310] and [429] it was shown how any existing MD5 collision, such as those originally presented by Xiaoyun Wang at the Crypto 2004 rump session, can be abused to mislead integrity checking software that uses MD5. A similar application of colliding executables, using freshly made collisions, was given on [527]. All these results use identical-prefix collisions and, similar to the colliding PostScript application mentioned earlier, differences in the colliding inputs are used to construct deviating execution flows.

However, digitally signed executables contain the signature itself, so in order to craft a collision for the hash input to the digital signature a slight modification is necessary. Didier Stevens [563] describes how this can be achieved for Microsoft Authenticode, which uses a special 'Attribute Certificate Table' section in the file containing the certificate chain and digital signatures. When computing the signature hash input one hashes the byte string of the executable file except for three parts that are cut out: a 4-byte checksum value, an 8-byte pointer to Attribute Certificate Table, and the Attribute Certificate table itself.

Chosen-prefix collisions allow a more elegant approach, because common operating systems ignore bit strings that are appended to executables: the programs will run unaltered, which is demonstrated in [572]. Furthermore, using tree-structured chosen-prefix collision appendages as in Section 7.3.8, any number of executables can be made to have the same MD5 hash value or MD5-based digital signature.

One can imagine two executables: a 'good' one (say Word.exe) and a 'bad' one (the attacker's Worse.exe). A chosen-prefix collision for those executables is computed, and the collision-causing bit strings are appended to both executables. The resulting altered file Word.exe, functionally equivalent to the original Word.exe, can be offered to a code-signing program such as Microsoft's Authenticode and receive an 'official' MD5-based digital signature. This signature will then be equally valid for the attacker's Worse.exe, and the attacker might be able to replace Word.exe by his Worse.exe (renamed to Word.exe) on the appropriate download site. This construction affects a common functionality of MD5 hashing and may pose a practical threat. It also allows people to get many executables signed at once at the cost of getting a single such executable signed, bypassing verification of any kind (e.g., authenticity, quality, compatibility, non-spyware, non-malware) by the signing party of the remaining executables.

7.3.10 Digital Forensics

In digital forensics, there are two main uses for cryptographic hash functions: file identification and integrity verification, for which MD5 is unsuitable. Contrary to the abundant evidence in the literature as described in this survey, a position paper has been published in 2019 that MD5 remains appropriate for integrity verification and file identification in the field of digital forensics by the Scientific Working Group on Digital Evidence (SWDGE) [523]. This document claims that

It is appropriate to use both MD5 and SHA1 for integrity verification provided the hash is securely stored or recorded in the examination documentation. This will prevent an individual from substituting a different file and its hash. This is true for all hash algorithms. Since there are no preimage attacks on any of the four hashing algorithms discussed, the only way to manipulate the evidence without detection is to do it before it is hashed.

It fails to acknowledge that collisions can be crafted in advance after which manipulation is undetectable with a non-collision resistant hash function. Note that actual manipulation of the evidence in custody may not be necessary for collisions to be problematic.

File Identification So-called hash sets are used to identify known files quickly. For example, when a hard disk is seized by law-enforcement officers, they may compute the hashes of all files on the disk, and compare those hashes to hashes in existing hash sets: a whitelist (for known harmless files such as operating system and other common software files) and a blacklist (for previously identified harmful files). Only files whose hashes do not occur in either hash set have to be inspected further. A useful feature of this method of recognising files is that the file name itself is irrelevant, since only the content of the file is hashed.

MD5 is a popular hash function for this application. Examples are NIST's National Software Reference Library Reference Data Set[6] and the US Department of Justice's Hashkeeper application.[7]

A conceivable, and rather obvious, attack on this application of hashes is to produce a harmless file (e.g., an innocent picture) and a harmful one (e.g., an illegal picture), and insert collision blocks that will not be noticed by common application software or human viewers. In a learning phase the harmless file might be submitted to the hash set and thus the common hash may end up on the whitelist. The harmful file will be overlooked from then on.

Integrity Verification When devices are seized by law-enforcement officers, they may compute the hashes of all files on the disk to digitally certify the contents. These can later be used in court to verify the chain of custody and

[6] http://www.nsrl.nist.gov/
[7] http://www.usdoj.gov/ndic/domex/hashkeeper.htm

to cryptographically verify the integrity of the evidence that it has not been tampered with.

Using document collisions, one can envision a potential criminal ensures all criminating files collide with harmless ones. He could try to dismiss the use of the incriminating files in court, by presenting the harmless files that match the hash values of the incriminating files and casting doubt on the chain of custody. For instance, in 2005 the Australian Hornsby Local Court dismissed a vehicle speeding case after the prosecutor could not provide evidence that the speed camera image hashed with MD5 had not been tampered with [511].

Another threat scenario of using colliding files, one harmless and one incriminating, is planting the harmless file on a victim's device later to be swapped by the incriminating one. Both cases may result in law enforcement presenting the incriminating file and the defendent presenting the harmless file. From a purely cryptographic hash point of view, these cases are indistinguishable when using a non-collision-resistant hash function such as MD5.

7.3.11 Peer-to-Peer Software

Hash sets such as for digital forensics are also used in peer-to-peer software. A site offering content may maintain a list of pairs (file name, hash). The file name is local only, and the peer-to-peer software uniquely identifies the file's content by means of its hash. Depending on which hash is used and how the hash is computed, such systems may be vulnerable to a chosen-prefix attack. Software such as eDonkey and eMule use MD4 to hash the content in a two-stage manner: the identifier of the content $c_1\|c_2\|\cdots\|c_n$ is $MD4(MD4(c_1)\|\cdots\|MD4(c_n))$, where the chunks c_i are about 9 MB each. One-chunk files, i.e., files not larger than 9 MB, are most likely vulnerable; whether multi-chunk files are vulnerable is open for research. Chosen-prefix collision attacks have not been presented against MD4. But as MD4 is a simpler version of MD5, an adaptation of MD5's attacks should result in an attacks on MD4 that are considerably faster compared with MD5's attacks.

7.3.12 Content-Addressed Storage

Content-addressed storage is a means of storing fixed content at a physical location of which the address is directly derived from the content itself. For example, a hash of the content may be used as the file name. See [485] for an example. Clearly, chosen-prefix collisions can be used by an attacker to fool such storage systems, e.g., by first preparing colliding pairs of files, by then storing the harmless-looking first one, and later overwriting it with the harmful second one.

7.3.13 Polyglots: Multi-Format Collisions

As file formats typically start with a specific fixed header that indicate the file format used, two colliding files of different file formats are difficult to construct using identical-prefix collisions. Using chosen-prefix collisions, one can take any two files and make them collide, one only has to correctly hide the collision bits for the file format's renders. The best-known example of four files with distinct file formats (PNG image, PDF document, MP4 video and PE executable) has been crafted by Ange Albertini.[8] In this case, it is a re-usable collision where the chosen prefixes contain only the file format headers, and do not cover any file content. A simple program by Albertini uses the pre-constructed four-way collision to convert given PNG, PDF, MP4 and PE files into four colliding files with the same hash.

7.3.14 Hashquines: Embedding the MD5 Hash in Documents

Another variant on the Nostradamus attack (see Section 7.3.6) that is feasible for short messages is as follows. The goal is to build a 2^d multi-collision crafted by d sequential identical-prefix collisions, where each of the 2^d files all having the same hash renders a different message. The precise method depends strongly on the file format being used. Later, after some event, one can choose which particular file out of the 2^d colliding files to show that matches the committed hash. One possible choice for the rendered message is the hash value itself; such documents that render their own hash are called hashquines. Hashquines have been demonstrated for the following file formats in the Poc‖GTFO journal [352]: GIF, PDF, Nintendo NES ROM and PostScript.

   In the case of executables it would be almost trivial to include these d sequential identical-prefix collisions, read out the chosen 'bit' b_i of collision i ($b_i = 0$ if collision data A is used, $b_i = 1$ if collision data B is used) and map the resulting d-bit string to a to-be-rendered message in the desired message space. After the complete file has been crafted one can commit to its hash.

   A more-common strategy that works for several file formats is to place collisions inside comment fields, where one of the message-block differences affects the comment length. In this manner, one can force different behaviours

---

[8] https://github.com/corkami/collisions#pileups-multi-collision

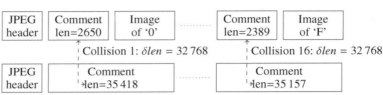

Figure 7.4 JPEG multi-collision: 16 different images, each prefixed with a comment whose length can be controlled with a collision. Only the first image that is not skipped over by a comment is rendered.

for parsers: either it parses a short comment followed by some content to render, or it parses a long comment that skips over the content. This idea for the JPEG file format is depicted in Figure 7.4, where 16 same-size JPEG images are placed into one file. Each of the 16 JPEG images renders to one of the 16 hexidecimal symbols $(0,\ldots,9,a,\ldots,f)$, and is prefixed by a comment whose length is controlled by a collision. In this case, long comments force the parser to skip over an image, and only the first short comment results in the next JPEG image data to be rendered. Any following JPEG images will be ignored. An MD5 hash consists of 32 hexidecimal symbols, so a PDF hashquine can be constructed by constructing 32 such JPEG images inside a PDF document. And once the PDF file MD5 hash is known, the 32 JPEG images can be altered (without changing the MD5 hash) such that together they render the MD5 hash.

7.4 MD5 Collision Attacks

In the sections below, we first give a brief mathematical description of MD5 itself. Then we discuss the details of several interesting chosen-prefix collision attacks: the currently best chosen-prefix collision attack [573] in Section 7.4.2, a single-block variant [573] in Section 7.4.3 and FLAME's attack by Nation States exposed by counter-cryptanalysis [187, 568] in Section 7.4.4.

7.4.1 MD5

The hash function MD5 was designed by Ron Rivest in 1992 with a 128-bit output hash and is built on the Merkle–Damgård framework by using its compression function *MD5Compress*.

(1) Padding. Pad the message with a '1'-bit followed by the minimal number

of '0'-bits to pad to a bit-length of 448 modulo 512. Append the bit-length of the original unpadded message as a 64-bit little-endian integer.

(2) Partitioning. The padded message is partitioned into k consecutive 512-bit blocks $M_1, \ldots, M_k$.

(3) Processing. MD5's chaining values CV_i are 128-bits long consisting of four 32-bit words a_i, b_i, c_i, d_i. For the initial value CV_0, these are fixed public values:

$$(a_0, b_0, c_0, d_0) = (\texttt{0x67452301}, \texttt{0xEFCDAB89}, \texttt{0x98BADCFE}, \texttt{0x10325476}).$$

For $i = 1, \ldots, k$, the chaining value is updated:

$$CV_i = (a_i, b_i, c_i, d_i) = MD5Compress(CV_{i-1}, M_i).$$

(4) Output. The resulting hash value is CV_k expressed as the concatenation of the hexadecimal byte strings of a_k, b_k, c_k, d_k (using little-endian).

MD5's compression function *MD5Compress* internally uses 32-bit words that are both used as integers modulo 2^{32} for addition (the least significant bit is the right-most bit), and as 32-bit strings for bitwise operations: AND ($\wedge$), OR ($\vee$), XOR ($\oplus$), NOT ($\overline{X}$), left/right rotation by n-bits (left: $X^{\lll n}$, right: $X^{\ggg n}$).

MD5Compress takes as input a 128-bit chaining value CV split into four words (a, b, c, d) and a 512-bit message block B split into 16 words $m_0 \| \cdots \| m_{15}$. We use an unrolled description of *MD5Compress* that initialises four state words with the chaining values $(Q_0, Q_{-1}, Q_{-2}, Q_{-3}) = (b, c, d, a)$ and performs 64 rounds for $t = 0, \ldots, 63$ in which it computes a next-state word Q_{t+1}:

$$F_t = f_t(Q_t, Q_{t-1}, Q_{t-2})$$
$$T_t = F_t + Q_{t-3} + AC_t + W_t$$
$$R_t = T_t^{\lll RC_t}$$
$$Q_{t+1} = Q_t + R_t,$$

where $AC_t = \lfloor 2^{32} \cdot |\sin(t+1)| \rfloor$ and W_t, $f_t(X, Y, Z)$ and RC_t are given in Table 7.7. At the end, *MD5Compress* returns $(a + Q_{61}, b + Q_{64}, c + Q_{63}, d + Q_{62})$.

7.4.2 MD5 Chosen-Prefix Collision Attack

The 2009 chosen-prefix collision construction used for the rogue CA certificate (cf. Section 7.3.2) remains the best-known attack. The attack has several parameters that can be adjusted. First, the maximum number of allowed near-collision blocks is denoted by r and can be used to trade-off between birthday-search time complexity and the cost of finding the r near-collision blocks.

Step	W_t	$f_t(X, Y, Z)$	RC_t
$0 \leq t < 16$	m_t	$(X \wedge Y) \oplus (\overline{X} \wedge Z)$	$(7, 12, 17, 22)_{[t \bmod 4]}$
$16 \leq t < 32$	$m_{(1+5t) \bmod 16}$	$(Z \wedge X) \oplus (\overline{Z} \wedge Y)$	$(5, 9, 14, 20)_{[t \bmod 4]}$
$32 \leq t < 48$	$m_{(5+3t) \bmod 16}$	$X \oplus Y \oplus Z$	$(4, 11, 16, 23)_{[t \bmod 4]}$
$48 \leq t < 64$	$m_{(7t) \bmod 16}$	$Y \oplus (X \vee \overline{Z})$	$(6, 10, 15, 21)_{[t \bmod 4]}$

Table 7.7 *MD5 round constants and Boolean functions.*

t	δQ_t	δF_t	δW_t	δT_t	δR_t	RC_t
31	$\mp 2^{p-10 \bmod 32}$					
32	0					
33	0					
34	0	0	$\pm 2^{p-10 \bmod 32}$	0	0	16
35–60	0	0	0	0	0	.
61	0	0	$\pm 2^{p-10 \bmod 32}$	$\pm 2^{p-10 \bmod 32}$	$\pm 2^p$	10
62	$\pm 2^p$	0	0	0	0	15
63	$\pm 2^p$	0	0	0	0	21
64	$\pm 2^p$ $+ \sum_{\lambda=0}^{w'} s_\lambda 2^{p+21+\lambda \bmod 32}$					

Table 7.8 *Family of partial differential characteristics using*
$\delta m_{11} = \pm 2^{p-10 \bmod 32}$, *where* $s_0, \ldots, s_{w'} \in \{-1, 0, +1\}$ *and* $w' = \min(w, 31 - p)$
for a fixed $w \geq 0$. *Interesting values for the parameter* w *are between 2 and 5.*

Second, k defines the birthday-search space (its size is $64 + k$) and the birthday-iteration function and can be used to trade-off between birthday-search time complexity, birthday-search memory complexity and average number of required near-collisions per birthday collision. Third, w defines the family of differential characteristics that can be used to construct the near-collision blocks and is the number of bit positions where arbitrary bit differences are allowed. It can be used to trade-off between the average number of required near-collision blocks per birthday collision and the cost of finding the r near-collision blocks.

The attack is based on the family of partial differential characteristics described in Table 7.8 using a single message-block bit difference $\delta m_{11} = \pm 2^i$ where both the sign and bit position i can be varied. This results in a difference $(\delta a, \delta b, \delta c, \delta d)$ to be added to the chaining-value difference, where $\delta a = 0$, $\delta d = \delta c = \pm 2^{i+10 \bmod 32}$, while δb consists of a fixed bit difference $\pm 2^{i+10 \bmod 32}$ as well as arbitrarily chosen differences on bit positions $i + 21, \ldots, i + 21 + w$ (modulo 32).

k = 0	w = 0			w = 1			w = 2			w = 3		
r	p	C_{tr}	M	p	C_{tr}	M	p	C_{tr}	M	p	C_{tr}	M
16	5.9	35.27	1 MB	1.75	33.2	1 MB	1.01	32.83	1 MB	1.	32.83	1 MB
15	7.2	35.92	1 MB	2.39	33.52	1 MB	1.06	32.86	1 MB	1.	32.83	1 MB
14	8.71	36.68	1 MB	3.37	34.01	1 MB	1.27	32.96	1 MB	1.04	32.84	1 MB
13	10.45	37.55	1 MB	4.73	34.69	1 MB	1.78	33.22	1 MB	1.2	32.93	1 MB
12	12.45	38.55	1 MB	6.53	35.59	1 MB	2.78	33.71	1 MB	1.66	33.16	1 MB
11	14.72	39.68	2 MB	8.77	36.71	1 MB	4.34	34.5	1 MB	2.61	33.63	1 MB
10	17.28	40.97	11 MB	11.47	38.06	1 MB	6.54	35.6	1 MB	4.18	34.42	1 MB
9	20.16	42.4	79 MB	14.62	39.64	2 MB	9.38	37.02	1 MB	6.46	35.56	1 MB
8	23.39	44.02	732 MB	18.21	41.43	21 MB	12.88	38.76	1 MB	9.52	37.09	1 MB
7	26.82	45.73	8 GB	22.2	43.43	323 MB	17.02	40.83	9 MB	13.4	39.02	1 MB
6	31.2	47.92	161 GB	26.73	45.69	8 GB	21.78	43.22	241 MB	18.14	41.4	20 MB
5	35.	49.83	3 TB	31.2	47.92	161 GB	27.13	45.89	10 GB	23.74	44.2	938 MB
4							34.	49.33	2 TB	30.19	47.42	81 GB

The columns p, C_{tr} and M denote the values of $-\log_2(p_{r,k,w})$, $\log_2(C_{tr}(r,k,w))$ and the minimum required memory M such that $C_{coll}(r,k,w,M) \leq C_{tr}(r,k,w)$, respectively.

Table 7.9 *Expected birthday-search costs for* $k = 0$.

Birthday Search As described in Section 7.2, the chosen-prefix collision attack starts with a tailored birthday search that results in a chaining value difference that can be eliminated by a sequence of near-collision attacks. In this case the search space V and iteration function f depend on the integer parameter $k \in \{0, 1, 2, \ldots, 32\}$; more precisely, first pad prefixes P and P' with arbitrary suffixes S_r and S'_r such that the total bit-lengths are equal, and $64 + k$ bits short of a multiple of the block size 512 bits. Then given $k \in \{0, \ldots, 32\}$, let B and B' be the last $512 - (64 + k)$ bits of padded prefixes $P\|S_r$ and $P'\|S'_r$, respectively. Then V and f are defined as follows:

$$V = \mathbb{Z}_{2^{32}} \times \mathbb{Z}_{2^{32}} \times \mathbb{Z}_{2^k},$$

$$f(x, y, z) = (a, c - d, c - b \bmod 2^k), \text{ where}$$

$$(a, b, c, d) = \begin{cases} MD5Compress(CV_{n-1}, B\|x\|y\|z) & \text{if } x \bmod 2 = 0; \\ MD5Compress(CV'_{n-1}, B'\|x\|y\|z) & \text{if } x \bmod 2 = 1. \end{cases}$$

The birthday collision, however, needs to satisfy several additional conditions that cannot be captured by V, f, or k: the prefixes associated with x and y in a birthday collision $f(x) = f(y)$ must be different, and the required number of pairs of near-collision blocks may be at most r when allowing differential characteristics with parameter w. The probability that a birthday collision satisfies all requirements depends not only on the choice of r and w, but also on the value for k, and is denoted by $p_{r,k,w}$. As a consequence, on average $1/p_{r,k,w}$ birthday collisions have to be found (see Table 7.9).

Assuming that M bytes of memory are available and that a single birthday trail requires 28 bytes of storage (namely 96 bits for the startpoint and endpoint each, and 32 for the length), this leads to the following expressions for the birthday-search costs [602]:

$$C_{tr}(r, k, w) = \sqrt{\frac{\pi \cdot 2^{64+k}}{2 \cdot p_{r,k,w}}}, \quad C_{coll}(r, k, w, M) = \frac{2.5 \cdot 28 \cdot C_{tr}(r, k, w)}{p_{r,k,w} \cdot M}.$$

For $M = 70/p_{r,k,w}$ as given in the last column of Table 7.9, the two costs are equal, and the overall expected birthday-search costs becomes $2C_{tr}(r, k, w)$. However, in practice it is advisable to choose M considerably larger. For $\epsilon \leq 1$, using $M = 70/(p_{r,k,w} \cdot \epsilon)$ bytes of memory results in $C_{coll} \approx \epsilon \cdot C_{tr}$ and the expected overall birthday-search cost is about $(1 + \epsilon) \cdot C_{tr}(r, k, w)$ MD5 compressions.

Near-Collision Construction Algorithm The birthday search results in a chaining-value difference δCV_n of the form $(0, \delta b, \delta c, \delta c)$. Let $\delta c = \sum_i k_i 2^i$ and $\delta b - \delta c = \sum_i l_i 2^i$, where $(k_i)_{i=0}^{31}$ and $(l_i)_{i=0}^{31}$ are signed binary digit expansions in non-adjacent form, denoted $NAF(a)$ for integer a (namely the unique signed binary digit expansion $(a_i)_{i=0}^{n}$ with $a_i \in \{-1, 0, 1\}$ and $a = \sum_i a_i 2^i$ where no two non-zero digits are adjacent). If $\delta c \neq 0$, let i be such that $k_i \neq 0$. Using a differential characteristic from Table 7.8 with $\delta m_{11} = -k_i 2^{i-10 \bmod 32}$ one can eliminate the difference $k_i 2^i$ in δc and δd and simultaneously change δb by

$$k_i 2^i + \sum_{\lambda=i+21 \bmod 32}^{i+21+w' \bmod 32} l_\lambda 2^\lambda,$$

where $w' = \min(w, 31 - i)$. Here one needs to be careful that each non-zero l_λ is eliminated only once in the case when multiple i-values allow the elimination of l_λ. Doing this for all k_i that are non-zero in the NAF of δc results in a difference vector $(0, \widehat{\delta b}, 0, 0)$ where $\widehat{\delta b}$ may be different from δb, and where the weight $w(NAF(\widehat{\delta b}))$ may be smaller or larger than $w(NAF(\delta b))$. More precisely, $\widehat{\delta b} = \sum_{\lambda=0}^{31} e_\lambda l_\lambda 2^\lambda$, where $e_\lambda = 0$ if there exist indices i and j with $0 \leq j \leq \min(w, 31 - i)$ such that $k_i = \pm 1$ and $\lambda = 21 + i + j \bmod 32$ and $e_\lambda = 1$ otherwise.

The bits in $\widehat{\delta b}$ can be eliminated as follows. Let $(\widehat{l_i})_{i=0}^{31} = NAF(\widehat{\delta b})$ and let j be such that $\widehat{l_j} = \pm 1$ and $j - 21 \bmod 32$ is minimal. Then the difference $\sum_{i=j}^{j+w'} \widehat{l_i} 2^i$ with $w' = \min(w, 31 - (j - 21 \bmod 32))$ can be eliminated from $\widehat{\delta b}$ using $\delta m_{11} = 2^{j-31 \bmod 32}$, which introduces a new difference $2^{j-21 \bmod 32}$ in δb, δc and δd. This latter difference is eliminated using $\delta m_{11} = -2^{j-31 \bmod 32}$, which then leads to a new difference vector $(0, \widetilde{\delta b}, 0, 0)$ with $w(NAF(\widetilde{\delta b})) < w(NAF(\widehat{\delta b}))$. The process is repeated until all differences have been eliminated. We refer to [570, 573] for details on the construction of the near-collision attacks themselves, including the differential characteristic construction algorithm and collision search algorithms, and to [566] for an implementation of the entire chosen-prefix collision attack.

Complete differential characteristics can be constructed with an average total complexity equivalent to roughly 2^{35} MD5 compressions. For small $w = 0, 1, 2$ and differential characteristics based on Table 7.8, finding a near-collision block pair requires on average roughly the equivalent of 2^{34} MD5 compressions. Combined with the construction of the differential characteristics, this leads to a rough overall estimate of about $2^{35.6}$ MD5 compressions to find a single pair of near-collision blocks for this chosen-prefix collision attack. The optimal parameters are given as $w = 2$, $k = 0$ and $r = 9$, for which the birthday-search cost is about 2^{37} MD5 compressions and constructing the $r = 9$ pairs of near-collision blocks costs about $2^{38.8}$ MD5 compressions, leading to the claimed total complexity of about $2^{39.1}$ MD5 compressions.

7.4.3 MD5 Single-Block Chosen-Prefix Collision Attack

It is possible to construct a chosen-prefix collision using only a single pair of near-collision blocks using a slightly different strategy [573]. Together with 84 birthday bits, the chosen-prefix collision-causing appendages are only $84 + 512 = 596$ bits long. This attack is based on a large family of differential characteristics that enables a corresponding large set of chaining-value differences to be eliminated by using a single pair of near-collision blocks.

Instead of using the family of differential characteristics based on $\delta m_{11} = \pm 2^i$, this attack re-uses the fastest-known collision attack for MD5 and varies the last few steps to find a large family of differential characteristics depicted in Table 7.10. Specifically, by varying those last steps and allowing the collision finding complexity to grow by a factor of about 2^{26}, this results in a set S of about $2^{23.3}$ different $\delta CV = (\delta a, \delta b, \delta c, \delta d)$ of the form $\delta a = -2^5$, $\delta d = -2^5 + 2^{25}$, $\delta c = -2^5 \bmod 2^{20}$ that can be eliminated. Such δCVs can be found using an 84-bit birthday search with step function $f : \{0, 1\}^{84} \to \{0, 1\}^{84}$ of the form

$$f(x) = \begin{cases} \phi(MD5Compress(CV, B\|x) + \delta\widehat{CV}) & \text{for } \tau(x) = 0 \\ \phi(MD5Compress(CV', B'\|x)) & \text{for } \tau(x) = 1, \end{cases}$$

where $\delta\widehat{CV}$ is of the required form, $\tau : x \mapsto \{0, 1\}$ is a balanced partition function and $\phi(a, b, c, d) \mapsto a\|d\|(c \bmod 2^{20})$. There are $2^{128-84} = 2^{44}$ possible δIHV values of this form, of which only about $2^{23.3}$ are in the allowed set S.

t	δQ_t	δF_t	δW_t	δT_t	δR_t	RC_t
26	-2^8					
27	0					
28	0					
29	0	0	2^8	0	0	9
30 − −33	0	0	0	0	0	.
34	0	0	2^{15}	2^{15}	2^{31}	16
35	2^{31}	2^{31}	2^{31}	0	0	23
36	2^{31}	2^{31}	0	0	0	4
37	2^{31}	2^{31}	2^{31}	0	0	11
38 − −46	2^{31}	2^{31}	0	0	0	.
47	2^{31}	2^{31}	2^8	2^8	2^{31}	23
48	0	0	0	0	0	6
49	0	0	0	0	0	10
50	0	0	2^{31}	0	0	15
51 − −59	0	0	0	0	0	.
60	0	0	2^{31}	2^{31}	-2^5	6
61	-2^5	0	2^{15}	2^{15}	2^{25}	10
62	$-2^5 + 2^{25}$	0	2^8	2^8	2^{23}	15
63	$-2^5 + 2^{25} + 2^{23}$	$2^5 - 2^{23}$	0	$2^5 - 2^{23}$	$2^{26} - 2^{14}$	21
64	$-2^5 + 2^{25} + 2^{23} + 2^{26} - 2^{14}$					

Partial differential characteristic for $t = 29, \ldots, 63$ using message differences $\delta m_2 = 2^8$, $\delta m_4 = \delta m_{14} = 2^{31}$, $\delta m_{11} = 2^{15}$. The probability that it is satisfied is approximately $2^{-14.5}$. It leads to a identical-prefix collision attack of approximated complexity 2^{16} MD5 compressions.

Table 7.10 *Partial differential characteristic for fast near-collision attack.*

It follows that a birthday collision $f(x) = f(x')$ has probability $p = 2^{23.3}/(2^{44} \cdot 2) = 2^{-21.7}$ to be useful, where the additional factor 2 stems from the fact that different prefixes are required, i.e., $\tau(x) \neq \tau(x')$.

A useful birthday collision can be expected after $\sqrt{\pi 2^{84}/(2p)} \approx 2^{53.2}$ MD5 compressions and requires approximately 400 MB of storage. The average complexity of finding the actual near-collision blocks is bounded by about $2^{14.8+26} = 2^{40.8}$ MD5 compressions and negligible compared to the birthday complexity. Thus the overall complexity is approximately $2^{53.2}$ MD5 compressions.

7.4.4 FLAME's MD5 Chosen-Prefix Collision Attack

As described in Section 7.3.3, by using counter-cryptanalysis all near-collision block pairs were recovered and thus the underlying differential characteristics could be reconstructed, proving that the used chosen-prefix collision was a yet-unknown variant attack [568]. After this initial discovery, Fillinger and Stevens performed an in-depth analysis wherein they reverse engineered this yet-unknown chosen-prefix collision attack [187]. The FLAME chosen-prefix collision attack differs significantly on several aspects from public research.

- The near-collision attack strategy to eliminate chaining-value differences $(\delta a, \delta b, \delta c, \delta d)$ using four blocks. Where δa and δd are essentially fixed and together have only three bit differences. Each of the four blocks focusses on a specific region of bit positions of δb and δc to cancel, while using the freedom to arbitrarily affect bit differences in bit position regions handled by later blocks. The last two blocks are able to cancel all bit differences in δb on bit positions $[0–4, 13–19, 27–31]$, but cancel only the last remaining bit difference in δc that is likely fixed in advance. The first two blocks cancel essentially all bit differences in δc, and they also affect δb but seemingly in a random manner: any difference added to δb is allowed as long as the last two blocks are still able to cancel them effectively.
- The differential characteristic construction method follows a variant meet-in-the-middle approach. Unlike the work of Stevens *et al.* [570] that uses many lower characteristics and many upper characteristics and then tries to combine pairs into a complete valid differential characteristic, the FLAME differential characteristics are fixed from δQ_6 up to δQ_{60}. This signifies that it uses only one upper characteristic and tries to connect many lower characteristics to it. The designers have chosen to have bit differences at all bit positions of Q_6, which may maximise the success probability. In the end, this appears to be significantly slower, as fixing δQ_6 in this manner implies a significant factor increase in the number of pairs that have to be tried, compared with the results in [570]. In comparison, the HashClash implementation is able to find replacement differential characteristics for FLAME's differential characteristics in a matter of seconds on a desktop computer, where the number of resulting bit conditions are also significantly lower: e.g., the replacement first characteristic has only 276 bit conditions versus 328 for FLAME's characteristic.
- The near-collision block search makes use of known collision-search speed-up techniques as actual bit values match many necessary bit conditions, which is unlikely to occur by chance. Noticable is that not all speed-up techniques that could be applied are visible in the actual blocks, as for some the

necessary bit conditions are not satisfied. This is not evidence that not all speed-up techniques are used. One possible explanation is that some speed-up techniques are used dynamically, depending on whether their necessary bit conditions are set. This would increase degrees of freedom, but only limited impact on the near-collision block-search complexity.

Fillinger and Stevens reconstruct a parameterised family of chosen-prefix collision attacks that should contain the FLAME collision attack. The precise parameters used for FLAME cannot be recovered with enough certainty. But it was possible to determine the minimal-cost parameters leading to estimated cost $C_{min} = 2^{46.6}$, and the minimal-cost parameters that are consistent with the observed data leading to estimated cost $C_{flame} = 2^{49.3}$. This is significantly higher than the estimated cost 2^{39} of the currently best-known MD5 chosen-prefix collision attack [573]. However, it was also observed that FLAME's attack is more suitable to be implemented on massively parallel hardware such as GPUs, and this has practical benefits.

PART II

IMPLEMENTATIONS

8

Efficient Modular Arithmetic

Joppe W. Bos, Thorsten Kleinjung and Dan Page

This chapter is concerned with one of the fundamental building blocks used in modern public-key cryptography: modular multiplication. Speed-ups applied to the modular multiplication algorithm or implementation directly translate in a faster modular exponentiation for RSA or a faster realisation of the group law when using elliptic-curve cryptography.

This chapter outlines one of the most commonly used modular multiplication method, known as 'Montgomery multiplication', for generic moduli, as well as different techniques when 'special' moduli of a particular shape are used. Moreover, we study approaches that might produce errors with a very small probability. Such faster 'sloppy reduction' techniques are especially beneficial in cryptanalytic settings. We look at this from a historical perspective as well as an applied implementation perspective. The best approach to implement modular multiplication on a modern 64-bit architecture with advanced single-instruction, multiple data-instruction-set extensions is, for example, quite different from the best approach on resource-constrained embedded devices.

Throughout this chapter we focus on the cryptographic setting unless we specifically discuss an algorithm for cryptanalysis. Contrary to many mathematical software applications, the running time of a cryptographic implementation (and hereby also the modular multiplication) should avoid secret-data-dependent branches and secretly indexed memory access. Such constant-time implementations are one of the basic countermeasures against timing attacks: advanced techniques that use information about the running time of the target algorithm to extract the used private key. Such attacks are part of a larger family of attacks known as side-channel attacks.

Throughout this chapter we represent a wn-bit non-negative integer X in the so-called radix-2^w representation $X = \sum_{i=0}^{n-1} x_i 2^{wi}$, where $0 \leq x_i < 2^w$. We denote x_i the ith word of the integer X.

8.1 Montgomery Multiplication

In order to accelerate the modular multiplication on modern computer platforms, Peter Montgomery introduced a modular reduction technique now known as Montgomery reduction [438]. The main idea behind this approach is to change the representatives of the residue classes and change the modular multiplication accordingly.

Let N be an odd wn-bit integer; here we assume w is the word size of the target computer platform (say 32- or 64-bit for modern architectures) and the modulus N can be represented by an array of n such computer words. More precisely, instead of computing the modular multiplication $A \cdot B \bmod N$ the Montgomery multiplication computes $\mathrm{MontMul}(A, B) = A \cdot B \cdot 2^{-wn} \bmod N$. In order to use this modular multiplication method, one needs to change the representation of the inputs. Given a wn-bit modulus N, we define the Montgomery form of an integer A to be $\tilde{A} = A \cdot 2^{wn} \bmod N$. This change of residue class ensures that the multiplication of two inputs in Montgomery form corresponds to the desired result in Montgomery form because

$$\mathrm{MontMul}(\tilde{A}, \tilde{B}) \equiv \tilde{A} \cdot \tilde{B} \cdot 2^{-wn} \equiv A \cdot 2^{wn} \cdot B \cdot 2^{wn} \cdot 2^{-wn} \equiv \widetilde{A \cdot B} \pmod{N}.$$

This change of representation is performed because computing the Montgomery multiplication can be done efficiently on modern computer architectures where multiplications and exact divisions by powers of two correspond to shifting the number to the left or right, respectively.

Montgomery multiplication uses the pre-computed value $\mu = -N^{-1} \bmod 2^{wn}$. Then, if we compute the Montgomery reduction of an integer C such that $0 \le C < N^2$, the idea is to add the multiple $N \cdot (\mu \cdot C \bmod 2^{wn})$ to C. Adding a multiple of N does not change the outcome modulo N and the computing of a reduction modulo 2^{wn} is for free on modern computer architectures: just take the wn least-significant bits. Adding this multiple of N ensures

$$C + N \cdot (\mu \cdot C \bmod 2^{wn}) \equiv C - N \cdot \left(N^{-1} \cdot C \bmod 2^{wn} \right)$$

$$\equiv C - C \equiv 0 \pmod{2^{wn}}.$$

Hence, $C + N \cdot (\mu \cdot C \bmod 2^{wn})$ is divisible by 2^{wn}, which can be computed by shifting wn positions to the right, avoiding an expensive division operation. Moreover, after this division by 2^{wn} the result has been reduced to, at most, $2N$ as

$$0 \le \frac{C + N \cdot (\mu \cdot C \bmod 2^{wn})}{2^{wn}} < \frac{N^2 + N \cdot 2^{wn}}{2^{wn}} < 2 \cdot N,$$

because $N < 2^{wn}$. This means a completely reduced result in $[0, N]$ can be computed with an additional conditional subtraction.

Algorithm 8.1 The radix-2^w interleaved Montgomery multiplication algorithm. Compute $A \cdot B \cdot 2^{-wn}$ modulo the odd modulus N using the pre-computed Montgomery constant $\mu = -N^{-1} \bmod 2^w$.

Input: $A = \sum_{i=0}^{n-1} a_i 2^{wi}, B, N$ such that N is odd,
$\quad 0 \le a_i < 2^w, \quad 0 \le A, B < 2^{wn}, \quad 2^{w(n-1)} \le N < 2^{wn}$.
Output: $C \equiv A \cdot B \cdot 2^{-wn} \pmod{N}$ such that $0 \le C < N$.
1: $C \leftarrow 0$
2: **for** $i = 0$ to $n - 1$ **do**
3: $\quad C \leftarrow C + a_i B$
4: $\quad q \leftarrow \mu C \bmod 2^w$
5: $\quad C \leftarrow (C + Nq)/2^w$
6: **if** $C \ge N$ **then**
7: $\quad C \leftarrow C - N$
8: **return** C

In many practical implementations of Montgomery multiplication the interleaved Montgomery multiplication algorithm is used. This approach merges the multiplication and reduction: after multiplying one computer word of A with the entire input of B the result is reduced modulo N. This has the advantage that the maximum size of the intermediate result remains significantly smaller: $n + 2$ computer words instead of the $2n + 1$ required when computing the full product $A \cdot B$ first and doing the Montgomery reduction next. This approach is outlined in Algorithm 8.1.

8.2 Arithmetic for RSA

Although embellishments such as padding (e.g., via OAEP [44], according to PKCS #1 [291]) are important from a security perspective, textbook RSA [501] can be described in terms of arithmetic in the multiplicative group $(\mathbb{Z}/N\mathbb{Z})^*$ as follows:

(1) generation of a public and private key pair: given a security parameter λ

$$
\text{KEYGEN}(\lambda) = \begin{cases}
\text{select random } \frac{\lambda}{2}\text{-bit primes } p \text{ and } q \\
\text{compute } N = p \cdot q \\
\text{compute } \varphi(N) = (p - 1) \cdot (q - 1) \\
\text{select random } e \in (\mathbb{Z}/N\mathbb{Z})^* \text{ such that } \gcd(e, \varphi(N)) = 1 \\
\text{compute } d = e^{-1} \pmod{\varphi(N)} \\
\text{return public key } (N, e) \text{ and private key } (N, d),
\end{cases}
$$

(2) encryption of a plaintext $m \in (\mathbb{Z}/N\mathbb{Z})^*$:

$$\text{ENC}((N, e), m) = m^e \pmod{N},$$

(3) decryption of a ciphertext $c \in (\mathbb{Z}/N\mathbb{Z})^*$:

$$\text{DEC}((N, d), c) = c^d \pmod{N}.$$

The description implies that efficiency of the underlying arithmetic and RSA itself are directly related. This fact could be viewed as an advantage, because improvement of the former will clearly yield improvement of the latter, or as a challenge: if the former cannot be efficient enough, RSA becomes impractical. Rivest, Shamir and Adleman themselves seem to have been well aware of this challenge. For example, preceding modern advice with respect to the co-design of cryptographic constructions and their implementation (see, e.g., the RWC 2015 invited talk by Bernstein [48, page 24]), their research paper [501] included overt focus on the latter. Section IV.A of [501], for example, discusses efficient realisation of encryption and decryption operations via an explicit left-to-right binary or 'square-and-multiply' algorithm (see, e.g., Knuth [338, Section 4.6.3] or Gordon [231, Section 2.1]) for modular exponentiation. As one of the first published public-key encryption schemes (cf. classified work by Cocks [548, Chapter 6]), it seems likely that one rationale for them to do so would be to rebut any claims of impracticality with respect to the technology landscape of that era. In fact, it remains challenging to implement efficient modular arithmetic and hence RSA on higher-end platforms (e.g., due to increasingly demanding workloads), and on lower-end platforms (which are, e.g., constrained with respect to computation and storage).

A vast range of literature has been dedicated to addressing such challenges more generally, so it seems reasonable to claim that RSA has acted as a driver for innovation with respect to modular arithmetic. Such innovation spans both software and hardware, of course, but, as noted as an aside in [501, Section IV.A], special-purpose hardware was (and still is) an attractive way to deliver efficiency. Rivest [499] describes an ASIC-based implementation by himself, Shamir and Adleman; as the first such implementation of RSA, it provides valuable insight into the state-of-the-art in efficient modular arithmetic of that era. The implementation is best described as a co-processor for multi-precision integer arithmetic, requiring $40\,000$ transistors within a single 42 mm^2 chip. The co-processor could be directed to compute various operations on operands stored in eight 512-bit registers. A limited set of operations, e.g., multiply-accumulate, were supported directly by a 512-bit ALU, and others via a 224-word micro-code program. Use of a micro-coded approach allowed more complex, number theoretic operations and, crucially, modular multiplication and exponentiation as required by RSA. Pre-dating techniques such as Montgomery multiplication (see Section 8.1), modular reduction would likely have been realised using integer division, which, in turn, likely used a shift-and-subtract approach [468, Section 13.1] supported by the ALU. Operating at 4 MHz, the co-processor was able to 'perform RSA encryption at rates in excess of 1200 bits/second'. Although the manufactured result never worked reliably, the design process is also remarkable in that it leveraged a purpose-built, LISP-style HDL, and motivated subsequent work by Rivest on the theory and practice of place-and-route [535].

Many facets of the technology landscape have changed since publication of RSA in 1978, not least the increased societal awareness, and importance, of cryptography in general. Ultimately, however, RSA has remained largely (i.e., with caveats per [74]) secure and so has been widely commercialised (see, e.g., [500]) and deployed during what is a rich, 40+ year history. This demonstrates that associated challenges with respect to efficient modular arithmetic have at least been mitigated if not solved, with any resulting innovations refined and capitalised on by more recent constructions. In this section we survey a very selective, limited subset of that history, focussing largely, but not exclusively, on RSA-specific approaches.

8.2.1 Capitalising on Special-Form Moduli

The selection of special-form parameters is a common, general optimisation strategy; doing so allows specialisation of associated algorithms, e.g., eliminating any overhead required to cope with the general case. Indeed, certain RSA parameters can be optimised in this way. The use of a short and low Hamming weight encryption exponent, e.g., $e = 65\,537$, is common: this replaces a general-purpose modular exponentiation with a short, fixed sequence of modular multiplications, for example. Equally, it is plausible to focus on the modulus N as a strategy for optimising the modular multiplications themselves. Where Montgomery multiplication is used per Section 8.1, imposing a special-form on the pre-computed value $\mu = -N^{-1} \bmod 2^w$ represents one such strategy. More specifically, a multiplication by μ (e.g., μC in line 4 of Algorithm 8.1) will be more efficient, even trivial, if we can select N such that $\mu = \pm 1$. Crucially, however, the impact of over-optimisation on the security properties of RSA must be carefully considered. Selecting too small a value of e, e.g., $e = 3$, should be avoided due to the broadcast attack by Håstad [262] (see Section 4.3.2) while a small d should be avoided due to an attack of Wiener [621] (see Section 4.4.1), for example. Likewise, any special structure in N could, intuitively, be exploited during an attempt to factor it. See Chapters 4 and 6 for further details on potential pitfalls.

Selecting a Special-Form Modulus

Consider three classes of special-form N, namely

$$N = p \cdot q = \begin{cases} f_{MSB} \, \| \, r & \Rightarrow \quad \text{(1) pre-determined MSBs} \\ r \, \| \, f_{LSB} & \Rightarrow \quad \text{(2) pre-determined LSBs} \\ f_{MSB} \, \| \, r \, \| \, f_{LSB} & \Rightarrow \quad \text{(3) pre-determined MSBs and LSBs,} \end{cases}$$

where f_{MSB} and f_{LSB} denote fixed (or constrained) values and r denotes a random (or unconstrained) value, all of suitable length: given a choice of f_{MSB} and/or f_{LSB}, these classes imply the most-significant bits (MSBs) and/or the least-significant bits (LSBs) of N are pre-determined (i.e., selected) during key generation. The question is, how?

Vanstone and Zuccherato [604] describe an approach for the first and second classes, i.e., selection of some t MSBs [604, Section 2] or LSBs [604, Section 7]. Citing a range of prior art, Lenstra [367] reviews an 'obvious and straight-forward trick' that affords both generalisation and improved efficiency by (a) supporting all three classes, and (b) avoiding the need for factorisation as a sub-step. Although Lenstra provides a high-level description of how the three classes of special-form N could be capitalised on in the context of RSA [367, Section 4], selection of LSBs, in particular, directly enables the strategy outlined above: for example, one simply selects the w LSBs such that $-1 \equiv N$ (mod 2^w), which then naturally implies $\mu = 1 \equiv -N^{-1}$ (mod 2^w) as required.

Both Vanstone and Zuccherato [604, Section 11] and Lenstra [367, Section 4] consider the security implications of selecting an associated N, framed within the context of contemporary factoring algorithms. A (very) informal summary would be that such special-form N have no negative impact on the security of RSA, provided that the fixed portions are not too large.

Scaling a General-Form Modulus

Rather than selecting a special-form modulus N outright, some alternative approaches attempt to construct and use $N' = s \cdot N$, i.e., the product of an existing modulus N and scaling factor s; the result, termed a scaled modulus, exhibits the special form required and thereby supports associated optimisations. Note that using N' rather than N potentially implies a larger modulus. As a result, use of a small enough s such that $2^{w \cdot (n-1)} \leq s \cdot N < 2^{w \cdot n}$, i.e., N' and N have the same number of radix-2^w words, might typically be preferred. Doing so avoids increasing the loop bound in line 2 of Algorithm 8.1, for example, thereby maximising the value of using N' by avoiding any additional overhead.

Quisquater multiplication [487], first presented at the EUROCRYPT 1990 rump session, uses an instance of this approach; we note that Walter [614] independently developed a similar approach around the same time. Following the presentation in [304], consider that the quotient and remainder stemming from division of some X by N can be computed as $Q = \lfloor X/N \rfloor$ and $R = X - Q \cdot N = X - \lfloor X/N \rfloor \cdot N$ respectively. The integer division required to compute Q is (relatively) inefficient, so use of an approximate quotient $\hat{Q}$ can be more attractive when (a) $\hat{Q}$ can be computed more efficiently than Q and (b) any error resulting from use of $\hat{Q}$ versus Q can be corrected via a small number of efficient sub-steps. Quisquater observed that Q is lower-bounded by

$$\hat{Q} = \left\lfloor \frac{X}{2^c \cdot 2^{w \cdot n}} \right\rfloor \cdot \left\lfloor \frac{2^c \cdot 2^{w \cdot n}}{N} \right\rfloor$$

and yields an approximate remainder $\hat{R} = X - \hat{Q} \cdot N$. If one selects a c such that $s = \lfloor (2^c \cdot 2^{w \cdot n})/N \rfloor$ then pre-computes and uses the modulus $N' = s \cdot N$, the required remainder can be computed as $\hat{R} = X - \lfloor X/2^{c+(w \cdot n)} \rfloor \cdot N'$. This is (more) efficient, because the integer division involved can be replaced with an appropriate shift. Note that, by construction, the c MSBs of the special-form modulus N' are equal to 1, but we did not need to select the existing, general modulus N with that property.

Hars [260, Section 5.4] describes an approach he terms 'tail tailoring', set within the context of Montgomery multiplication. The idea is to set $s = \mu = -N^{-1}$ (mod 2^w), i.e., the pre-computed value related to N: doing so implies

$$N' = s \cdot N \equiv -N^{-1} \cdot n_0 \equiv -n_0^{-1} \cdot n_0 \equiv -1 \quad (\text{mod } 2^w),$$

recalling that n_0 denotes the 0th or least-significant word of N, and so the pre-computed value related to N' is $\mu' = -N'^{-1} \equiv 1$ (mod 2^w) as required to enable the strategy outlined above. Note that this time, by construction, the w LSBs of the special-form modulus N' are equal to 1, but we again did not need to select the existing, general modulus N with that property.

8.2.2 Capitalising on Parallelism

It should be obvious that, in the period since publication of the RSA cryptosystem [501], almost any platform on which one might expect RSA to be implemented has vastly improved with respect to computational, communication and storage capabilities. It is interesting to highlight a symbiotic relationship between study of the algorithms for modular arithmetic, and platforms on which any associated implementation is then developed. A common theme is that algorithms and platforms co-evolve to some extent, so that, e.g., an improvement in the latter might be capitalised on by the former and vice versa.

Support for parallel computation has, in particular, evolved from a once niche to now commodity feature in most platforms; it represents a central means of addressing the limitations on scaling (e.g., clock frequency) that stem from Moore's law [441]. Flynn's taxonomy [190] offers a structured way to reason about the use of parallelism, describing instances as either single (i.e., scalar) or multiple (i.e., parallel) along two dimensions, namely instructions and data: the four classes are Single Instruction Single Data (SISD), Single Instruction Multiple Data (SIMD), Multiple Instruction Single Data (MISD) and Multiple Instruction Multiple Data (MIMD). Given the computational demands of modular arithmetic, it is unsurprising that one can identify work attempting to capitalise on any class outlined above as a way to deliver improved efficiency. However, for various reasons, SIMD is arguably the most interesting. For example, it has wide-scale support through instances of the Intel MMX/SSE/AVX, ARM NEON, PowerPC AltiVec, and AMD 3DNow! families, and aligns with genuine vector-like support from, for example ARM SVE, and the RISC-V V extension. In addition, it forces lower-level relationship between algorithms and platform, in the sense the ISA dictates the exact form of SIMD support.

Intra-Multiplication Parallelism

Bos *et al.* [96] adopt an intra-multiplication (i.e., within each multiplication) approach to parallelisation, splitting Algorithm 8.1 into two threads, which can then be computed in parallel. We continue to use this term according to [96], but note that lanes might be a better choice: the idea is to compute in two SIMD-based lanes within one execution context, not two execution contexts (i.e., a multi-threaded implementation). Their approach is based on two main ideas. First, lines 3, 4, and 5 of Algorithm 8.1 capture $1 \times n$, 1×1, and $1 \times n$ word multiplications respectively; these constitute the computational core, but are difficult to parallelise owing to dependencies between them. Notice, however, that computation of $q = \mu C \bmod 2^w$ requires c_0 alone: this fact implies it is possible to compute the $1 \times n$ word multiplications in parallel, if one duplicates the computation of c_0 in each thread so as to eliminate the problematic dependency. Second, rather than use the Montgomery constant $\mu = -M^{-1} \bmod 2^w$ as is, the sign is flipped to yield $\mu = M^{-1} \bmod 2^w$ instead. Doing so yields two outcomes, namely that (1) both D and E are bounded by M, meaning they can be represented in n (versus $n + 1$) words (this fact simplifies management of carries), and (2) the conditional final subtraction in Algorithm 8.1 becomes a conditional final addition.

Algorithm 8.2 Bos *et al.* [96, Algorithm 2]: a parallel radix-2^w interleaved (SIMD-friendly) Montgomery multiplication algorithm.

Input: A, B, M, and μ, such that $A = \sum_{i=0}^{n-1} a_i 2^{wi}$, $B = \sum_{i=0}^{n-1} b_i 2^{wi}$, $M = \sum_{i=0}^{n-1} m_i 2^{wi}$, $0 \leq A, B < M$, $2^{w \cdot (n-1)} \leq M < 2^{w \cdot n}$, $2 \nmid M$, $\mu = M^{-1} \bmod 2^w$.

Output: $C \equiv A \cdot B \cdot 2^{-w \cdot n} \bmod M$ such that $0 \leq C < M$.

1: **parallel**
2:     **thread**
3:         $D \leftarrow 0$, i.e., $d_i = 0$ for $0 \leq i < n$
4:         **for** $j = 0$ to $n - 1$ **do**
5:             $q \leftarrow ((\mu \cdot b_0) \cdot a_j + \mu \cdot (d_0 - e_0)) \bmod 2^w$
6:             $t_0 \leftarrow a_j \cdot b_0 + d_0, t_0 \leftarrow \lfloor t_0 / 2^w \rfloor$
7:             **for** $i = 1$ to $n - 1$ **do**
8:                 $p_0 \leftarrow a_j \cdot b_i + t_0 + d_i, t_0 \leftarrow \lfloor p_0 / 2^w \rfloor, d_{i-1} \leftarrow p_0 \bmod 2^w$
9:             $d_{n-1} \leftarrow t_0$
10:     **end thread**
11:     **thread**
12:         $E \leftarrow 0$, i.e., $e_i = 0$ for $0 \leq i < n$
13:         **for** $j = 0$ to $n - 1$ **do**
14:             $q \leftarrow ((\mu \cdot b_0) \cdot a_j + \mu \cdot (d_0 - e_0)) \bmod 2^w$
15:             $t_1 \leftarrow q \cdot m_0 + e_0, t_1 \leftarrow \lfloor t_1 / 2^w \rfloor$
16:             **for** $i = 1$ to $n - 1$ **do**
17:                 $p_1 \leftarrow q \cdot m_i + t_1 + e_i, t_1 \leftarrow \lfloor p_1 / 2^w \rfloor, e_{i-1} \leftarrow p_1 \bmod 2^w$
18:             $e_{n-1} \leftarrow t_1$
19:     **end thread**
20: **end parallel**
21: $C \leftarrow D - E$
22: **if** $C < 0$ **then**
23:     $C \leftarrow C + M$
24: **return** C

Algorithm 8.2 reproduces [96, Algorithm 2], thus capturing the approach for completeness. Notice that the threads detailed in lines 2–10 and 11–19 adhere to a SIMD computational model, in the sense they perform the same operations (e.g., line 8 versus 17) with different data (respectively D versus E). The only caveat is perhaps lines 5 and 14, which compute q. Since both require d_0 and e_0, some synchronisation is required; in reality, it is likely easier to compute q sequentially then distribute the result to both threads. The fact that there are two threads somewhat specialises the approach to ISAs with support for 2-way (or

2-lane) SIMD; the authors focus on $w = 32$ specifically, in line with concrete ISAs providing such support. Compared with a sequential 32-bit implementation, [96, Section 4] reports improvement by a factor of 1.68 to 1.76 using a 2048-bit modulus on platforms enabled with the Intel SSE (e.g., Intel Xeon) and ARM NEON (e.g., ARM Cortex-A9) instruction set architectures (ISAs). However, compared with a sequential 64-bit implementation the results are not as positive: the parallel approach is limited by the form of SIMD supported, in the sense the ISAs allow (32×32)-bit multiplication only. Constraints imposed by, e.g., non-orthogonality of the ISA, are a common challenge for SIMD-based implementations. For example, in early work Acar [2, Section 5.4.1] noted the difficulty of using MMX for modular arithmetic owing to a lack of suitable unsigned multiplication instructions. This issue has arguably improved over time, as ISAs have evolved away from their media-oriented origins and toward support for general-purpose workloads. Either way, use of intra-multiplication parallelism is advantageous in the sense that improvement could be harnessed by any workload based on modular multiplication: use of modular exponentiation in RSA is one example, but, equally, others can (transparently) benefit.

Inter-Multiplication Parallelism

Within the context of cryptographic implementation, the term bit-slicing is normally attributed to Biham [61]: it refers to a non-standard representation of data, plus a non-standard implementation of functions that operate on instances of said representation. Consider a w-bit word x, where x_i denotes the ith bit for $0 \le i < w$. Computing the result of operations on such words, e.g., $x \oplus y$, the XOR of words x and y, is rendered efficient by native support in the underlying processor. However, computing the result of $x_i \oplus x_j$, i.e., the XOR of bits within x, is more difficult due to a lack of the same (native) support. Where such operations are common, the use of bit-slicing can be attractive: it transforms such an x into w slices, say $\hat{x}[i]$ for $0 \le i < w$, such that $\hat{x}[i]_k = x_i$ (i.e., the ith bit of x) for some k. Put another way, the ith bit of x is placed at the kth index within the ith slice. Visually, this transformation is described by

$$x = \langle x_0, x_1, \ldots, x_{w-1} \rangle \mapsto \begin{cases} \hat{x}[0] & = \langle \ \ldots, & x_0, & \ldots \ \rangle \\ \hat{x}[1] & = \langle \ \ldots, & x_1, & \ldots \ \rangle \\ & \vdots \\ \hat{x}[w-1] & = \langle \ \ldots, & x_{w-1}, & \ldots \ \rangle. \end{cases}$$

Under such a representation, the original operation can again be efficient: we can (natively) compute $\hat{x}[i] \oplus \hat{x}[j]$, because x_i and x_j are at the same index in those slices. More generally, any function previously used as $r = f(x)$ must be transformed into an alternative $\hat{r} = \hat{f}(\hat{x})$ in order to process bit-sliced operands. This implies two disadvantages: (1) there is an overhead related to the conversion of x into $\hat{x}$ and $\hat{r}$ into r, and (2) many operations must be translated into a 'software circuit', composed of Boolean operations on the slices, within $\hat{f}$; although one can (natively) compute $x + y$, for example, the same is not true of $\hat{x}$ and $\hat{y}$. Crucially, however, if each slice is itself represented as a w-bit word, then it is possible to compute w instances of $\hat{f}$ in parallel on suitably packed $\hat{x}$. A common analogy is that of bit-slicing transforming the w-bit, 1-way scalar processor into a 1-bit, w-way SIMD processor, thus yielding upto a w-fold improvement which acts to compensate for the disadvantages.

An analogous technique can be applied to (modular) integer arithmetic using a radix-2^w representation: the idea is again to slice an $X \in \mathbb{Z}$ into n slices, say $\hat{X}[i]$ for $0 \leq i < n$, such that $\hat{X}[i]_k = x_i$ (i.e., the ith word of X) for some k. Visually, this transformation is described by

$$X = \langle x_0, x_1, \ldots, x_{n-1} \rangle \quad \mapsto \quad \begin{cases} \hat{X}[0] &= \langle \ \ldots, \quad x_0, \quad \ldots \ \rangle \\ \hat{X}[1] &= \langle \ \ldots, \quad x_1, \quad \ldots \ \rangle \\ &\vdots \\ \hat{X}[n-1] &= \langle \ \ldots, \quad x_{n-1}, \quad \ldots \ \rangle \end{cases}$$

As with bit-slicing this can make it easier to combine words within X, say x_i and x_j, e.g., to deal with carries between words. This technique, termed word-slicing, seems to have independent origins from bit-slicing itself. More specifically, Montgomery [439] originally observed that the vectorisation of modular arithmetic 'horizontally' (or intra-operation) was more difficult than 'vertically' (or inter-operation) on a vector-based Cray Y-MP. To solve this problem he adopted word-slicing, vectorising an ECM [388] implementation 'vertically' to allow multiple trials in parallel (versus the 'horizontal' parallelisation of one trial). Page and Smart [465] have rediscovered the technique in adopting an inter-multiplication approach to parallelisation. Their idea is to support parallel computation of $R[k] = X[k]^e \pmod{N[k]}$ for $0 \leq k < 4$, i.e., four exponentiations using different bases and moduli but the same exponent; note that use of the same exponent implies uniform control-flow within each exponentiation, i.e., they follow SIMD-style computation. Set within the context of RSA this permits a form of 'batch' encryption or decryption, e.g., by a server dealing with multiple clients. Implementing the parallel exponentiation reduces to implementing parallel Montgomery multiplication, namely Algorithm 8.1, through use of word-slicing: one simply slices and packs words in

$$
\begin{aligned}
R[k] &= \langle \quad r[k]_0, \quad r[k]_1, \quad \ldots, \quad r[k]_{n-1} \quad \rangle \\
N[k] &= \langle \quad n[k]_0, \quad n[k]_1, \quad \ldots, \quad n[k]_{n-1} \quad \rangle \\
X[k] &= \langle \quad x[k]_0, \quad x[k]_1, \quad \ldots, \quad x[k]_{n-1} \quad \rangle
\end{aligned}
$$

for $0 \le k < 4$ to yield

$$
\begin{aligned}
\hat{R}[i] &= \langle \quad r[0]_i, \quad r[1]_i, \quad r[2]_i, \quad r[3]_i \quad \rangle \\
\hat{X}[i] &= \langle \quad x[0]_i, \quad x[1]_i, \quad x[2]_i, \quad x[3]_i \quad \rangle \\
\hat{N}[i] &= \langle \quad n[0]_i, \quad n[1]_i, \quad n[2]_i, \quad n[3]_i \quad \rangle
\end{aligned}
$$

for $0 \le i < n$, then implements a suitable $\hat{f}$ such that $\hat{R} = \hat{f}(\hat{X}, e, \hat{N})$ computes the required exponentiations. The only arithmetic complication is the conditional final subtraction step in lines 6–7 of Algorithm 8.1, as used to produce the least residue modulo N such that $0 \le C < N$ as output; without it, the output would satisfy $0 \le C < 2 \cdot N$. As observed by Walter [613, 615] (and others [250, 253]), where Algorithm 8.1 is used iteratively (where the output is reused as a subsequent input, e.g., in an exponentiation) the subtraction step can be eliminated whenever $4 \cdot N < 2^{wn}$ because then all input and output to the Montgomery multiplication are bounded by $2 \cdot N$ and represented in a redundant Montgomery form. Having removed the conditional subtraction, a Montgomery multiplication can be computed with no data-dependent control-flow; the (minor) trade-off is a requirement to cater for 1-word larger operands when the length of the modulus is close to a multiple of w (which for standard RSA parameters, it will be). Compared with a sequential 32-bit implementation, [465, Section 3.3] reports improvement by close to a factor of 2 using a 2048-bit modulus on platforms enabled with the Intel SSE (Intel Pentium 4) ISA. Of course, realising the improvement in practice assumes a usable batch of exponentiations is available; where such a batch cannot be guaranteed, use of an intra-multiplication approach [96] could be a more sensible use of the computational resources.

8.2.3 Dealing with Large Moduli

Set against the context of contemporary factoring algorithms (as discussed in [501, Section IX]), Rivest, Shamir and Adleman 'recommend[ed] using 100-digit (decimal) prime numbers p and q, so that [N] has 200 digits' [501, Section VII] to ensure the security of RSA; this means a 665-bit N. However, the technology landscape has obviously evolved since then (see Chapter 3). This has meant improvement in factoring algorithms, their implementation, the platforms they are executed on, and therefore, ultimately, their efficiency: all these

factors have contributed to a significant increase in the length of plausibly factorable moduli, and hence recommendations for secure parameterisation.

For example, in 2001 Lenstra and Verheul [379] developed a methodology
and analysis of parameter selection for a range of cryptographic constructions,
including RSA. They conclude that 'RSA keys that are supposed to be secure
until 2040 are about three times larger than the popular 1024-bit RSA keys that
are currently secure' noting an impact on the efficiency of associated modular
arithmetic: use of RSA will be '9–27 times slower'. ENISA [549, Section 3.6]
offer a longer-term perspective, recommending 15360-bit moduli to ensure 'security [of RSA] for thirty to fifty years'. Resources such as

$$\texttt{https://www.keylength.com}$$

offer a useful summary of recommendations, over time and from different
sources (see also Chapter 11). Even viewing such recommendations as approximate, the obvious challenge is how to scale, i.e., how to mitigate the impact
of increased moduli lengths on the efficiency of associated modular arithmetic
and hence RSA.

'Double-Length' Arithmetic

Consider the Karatsuba–Ofman [315] technique, which allows the computation of a product $R = X \cdot Y$ by decomposition. For n-bit integers X and Y,
assuming for simplicity that n is even, it decomposes

$$
\begin{aligned}
X &= X_1 \cdot 2^{n/2} + X_0 \\
Y &= Y_1 \cdot 2^{n/2} + Y_0,
\end{aligned}
$$

where X_i and Y_i are $(n/2)$-bit integers, then computes

$$R = R_2 \cdot 2^n + R_1 \cdot 2^{n/2} + R_0,$$

where

$$
\begin{aligned}
R_0 &= T_0 \\
R_1 &= T_1 - T_0 - T_2 \\
R_2 &= T_2
\end{aligned}
$$

$$
\begin{aligned}
T_0 &= X_0 \cdot Y_0 \\
T_1 &= (X_0 + X_1) \cdot (Y_0 + Y_1) \\
T_2 &= X_1 \cdot Y_1.
\end{aligned}
$$

Put another way, this technique realises an n-bit integer multiplication by using (a) three $(n/2)$-bit multiplications plus (b) some auxiliary additions and
subtractions. In the context of hardware implementation, for example, this is

attractive because it enables various trade-offs including between area and latency: it implies reduced area of a $(n/2)$-bit versus n-bit multiplier, as a trade-off against increased latency with respect to their reuse when computing T_0, then T_1, then T_2.

One might consider scaling RSA using a conceptually similar technique: if we have efficient modular arithmetic for an n-bit modulus (e.g., in the form of dedicated hardware, or co-processor), the idea would be to somehow leverage it in delivering what we want, i.e., arithmetic for a $(2 \cdot n)$-bit or 'double-length' modulus. Paillier [467, Section 1] succinctly outlines two research challenges of this type, namely '[h]ow to optimally implement nk-bit modular operations using k-bit modular operations' and, more specifically, '[h]ow to implement an nk-bit modular multiplication using k-bit modular operations with a minimal number of k-bit multiplications', leading to various associated work (see, e.g., [124, 630]). As a representative example, consider the solution of Fischer and Seifert [189], which assumes an API that provides an operation

$$(Q, R) = \texttt{MultModDiv}(X, Y, M)$$

for an n-bit M, where

$$
\begin{aligned}
Q &= (X \cdot Y) \bmod M \\
R &= \lfloor (X \cdot Y)/M \rfloor.
\end{aligned}
$$

The simpler of two algorithms is presented in [189, Section 3.1]: it realises a $(2 \cdot n)$-bit multiplication modulo N, by using seven invocations of $\texttt{MultModDiv}$ (noting that some can be executed in parallel, if/when possible), plus some auxiliary additions and subtractions. We reproduce the approach in Algorithm 8.3 for completeness, but omit the proof of correctness provided in detail by [189, Section 3.1].

'Approximate-then-Correct' Arithmetic

Using the long-term ENISA [549, Section 3.6] recommendation as motivation, Bentahar and Smart [46] explore the efficiency of algorithms for arithmetic modulo a 15 360-bit N, via both analytical and experimental approaches: their premise is that asymptotically efficient yet concretely unsuitable algorithms, for smaller moduli, become 'in scope' for such larger moduli.

The use of 'wooping' is outlined in [46, Section 2.4]; this is a technique attributed to Bos [84, Chapter 6] (see also [186, Section 15.1.1]) as a means of error correction. Imagine one computes the integer product $R = X \cdot Y$. To (probabilistically) verify whether R is correct, one selects a random prime p,

Algorithm 8.3 Fischer and Seifert [189, Section 3.1]: 'basic doubling' algorithm.

$$\textbf{Input:} \quad \left. \begin{array}{l} N = N_t \cdot 2^n + N_b \\ X = X_t \cdot 2^n + X_b \\ Y = Y_t \cdot 2^n + Y_b \end{array} \right\} \text{ such that } 0 \le N_b, X_b, Y_b < 2^n$$

Output: $R = X \cdot Y \pmod{N}$

1: $(Q_1, R_1) \leftarrow \texttt{MultModDiv}(Y_t, 2^n, N_t)$
2: $(Q_2, R_2) \leftarrow \texttt{MultModDiv}(Q_1, N_b, 2^n)$
3: $(Q_3, R_3) \leftarrow \texttt{MultModDiv}(X_t, R_1 - Q_2 + Y_b, N_t)$
4: $(Q_4, R_4) \leftarrow \texttt{MultModDiv}(X_b, Y_t, N_t)$
5: $(Q_5, R_5) \leftarrow \texttt{MultModDiv}(Q_3 + Q_4, N_b, 2^n)$
6: $(Q_6, R_6) \leftarrow \texttt{MultModDiv}(X_t, R_2, 2^n)$
7: $(Q_7, R_7) \leftarrow \texttt{MultModDiv}(X_b, Y_b, 2^n)$
8: $Q \leftarrow R_3 + R_4 - Q_5 - Q_6 + Q_7$
9: $R \leftarrow R_7 - R_6 - R_5$
10: **return** $Q \cdot 2^n + R \pmod{N}$

computes

$$\begin{array}{rcll} \hat{X} & = & X & \pmod{p} \\ \hat{Y} & = & Y & \pmod{p} \\ \hat{R} & = & \hat{X} \cdot \hat{Y} & \pmod{p} \end{array}$$

then tests whether $R \stackrel{?}{=} \hat{R} \pmod{p}$. If the equality does not hold we infer R is incorrect; if the equality holds, the probability of a false-positive is $1/p$. $\hat{R}$ can be viewed as an arithmetic checksum, which (a) can be efficiently computed if p is small (e.g., a single word, meaning w bits; this is particularly true when also using special-form p), and (b) offers a low false-positive probability if p is large enough, or multiple such checksums are used. This technique is then used [46, Section 3] to yield efficient algorithms for Montgomery [438] and Barrett [41] reduction. The central idea is to approximate certain full products using (more efficient) half products, then correct said approximations via wooping.

8.3 Arithmetic for ECC

Besides RSA (see Section 8.2), the other popular approach to realise public-key cryptography is based on the algebraic structure of elliptic curves over finite fields. This elliptic-curve cryptography (ECC) [340, 433] has enjoyed increasing popularity since its invention in the mid 1980s and has become the preferred alternative to RSA due to the attractiveness of smaller key-sizes [368, 379].

As opposed to RSA, the cryptographic standard includes many parameters that determine the exact choice of curve used and over which finite field. Since the initial standardisation of elliptic-curve cryptography [283] there has been significant progress: this includes taking into account various new types of attacks (such as side-channel attacks) and performance improvements. The latter area includes using different curve models, which lowers the total cost of elliptic-curve group operations. However, all the curve operations have to be implemented by using a sequence of operations in the underlying finite field. There have been advances in the choice of the shape of these primes in order to increase the performance of the modular arithmetic.

In this section we will describe the choices made for the prime shapes in the ECC standard by the National Institute of Standards and Technology (NIST) [555], and will ask why these primes result in particularly efficient modular reduction implementations and what caused the choice of slightly different prime shapes with the new elliptic curves.

8.3.1 Generalised Mersenne Numbers

It is well known that reduction modulo Mersenne primes is very efficient. A Mersenne prime is a prime number that is one less than a power of two: $2^x - 1$. When computing $a \cdot b$ modulo $2^x - 1$ with $0 \le a, b < 2^x - 1$, one can compute the reduction without multiplications as

$$c = a \cdot b = c_1 \cdot 2^x + c_0 \equiv c_1 + c_0 \pmod{2^x - 1},$$

where $0 \le c_0, c_1 < 2^x$, because $2^x \equiv 1 \bmod 2^x - 1$. However, for even more practical convenience, one would like x to be close to a multiple of 32 or 64 since this matches (a multiple of) the word size of virtually all modern computer architectures.

Unfortunately, there are not many Mersenne primes in the range where this is of interest for elliptic-curve cryptography. For example, when looking at the range $100 < x < 1000$ only four such values are available: $x \in \{107, 127, 521, 607\}$. The modulus $2^{127} - 1$ has been proposed for usage with hyperelliptic curve cryptography in genus 2 (cf. [47, 92, 211]) where it offers sufficient security while $2^{521} - 1$ is used in the NIST standard for elliptic curve (genus 1) cryptography.

One direction to generalise this idea is to use prime moduli of the form $2^x - c$, where c is small enough to fit in a computer word. These are sometimes referred to as Crandall numbers [145] and will be explored in more detail in Section 8.3.2. Another direction to generalise was studied by Solinas in [555] and later adopted by NIST. A potential reason to go for these generalised Mersenne numbers over Crandall numbers might be patent-related (cf. [145]).

Solinas studied efficient reductions, e.g., reductions that do not require multiplications, modulo polynomials of the form

$$f(t) = t^d + \sum_{i=0}^{d-1} c_i \cdot t^i, \tag{8.1}$$

where $c_i \in \{-1, 0, 1\}$. For selected integers k and d these techniques can then be used for efficient reduction modulo $f(2^k)$, as will be shown in more detail in this section. The five selected generalised Mersenne primes of this shape that are specified in the NIST standard are

$$p_{192} = 2^{192} - 2^{64} - 1$$
$$p_{224} = 2^{224} - 2^{96} + 1$$
$$p_{256} = 2^{256} - 2^{224} + 2^{192} + 2^{96} - 1$$
$$p_{384} = 2^{384} - 2^{128} - 2^{96} + 2^{32} - 1$$
$$p_{521} = 2^{521} - 1.$$

It should be noted that for the first four generalised Mersenne primes all the exponents are a multiple of 32. This makes implementations on 32-bit platforms significantly easier and faster. The exception is for the Mersenne prime p_{521}; here the choice was clearly made to go for a Mersenne prime instead of a generalised Mersenne prime such as $2^{512} - 2^{32} - 1$, $2^{512} - 2^{32} + 1$, or $2^{512} - 2^{288} + 1$.

Using the Prime Shape Directly

Let us give an example of how this efficient modular reduction works. We will use the modulus p_{256} as an example because this is arguably the most frequently used prime out of these five as it targets the 128-bit security level when used with an appropriate elliptic curve for usage in cryptography. Similar to the approach used for Mersenne primes, one writes large powers of two (where the exponent is ≥ 256) in terms of smaller powers of two. For example, $2^{256} \equiv 2^{224} - 2^{192} - 2^{96} + 1$ by the definition of p_{256}. For larger exponents this can be done similarly and one obtains

$$2^{256} \equiv 2^{224} - 2^{192} - 2^{96} + 1,$$
$$2^{288} \equiv -2^{192} - 2^{128} - 2^{96} + 2^{32} + 1,$$
$$2^{320} \equiv -2^{224} - 2^{160} - 2^{128} + 2^{64} + 2^{32},$$
$$2^{352} \equiv -2^{224} - 2^{160} + 2 \cdot 2^{96} + 2^{64} - 1,$$
$$2^{384} \equiv -2^{224} + 2 \cdot 2^{128} + 2 \cdot 2^{96} - 2^{32} - 1,$$
$$2^{416} \equiv -2^{224} + 2^{192} + 2 \cdot 2^{160} + 2 \cdot 2^{128} + 2^{96} - 2^{64} - 2^{32} - 1,$$
$$2^{448} \equiv 3 \cdot 2^{192} + 2 \cdot 2^{160} + 2^{128} - 2^{64} - 2^{32} - 1,$$
$$2^{480} \equiv 3 \cdot 2^{224} + 2 \cdot 2^{192} + 2^{160} - 2^{96} - 2^{64} - 2^{32}.$$

Hence, after an initial multiplication $c = a \cdot b$ for $0 \le a, b < p_{256}$ the modular reduction can be done efficiently by substituting the powers of two in $c = \sum_{i=0}^{15} c_i 2^{32i}$ with the congruent values from above. When grouping terms together things boil down to

$$c = a \cdot b = \sum_{i=0}^{15} c_i 2^{32i} \equiv s_1 + 2s_2 + 2s_3 + s_4 + s_5 - s_6 - s_7 - s_8 - s_9 \bmod p_{256},$$

where the 32-bit coefficients of the s_i are defined in terms of the coefficients c as follows.

	2^{224}	2^{192}	2^{160}	2^{128}	2^{96}	2^{64}	2^{32}	2^0
s_1	c_7	c_6	c_5	c_4	c_3	c_2	c_1	c_0
s_2	c_{15}	c_{14}	c_{13}	c_{12}	c_{11}	0	0	0
s_3	0	c_{15}	c_{14}	c_{13}	c_{12}	0	0	0
s_4	c_{15}	c_{14}	0	0	0	c_{10}	c_9	c_8
s_5	c_8	c_{13}	c_{15}	c_{14}	c_{13}	c_{11}	c_{10}	c_9
s_6	c_{10}	c_8	0	0	0	c_{13}	c_{12}	c_{11}
s_7	c_{11}	c_9	0	0	c_{15}	c_{14}	c_{13}	c_{12}
s_8	c_{12}	0	c_{10}	c_9	c_8	c_{15}	c_{14}	c_{13}
s_9	c_{13}	0	c_{11}	c_{10}	c_9	0	c_{15}	c_{14}

Computing the reduction modulo p_{256} can be done using six modular additions and four modular subtractions. Computing the modular addition $c = a + b \bmod p_{256}$ in constant running time (see the beginning of this chapter) can be done at the cost of approximately one addition and one subtraction by using

$$c_1 \leftarrow a + b, \quad c_2 \leftarrow c_1 - p_{256}, \quad c \leftarrow \text{select}(\lfloor c_1/2^{256} \rfloor, c_1, c_2),$$

where the function $\text{select}(x, a, b)$ returns a if $x = 0$ or b otherwise. Such a selection can be implemented efficiently in constant time by masking out the correct results. Ergo, the total cost of a constant-time modular reduction is ten 256-bit additions and ten 256-bit subtractions.

Using Montgomery Reduction

Another approach to perform arithmetic modulo generalised Mersenne numbers is by using Montgomery arithmetic (see Section 8.1). Recall that radix-2^w Montgomery reduction requires a multiplication with the pre-computed Montgomery constant $\mu = -N^{-1} \bmod 2^w$ and a multiplication with the modulus N. Similar to the observation made in Section 8.2.1 in the setting of RSA, the multiplication with μ can be avoided whenever $N \equiv \pm 1 \bmod 2^w$ because then $\mu \equiv \mp 1 \bmod 2^w$. This observation has been made and used in various cryptographic applications before, see for example Lenstra [367], Acar and Shumow [3], Knežević, Vercauteren and Verbauwhede [336], Hamburg [255] and Bos, Costello, Hisil and Lauter [92, 93].

Turning to our running example when using p_{256}, this multiplication can be omitted since $-p_{256}^{-1} \bmod 2^x = 1$ for all positive integers $x \leq 96$. Let us present an approach where one uses a Montgomery-radix of 2^{64} for a 32-bit platform when computing $c = A \cdot B = \sum_{i=0}^{3} a_i 2^{64i} \cdot B \bmod p_{256}$. After computing and accumulating the product of a_i with B as $c = c + a_i \cdot B$ the Montgomery reduction can be simplified as

$$
\begin{aligned}
c &= (c + p_{256} \cdot (c_1 \cdot 2^{32} + c_0))/2^{64} \\
&= \left(c + \left(c_0 \cdot 2^{256} - c_0 \cdot 2^{224} + c_0 \cdot 2^{192} + c_0 \cdot 2^{96} - c_0\right) + \right. \\
&\quad \left. \left(c_1 \cdot 2^{288} - c_1 \cdot 2^{256} + c_1 \cdot 2^{224} + c_1 \cdot 2^{128} - c_1 \cdot 2^{32}\right)\right)/2^{64} \\
&= \left(\sum_{i=2}^{9} c_i \cdot 2^{32(i-2)}\right) - \left(c_1 \cdot 2^{192} + c_0 \cdot 2^{160}\right) + \\
&\quad \left(c_1 \cdot 2^{224} + c_0 \cdot 2^{192} + c_1 \cdot 2^{160} + c_0 \cdot 2^{128} + c_1 \cdot 2^{64} + c_0 \cdot 2^{32}\right)
\end{aligned}
$$

when using the special shape of p_{256}. This process reduces the size of the result by 64 bits and needs to be performed four times in total to compute a full Montgomery reduction. This results in the following interleaved Montgomery multiplication routine for p_{256} on a 32-bit platform.

$c \leftarrow 0$
for $i = 0$ to 3 **do**
$\quad c \leftarrow c + a_i \cdot B$
$\quad t_1 \leftarrow c_1 \cdot 2^{224} + c_0 \cdot 2^{192} + c_1 \cdot 2^{160} + c_0 \cdot 2^{128} + c_1 \cdot 2^{64} + c_0 \cdot 2^{32}$
$\quad t_2 \leftarrow c_1 \cdot 2^{192} + c_0 \cdot 2^{160}$
$\quad c \leftarrow \left\lfloor \frac{c}{2^{64}} \right\rfloor + t_1 - t_2$

Hence, the Montgomery reduction modulo p_{256} takes only four 256-bit additions and four 256-bit subtractions in total. When reduction between zero and p_{256} is required, a conditional subtraction needs to be computed, which can be done by using one additional 256-bit addition and one additional 256-bit subtraction. This approach is more efficient compared with the direct approach described previously, and allows for a significantly simpler implementation.

8.3.2 Arithmetic for Curve25519

Since the standardisation of the NIST curves, together with the prime used to define the finite field, a significant amount of progress has been made in the field of elliptic curve cryptography with respect to the performance and security properties. This led Bernstein to introduce an alternative elliptic curve for usage in public-key cryptography [49]. This curve is denoted Curve25519, referring to the shape of the prime used to define the finite field: $2^{255} - 19$. From a modular arithmetic point of view this approach deviates from the generalised Mersenne approach used by NIST. Bernstein proposes to use a Crandall number [145] of the form $2^x - c$, where c is small compared with the word size used instead.

Bernstein proposes to represent integers modulo $2^{255} - 19$ as elements of the ring of polynomials $\sum_{i=0}^{9} u_i x^i$, where u_i is an integer multiple of $2^{\lceil 25.5i \rceil}$ such that $a_i / 2^{\lceil 25.5i \rceil} \in \{-2^{25}, -2^{25} + 1, \ldots, -1, 0, 1, \ldots, 2^{25} - 1, 2^{25}\}$ are a reduced-degree reduced-coefficient polynomial and represent elements of $\mathbb{Z}/(2^{255} - 19)\mathbb{Z}$ where each polynomial represents its value at one. Hence, the idea is to represent a 255-bit integer using ten 26-bit pieces. This approach is motivated by using the fast floating-point operations available on most modern processors.

Although this floating-point approach is of interest by itself, it is not the preferred approach used by many of the implementations in cryptographic libraries. In practice, either a representation that allows a non-unique representation of the elements to avoid carry-propagation is used or a redundant representation is used. Both approaches are outlined below.

One common approach to implement efficient multi-precision arithmetic is to use a non-unique representation where the coefficients have sufficient space to grow without the need for carry propagation. This allows efficient accumulation of results during the multiplication and reduction step. For example, in the implementation approach of EdDSA [50], which targets 64-bit Intel and AMD CPUs, a radix-2^{64} can cause a bottleneck when dealing with carries. Therefore an element $x \in \mathbb{Z}/(2^{255} - 19)\mathbb{Z}$ is represented as $x = \sum_{i=0}^{4} x_i 2^{51i}$. A unique representation would require that each $0 \le x_i < 2^{51}$, however, in practice the coefficients x_i are stored in 64-bit computer words that avoid carry propagation. The idea is that after multiplication $c = a \cdot b$ for $0 \leq a, b < 2^{255} - 19$ the integer $c = \sum_{i=0}^{9} c_i 2^{51i}$ can be efficiently reduced by computing

$$(x_i + 19x_{i+5})\, 2^{51i}, \quad \text{for } 0 \leq i \leq 4.$$

Moreover, when the coefficients x_i are $0 \leq x_i < 2^{51+\delta}$ for a positive integer $\delta > 0$ then $(x_i + 19x_{i+5})\, 2^{51i} < 2^{56+\delta}$. In practice, the multiplication accepts inputs with each limb having up to 54 bits (i.e., $\delta = 3$). The disadvantage of this approach is that comparisons in this non-unique representation become more cumbersome and, more importantly, more registers are needed to represent an element.

When the number of available registers is limited it might be more efficient to work with a redundant representation and compute with elements from $\mathbb{Z}/(2(2^{255} - 19))\mathbb{Z} = \mathbb{Z}/(2^{256} - 38)\mathbb{Z}$. For example, Düll *et al.* [170] outline the most efficient approach for the embedded platforms AVR ATmega (8-bit), MSP430X (16-bit), and the ARM Cortex M0 (32-bit). This approach uses the 'straightforward' approach, as also considered by Bos [85]. Given two integers a, b such that $0 \leq a, b < 2^{256} - 38$, first compute the multiplication $c = a \cdot b$. An initial reduction step can be computed with

$$c = c_1 2^{256} + c_0 \equiv c_1 38 + c_0 \bmod (2^{256} - 38)$$

by using a single 6×256 bits multiplication where $0 \leq c_0, c_1 < 2^{256}$. The resulting integer $d = c_1 38 + c_0$ can be further reduced similarly $d = d_1 2^{256} + d_0 \equiv d_1 38 + d_0 \bmod (2^{256} - 38)$ by using a 6×6 bits multiplication. The resulting $d_1 38 + d_0 < 38^2 + 2^{256}$ and a final conditional subtraction is needed to reduce the input to the range $[0, \ldots, 2^{256} - 38 - 1]$.

8.4 Special Arithmetic

Apart from RSA and ECC there are many more situations in cryptology where efficient modular multiplication is needed. As in the case of RSA and ECC, most or all cryptosystems allow specialising parameters in order to speed up their arithmetic. Instead of enumerating all improvements of modular multiplication in special situations we focus on two examples, namely sloppy reduction [91], which speeds up modular reduction for certain moduli at the expense of correctness, and a method for arithmetic operations modulo Mersenne numbers [90] whose efficiency depends on the available arithmetic instructions.

8.4.1 Sloppy Reduction

As already alluded to at the end of the previous section, modular reduction is quite easy if the modulus is an integer of the form $N = 2^n - r$ (Crandall number), where r is a small positive integer, say, $r^2 + r \leq 2^n$. In this case it is advantageous to extend the range of operands and results from $[0, \ldots, N - 1]$ to $[0, \ldots, R - 1]$ with $R = 2^n$. The modular reduction of a product $c = a \cdot b$, $0 \leq a, b < R$ can be computed by a repeated application of the following simple reduction step $\mathcal{R}$. For an integer d with $0 \leq d < R^2$, and $d = d_0 + d_1 R$ with $0 \leq d_0, d_1 < R$ define $\mathcal{R}(d) = d_0 + d_1 r$. Equivalently, one has $\mathcal{R}(d) = d - d_1 N$, which implies $\mathcal{R}(d) \equiv d \pmod{N}$. Moreover, $\mathcal{R}(d) < d$ holds if and only if $d \geq R$. Thus, by applying $\mathcal{R}$ sufficiently often to the product c it will be reduced to an integer below R and a final conditional subtraction reduces it to the range $[0, \ldots, N - 1]$ if so desired. It is easy to show that, for $r^2 \leq R$, three reduction steps are always sufficient; for $r = 1$ two steps suffice.

Because the time spent in a reduction step is not negligible, the idea of sloppy reduction is to skip the third reduction step and output $\mathcal{R}(\mathcal{R}(d)) \bmod R$ as the modular reduction, leading to the following reduction routine.

> **for** $i = 1$ to 2 **do**
> $\quad d_0 \leftarrow d \bmod R$
> $\quad d_1 \leftarrow \lfloor \frac{d}{R} \rfloor$
> $\quad d \leftarrow d_0 + d_1 r$
> $d \leftarrow d \bmod R$

This can produce a wrong result, so sloppy reduction is useful only if the probablility of an incorrect result is sufficiently low. In the case that the result can be verified by other means, the probablility must be low enough such that the cost of verification and possible recomputation does not exceed the saving from sloppy reduction. For example, in cryptanalysis it is often the case that the results of only a small fraction of many computations are used and that these results can be checked quickly. If no verification of the result is possible, sloppy reduction may still be used, e.g., if the probablility of an incorrect result is much smaller than the probablility of errors from other sources (algorithmic or hardware).

Analysis of Incorrectness Let d be an integer with $0 \leq d < R^2$. In the following, equivalent conditions for $\mathcal{R}(\mathcal{R}(d)) \geq R$, i.e., failure of sloppy reduction for d, will be derived. Writing $d = x + yN = x' + (y - z)R$ with $0 \leq x < N$ and $0 \leq x' < R$ one gets $x' = x - yr + zR$ and thus $0 \leq z \leq r + 1$ (as $y \leq R + r$, a consequence of $r^2 + r \leq R$). Therefore it follows from $\mathcal{R}(d) = x + zN = (x - zr) + zR$

that

$$R(R(d)) = \begin{cases} x & x \geq zr \\ x + N & x < zr \end{cases}$$

holds. Thus sloppy reduction fails if and only if $r \leq x < zr$ holds. In order to express this statement on a more managable quantity than z, write $x = u + vr$ with $0 \leq u < r$; from the above one also has $1 \leq v \leq z - 1 \leq r$. Because $x' = x - yr + zR < R$ implies $yr > x + zR - R \geq x + vR$, one obtains

$$d = x + yN > \frac{xr + xN + vRN}{r} = \frac{R(x + vN)}{r} = \frac{R(u + vR)}{r}.$$

Conversely, this inequality implies that $x < zr$ holds. Because $\frac{R(u+vR)}{r} \geq R^2$ for $v \geq r$, the bound on v can be improved to $1 \leq v < r$ (this also implies $R(R(d)) < N + r^2 < 2N$ so that a third reduction step will always result in a correct modular reduction).

Summarising, one obtains the following

Fact Let d be an integer with $0 \leq d < R^2$. Then $R(R(d)) \geq R$ if and only if there exist $0 \leq u < r$ and $1 \leq v < r$ such that $d \equiv u + vr \bmod N$ and $d > \frac{R(u+vR)}{r}$ hold.

From these conditions one easily obtains an estimate of the number of $0 \leq d < R^2$ with $R(R(d)) \geq R$. Indeed, it is

$$\sum_{u=0}^{r-1} \sum_{v=1}^{r-1} \# \left\{ d \mid d \equiv u + vr \bmod N \quad \text{and} \quad \frac{R(u + vR)}{r} < d < R^2 \right\},$$

which can be approximated by

$$\sum_{u=0}^{r-1} \sum_{v=1}^{r-1} \frac{rR^2 - R(u + vR)}{Nr} = \sum_{v=1}^{r-1} \frac{(r - v)R^2}{N} - \sum_{u=0}^{r-1} \frac{(r - 1)uR}{Nr} = \frac{R(rR - r + 1)(r - 1)}{2N}$$

with an error of at most $r(r - 1)$. Thus, if sloppy reduction is applied to a randomly chosen number in the interval $[0, R^2 - 1]$, the probability of an incorrect result is about $\frac{r(r-1)}{2N}$.

However, sloppy reduction is usually used after a multiplication of two uniformly distributed factors in the interval $[0, R - 1]$, possibly followed by an addition or a subtraction of a uniformly distributed integer in the same interval, so that the input of sloppy reduction is not uniformly distributed in $[0, R^2 - 1]$. In these cases the analysis is more difficult; the analysis given below will rely on heuristic assumptions.

Consider first the case $c = a \cdot b + a'$ with $0 \leq a, b < R$ and $|a'| < R$. For a rough approximation of the failure probability one discards a' and assumes that the condition $c \equiv u + vr \bmod N$ is met with probability $\frac{1}{N}$ for given u and

v. The number of pairs (a, b) satisfying $a \cdot b > \frac{R(u+vR)}{r}$ can be approximated by the integral

$$\int_{\frac{(u+vR)}{r}}^{R} \left(R - \frac{R(u+vR)}{ar} \right) da = R \left(R - \frac{(u+vR)}{r} \right) - \frac{R(u+vR)}{r} \log \frac{rR}{u+vR}. \quad (8.2)$$

By approximating $u + vR$ with vR and summing over u and v one gets the following estimate for the probability of an incorrect result

$$\frac{1}{NR^2} \sum_{u=0}^{r-1} \sum_{v=1}^{r-1} R \left(R - \frac{vR}{r} \right) - \frac{vR^2}{r} \log \frac{r}{v} = \frac{1}{N} \left(\frac{r(r-1)}{2} - \sum_{v=1}^{r-1} v \log \frac{r}{v} \right).$$

The case $c = a \cdot b$ becomes more involved because the number and properties of solutions of $ab \equiv u + vr \bmod N$ depend on the greatest common divisor of $u + vr$ and N. For example, if $u + vr$ and N are even, at least one of a and b must be even too.

Given $0 \le u < r$ and $1 \le v < r$ let $w = \gcd(u + vr, N)$ and consider the solutions (a, b) of $ab \equiv u + vr \bmod N$. These solutions satisfy $\gcd(a, N) = \gcd(a, w)$. For a fixed divisor $g \mid w$ the number of residue classes of a modulo N satisfying $\gcd(a, N) = \gcd(a, w) = g$ is $\varphi \left(\frac{N}{g} \right)$ and for each such residue class b is determined uniquely modulo $\frac{N}{g}$. Thus the number of residue classes of pairs (a, b) satisfying $ab \equiv u + vr \bmod N$ is $\sum_{g \mid w} \varphi \left(\frac{N}{g} \right) g$.

Under the heuristic assumption that this expression divided by N^2 approximates the probability that a pair (a, b) with $a \cdot b > \frac{R(u+vR)}{r}$ also satisfies $ab \equiv u + vr \bmod N$, the analysis for the case $c = a \cdot b + a'$ can be redone by adjusting Eq. (8.2) with this probability. This leads to a long formula, which is not displayed here but which is discussed in a few examples.

- Assume that the smallest prime factor of N is bigger than r^2. This applies for example to $r = 3$, $N = 2^n - 3$ with $n \not\equiv 3 \bmod 4$. Then $w = 1$ for all u, v as above so that the probability of failure is approximated by $\frac{\varphi(N)}{N^2} \left(\frac{r(r-1)}{2} - \sum_{v=1}^{r-1} v \log \frac{r}{v} \right)$. In the case $r = 3$ this simplifies to $\frac{\varphi(N)(3 - \log \frac{27}{4})}{N^2}$, which is less than $\frac{1.1}{N}$. Notice that this is much lower than the probability $\frac{3}{N}$ for inputs uniformly distributed in $[0, R^2 - 1]$.
- Consider the case $r = 38$, $n = 256$ so that $N = 2p$ where p is the prime used in Curve25519 (cf. final part of Section 8.3.2). Then w can only take the values 1 and 2 with the former occuring if u is odd and the latter if u is even. Thus the probability that $ab \equiv u + vr \bmod N$ is satisfied becomes $\frac{\varphi(N)}{N^2} = \frac{p-1}{N^2} \approx \frac{1}{2N}$ if u is odd and $\frac{\varphi(N) + 2\varphi(\frac{N}{2})}{N^2} = \frac{3(p-1)}{N^2} \approx \frac{3}{2N}$ if u is even. Because $u + vR$ can be approximated with vR and the sum over u contains as many even as odd values of u, one obtains the same result as in the analysis

of the case $c = a \cdot b + a'$, namely

$$\frac{1}{N}\left(703 - \sum_{v=1}^{37} v \log \frac{38}{v}\right) \approx \frac{343}{N}$$

as an approxmiation of the probability of an incorrect result. Again this is lower than in the case of inputs uniformly distributed in $[0, R^2 - 1]$.

8.4.2 Special Arithmetic for Mersenne Numbers

In the case of a Mersenne number $N = 2^n - 1$, modular reduction is fairly simple because two reduction steps $\mathcal{R}$ are sufficient and, moreover, the second reduction step consists of adding either 0 or 1. However, if n is not a multiple of the word size, this involves many shifts and in some cases it may be advantageous to interweave these shifts with operations of the preceding multiplication. For example, this is done in [90] and will be outlined in the following subsections.

Let w denote the word size of the underlying architecture and assume that w-bit additions and subtractions, preferably with carry propagation, are available as well as a $w/2 \times w/2$ to w-bit multiplication. Furthermore, assume that the number of words of N is in a range where one can benefit from using the Karatsuba–Ofman technique (or similar ones). Since branches are usually expensive, especially in a SIMD-setting (as in [90]), it is advantageous to adapt the word size in the algorithm so that carries in intermediate results of the Karatsuba–Ofman technique can be avoided. More precisely, let $s < w/2$ be the adapted word size, let $S = 2^s$ be the radix and let m be the number of s-bit words used to represent integers modulo N. Thus up to $w/2 - s$ levels of the Karatsuba–Ofman technique can be applied. If an (unsigned) radix-2^s representation is used, the intermediate coefficients $d_k = \sum_{i,j=0,i+j=k}^{m-1} a_i b_j$ in the product $\left(\sum_{i=0}^{m-1} a_i S^i\right) \cdot \left(\sum_{j=0}^{m-1} b_j S^j\right)$ are bounded by mS^2. This bound can be halved by using a signed radix-2^s representation, i.e., writing $A = \sum_{i=0}^{m-1} a_i S^i$ with $-2^{s-1} \le a_i < 2^{s-1}$ which allows to uniquely represent any integer in the interval $[-2^{ms-1}, 2^{ms-1} - 1]$. By using this representation the intermediate coefficients d_k fit in $\lceil \log_2 m \rceil + 2s - 1$ bits and it is assumed that s and m are chosen such that $\log_2 m + 2s - 1 \le w$ holds. For such a signed radix-2^s representation the arithmetic operations, i.e., modular addition, subtraction and multiplication as well as conversions between signed radix-2^s and (unsigned) radix-2^w, are described in the following subsections. For $* \in \{+, -, \cdot\}$ the operands of $C \equiv A * B \bmod N$ are written as $A = \sum_{i=0}^{m-1} a_i S^i$, $B = \sum_{i=0}^{m-1} b_i S^i$ and $C = \sum_{i=0}^{m-1} c_i S^i$.

Modular Addition and Subtraction

These operations are done coefficient wise with a subsequent transformation of
the result modulo N into a signed radix-2^s representation. The transformation
is facilitated by adding the constant $C_0 = 2^{s-1} \sum_{i=0}^{m-1} S^i$ to A and subtracting it
from the result. The steps are as follows.

(1) Compute $a_i' = a_i + 2^{s-1}$ for $0 \le i < m$ (adding C_0).

(2) Set $c_i = a_i' + b_i$ in the case of an addition or $c_i = a_i' - b_i$ in the case of a
    subtraction for $0 \le i < m$.

(3) Set $c = 0$. For $i = 0$ to $m - 1$ first replace c by $c + c_i$, then set c_i to $c \bmod S$
    and finally replace c by $\lfloor \frac{c}{S} \rfloor$.

(4) Add $c \cdot 2^\alpha$ to c_β where α and β are integers such that $sm - n = \alpha + s\beta$ with
    $0 \le \alpha < s$. If $c_\beta \ge S$, propagate the carry as in the previous step until $c = 0$.
    If handling c_{m-1} produces a carry, repeat this step.

(5) Set $c_i = c_i - 2^{s-1}$ for $0 \le i < m$ (subtracting C_0).

In the first step the constant C_0 is added to A, then A and B are added or subtracted in the second step. The third step normalises the coefficients of C to
the range $[0, S - 1]$ while producing a carry c, which corresponds to $c \cdot S^m$ and
which is taken care of in the fourth step thus producing a radix-2^s representation of $C + C_0$. In the fifth step C_0 is subtracted so that C is in a signed radix-2^s
representation.

Modular Multiplication

This is done in two phases. In the first phase, the intermediate coefficients
$d_k = \sum_{i,j=0,i+j=k}^{m-1} a_i b_j$, $0 \le k < 2m - 1$ are computed so that $\sum_{k=0}^{2m-2} d_k S^k$ modulo
N is the result C. The second phase handles the reduction of this sum modulo
N and its transformation into a signed radix-2^s representation. In the first phase
the Karatsuba–Ofman technique (for polynomials) can be used. The reduction
of $\sum_{k=0}^{2m-2} d_k S^k$ modulo N proceeds in two sub-steps. Firstly, the d_k are made
non-negative by adding a constant which is 0 modulo N. Next, the d_k with
$k \ge m$ are split into two $w/2$-bit words that are added with an appropriate shift
to the appropriate $d_{k'}$ with $k' < m$. The resulting sum $\sum_{k=0}^{m-1} d_k S^k$ satisfies $0 \le$
$d_k < 2^w$. The transformation into a signed radix-2^s representation is essentially
the same as in the addition method above, i.e., first the constant C_0 is added,
next the carries are propagated, and finally the constant C_0 is subtracted. This
leads to the following steps.

(1) Compute for $0 \le k < 2m - 1$ the intermediate coefficients $d_k = $
    $\sum_{i,j=0,i+j=k}^{m-1} a_i b_j$ of the product $A \cdot B$ by using at most $w/2 - s$ levels of the
    Karatsuba–Ofman technique.

(2) Set $d_0 = d_0 + 2^{w-1}$, $d_i = d_i + 2^{w-1} - 2^{w-1-s}$ for $1 \leq i < 2m-1$ and $d_\beta = d_\beta - 2^\alpha$ where α and β are integers such that $s(2m - 2) + w - 1 - n = \alpha + s\beta$ with $0 \leq \alpha < s$.

(3) Set $d_i = d_i + 2^{s-1}$ for $0 \leq i < m$ (adding C_0).

(4) For $m \leq i < 2m-1$ do the following operations. First write $d_i = d_i' + d_i'' \cdot 2^{w/2}$, then add $d_i' \cdot 2^{\alpha'}$ to $d_{\beta'}$ and $d_i'' \cdot 2^{\alpha''}$ to $d_{\beta''}$ where $\alpha', \beta', \gamma', \alpha'', \beta'', \gamma'' \in \mathbb{Z}$ such that $is = \alpha' + s\beta' + n\gamma'$ and $is + w/2 = \alpha'' + s\beta'' + n\gamma''$ with $0 \leq \alpha', \alpha'' < s$ and $0 \leq \alpha' + s\beta', \alpha'' + s\beta'' < n$.

(5) Perform the third and fourth step of the addition method to d_i, $0 \leq i < m$.

(6) Set $c_i = d_i - 2^{s-1}$ for $0 \leq i < m$ (subtracting C_0).

Notice that the m additions of the third step can be combined with corresponding additions in the second step.

Conversions

Usually, the input of a long calculation is available in a radix-2^w representation and this is also the desired format for the result. Therefore, a few conversions between signed radix-2^s and radix-2^w representations are needed at the beginning and at the end. Although in most cases a simple and slow conversion is sufficient, it is possible that during the course of the calculation many conversions are needed so that it must be more efficient. With some of the tricks used above, reasonably efficient conversion routines can be built which are sketched below. These can be further optimised by combining some of their steps with steps of the preceding or subsequent modular operation, e.g., the fourth step of the following conversion is the inverse of the first step of modular addition. As above, let $A = \sum_{i=0}^{m-1} a_i S^i$ be the signed radix-2^s representation and $\tilde{A} = \sum_{i=0}^{\tilde{m}-1} \tilde{a}_i 2^{wi}$ the radix-2^w representation of the same residue class modulo N with $0 \leq \tilde{A} < N$ and an appropriate $\tilde{m}$.

The conversion from radix-2^w to signed radix-2^s proceeds as follows.

(1) Add the constant $C_0 = 2^{s-1} \sum_{i=0}^{m-1} S^i$ to $\tilde{A}$ by using the standard addition technique with carry propagation.

(2) Set $a_i' = 0$ for $0 \leq i < m$. For $0 \leq i \leq \tilde{m}$ split $\tilde{a}_i$ appropriately and add the appropriately shifted parts to the appropriate a_i'.

(3) Perform the third and fourth step of the addition method to a_i', $0 \leq i < m$.

(4) Set $a_i = a_i' - 2^{s-1}$ for $0 \leq i < m$ (subtracting C_0).

With the pre-computed constant C_1, where $C_1 \equiv -C_0 \bmod N$ and $0 \leq C_1 < N$, the conversion from signed radix-2^s to radix-2^w proceeds as follows.

(1) Set $a_i = a_i + 2^{s-1}$ for $0 \leq i < m$ (adding C_0).

(2) Initialise the $\tilde{a}_j$ with the pre-computed constant C_1.

(3) For $0 \le i \le m$ split $a_i \cdot 2^{\alpha_i}$ into (at most) two w-bit words where $0 \le \alpha_i < w$ and $is - n \left\lceil \frac{is}{n} \right\rceil \equiv \alpha_i \bmod w$. Then add these words to the appropriate $\tilde{a}_j$ (in order to minimise carry propagation, perform these additions in increasing order of j).

(4) Reduce $\sum_{i=0}^{\tilde{m}-1} \tilde{a}_i 2^{wi}$ modulo N; usually this requires only a few shifts and additions.

9
Arithmetic Software Libraries

Victor Shoup

9.1 Introduction

This chapter discusses a library for doing number theory (NTL) as well as its relation to a few other libraries.

NTL is a high-performance, portable C++ library providing data structures and algorithms for manipulating signed, arbitrary-length integers, and for vectors, matrices and polynomials over the integers and over finite fields. It is maintained and mostly written by the author, although a number of contributions have been made by others.

9.1.1 NTL, LIP and GMP

Work on NTL started around 1990, when the author wanted to implement some new algorithms for factoring polynomials over finite fields and related problems. It seemed that none of the publicly available software was adequate for this task, mainly because the code for polynomial arithmetic offered by this software was too slow. More precisely, many papers by the author and others working in this field would give big-O estimates for algorithms under the assumption of fast (i.e., quasi-linear) polynomial multiplication algorithms; however, the publicly available software at the time did not implement any of these fast multiplication algorithms. Thus, to turn these theoretical results into practically useful implementations, the author began working on software for fast polynomial arithmetic over finite fields. This software eventually grew into a full-fledged library, and eventually led to the public release of NTL version 1.0 in 1997. NTL has evolved quite a bit since then, with a lot of new functionality and algorithmic improvements, as well as features such as thread-safety, and exploitation of multi-core and SIMD hardware.

The starting point for NTL was Arjen Lenstra's LIP (long-integer package) for long (i.e., multi-precision) integer arithmetic, which was written in C. LIP was originally written by Arjen Lenstra in the late 1980s. Later, in 1995, a version of LIP called FreeLIP was released and maintained by Paul Leyland. It soon became clear that using C++ instead of C would be much more productive and less prone to errors, mainly because of C++'s constructors and destructors, which allowed memory management to be automated. Using C++ had other benefits as well, such as function and operator overloading, which makes for more readable code. So an essential part of NTL's original design was a lightweight C++ wrapper around LIP to manage the storage of long integers. For example, suppose one wanted to write a program that input integers a, b, c, d and output $ab + cd$. This could be written in LIP as follows.

```
verylong a=0, b=0, c=0, d=0, t1=0, t2=0;
zread(&a); zread(&b); zread(&c); zread(&d);
zmul(&t1, a, b); zmul(&t2, c, d);
zadd(&t1, t1, t2); zwriteln(t1);
zfree(&a); zfree(&b); zfree(&c); zfree(&d);
zfree(&t1); zfree(&t2);
```

However, in NTL, this could be written as follows.

```
ZZ a, b, c, d;
cin >> a >> b >> c >> d;
cout << (a*b + c*d) << "\n";
```

Among other things, LIP was designed to be used in widely distributed integer factorisation projects, and so portability, without sacrificing too much performance, was a major design feature of LIP. Indeed, quoting from the LIP documentation itself: 'This very long int package is supposed to be easy to use, portable, and not too slow'. This design principle was adopted for NTL as well, for a number of reasons, not the least of which was that the author's computing environment kept changing whenever he changed jobs (which was unfortunately happening quite frequently in those days), and so portability was a real issue. Achieving portability eventually became easier as standards, such as IEEE floating-point [223], got widely adopted, as the definition of and implementations of the C++ language became standardised and more stable, and as important language extensions, such as double-word integer arithmetic, became more widely available.

When work started on NTL, besides LIP, there were not that many good, portable long-integer packages around. Shortly thereafter, Torbjörn Granlund released (in 1991) the first version of GMP (the GNU Multiple Precision arithmetic library) [247]. GMP itself was largely compatible with the earlier

Berkeley MP library, but it was quite a bit more efficient. From the documentation of one of the first public releases of GMP (v1.1, 29 Sept. 1991): 'The speed of GNU MP is about 5 to 100 times that of Berkeley MP for small operands. The speed-up increases with the operand sizes for certain operations, where GNU MP has asymptotically faster algorithms'. At that time, it was not clear that GMP was going to be developed and maintained as well as it ultimately proved to be, so NTL continued to use LIP for its long-integer arithmetic. However, as time went on, GMP became much faster relative to LIP, and was very well supported across many platforms, and so NTL was eventually (in 2000) restructured to use GMP instead of LIP. For the most part, this restructuring was invisible to NTL users, as these implementation details are all hidden behind a layer of abstraction. Moreover, the current version of NTL can still be built without GMP, in which case it reverts to logic that is derived from LIP (although by now, even that logic has been almost entirely restructured). LIP itself has apparently not been supported for many years.

9.1.2 A Quick Tour of NTL

As already mentioned above, the initial purpose for working on NTL was to implement some new algorithms for factoring polynomials over finite fields. The author's early work in this area was reported in [540]. Besides the class ZZ mentioned above, which represents multi-precision integers, NTL also provides a class ZZ_pX, which represents the ring of univariate polynomials $\mathbb{Z}_p[X]$ over the quotient ring $\mathbb{Z}_p = \mathbb{Z}/p\mathbb{Z}$, where p is a multi-precision integer (not necessarily prime). NTL also provides a corresponding class zz_pX, optimised for single-precision p, and a class GF2X, optimised for the important special case of $p = 2$. In addition, NTL provides classes for the ring of univariate polynomials $E[X]$ over the quotient ring $E = \mathbb{Z}_p[T]/(m(T))$, where $m(T) \in \mathbb{Z}_p[T]$. These classes are called ZZ_pEX, zz_pEX and GF2EX, respectively, depending on whether p is multi-precision, single-precision or equal to 2. For each of these polynomial classes, NTL provides algorithms for all of the basic arithmetic operations and when the coefficient domain is a field, for other operations, such as GCD, factoring, minimal polynomial calculation and generating irreducible polynomials.

In addition to polynomials over finite rings and fields, NTL provides the class ZZX, which represents the ring $\mathbb{Z}[X]$ of univariate polynomials over the integers, where coefficients are of arbitrary size. Naturally, NTL provides algorithms for all of the basic arithmetic operations over $\mathbb{Z}[X]$, as well as GCD and factoring algorithms.

Unfortunately, at the present time, NTL does not provide any support for multivariate or even bivariate polynomials.

NTL provides support for matrices over finite fields and rings, as well as over the integers. For integer matrices, NTL provides implementations of various lattice-basis reduction algorithms, including (fast, heuristic) algorithms that are based on floating-point arithmetic. In support of these floating-point lattice-basis reduction routines, and for other applications as well, NTL also provides specialised floating-point classes.

NTL is licenced under the GNU LGPL licence, and is available at `www.shoup.net/ntl`.

9.1.3 Outline

In Section 9.2, we describe some of the technical details of how LIP and NTL originally implemented long-integer arithmetic, and compare the performance of these techniques with GMP's performance. In Section 9.3, we describe NTL's implementation techniques for the Number-Theoretic Transforms (NTT), i.e., the Fast Fourier Transform (FFT) for small primes. As we shall see, the NTT plays a central role in a number of NTL's other arithmetic modules. In Section 9.4, we discuss how NTL implements arithmetic in $\mathbb{Z}_p[X]$ for multi-precision p. In Section 9.5, we discuss how NTL implements arithmetic in $\mathbb{Z}_p[X]$ for single-precision p. In Section 9.6, we discuss how NTL implements matrix arithmetic over $\mathbb{Z}_p$. In Section 9.7, we discuss how NTL implements polynomial and matrix arithmetic over other finite rings. Finally, in Section 9.8, we discuss how NTL implements polynomial and matrix arithmetic over the integers, including a brief discussion of NTL's lattice-basis reduction algorithms. In Section 9.9 we conclude with a brief look into the future of NTL.

9.2 Long-Integer Arithmetic

In this section, we describe some of the technical details of how LIP and NTL originally implemented long-integer arithmetic. These details are mostly of only historical interest, as NTL now mainly relies on GMP for its long-integer arithmetic.

As mentioned above, LIP was designed to be highly portable, but not too slow, and NTL adopted this philosophy as well. One way in which LIP initially achieved a good balance between portability and performance was to avoid the use of double-word integer arithmetic, and instead, to use double-precision floating-point to achieve similar functionality.

By 'double-word integer arithmetic', we mean some mechanism that allows one to multiply two single-word integers to get the double-word product, as well as related operations such as adding a single- or double-word integer to a double-word integer. At the time NTL was initially developed, some hardware simply did not have any implementation at all of double-word integer arithmetic, and even if it did, its implementation could often be fairly inefficient compared to its floating-point implementation; moreover, programming languages and compilers often did not give the programmer access to this hardware, and so exploiting such hardware would require writing and maintaining a lot of assembly code (which, among other things, is exactly what GMP does).

It might seem surprising to use floating-point arithmetic, which is inherently subject to rounding errors, to implement long-integer arithmetic, which does not tolerate any errors. Nevertheless, under reasonable assumptions, this can indeed be done. The assumptions we need are simply bounds on the relative error introduced when performing the basic arithmetic operations of addition, subtraction, multiplication and division. These assumptions are certainly implied by the IEEE floating-point standard, which was already being quickly adopted at the time of NTL's initial development.

On 32-bit machines (which were ubiquitous in the early days of NTL development), LIP represents long-integers in base R, where $R = 2^{30}$. Thus, a 'base-R digit' is an integer in the interval $[0, R)$. A basic operation that is used in several algorithms, including long-integer multiplication, is *addmulp*, which takes as input base-R digits a, b, d and t, computes the two base-R digits representing $b \cdot d + a + t$, and stores the high-order digit back in t and the low-order digit back in a.

The original implementation of *addmulp* in LIP was essentially that presented in Algorithm 9.1.[1] We are assuming that the types `long` and `unsigned long` represent 32-bit integers. To see why this works, let lo and hi denote the low-order and high-order digit, respectively, of $b \cdot d + a + t$. Observe that the value of $t3$ equals lo, since unsigned arithmetic is guaranteed to be done correctly modulo 2^{32}. Also observe that

$$hi = (b \cdot d + a + t - lo)/R.$$

Now, the value of $d1$ is

$$bd(1 + \theta_1),$$

---

[1] Actually, the code in Algorithm 9.1 is somewhat more robust and efficient than the original LIP implementation, but it is still in the same spirit.

Algorithm 9.1 A floating-point implementation of *addmulp*.

```
typedef long L;
typedef unsigned long U;
typedef double D;
const int NBITS = 30;
const U R = U(1) << NBITS;

inline void addmulp(U& a, U b, U d, U& t)
{
    U t1 = b * d;
    U t2 = a + t;
    U t3 = (t1+t2) & (R-U(1));

    double d1 = D(L(b))* D(L(d));
    double d2 = d1 + D( L(t2) - L(t3) + L(R/2) );
    double d3 = d2 * (1.0 / D(L(R)) );

    t = L(d3);
    a = t3;
}
```

---

where $|\theta_1| \leq \epsilon := 2^{-53}$ on a machine that correctly implements the IEEE floating-point standard. The value of $d2$ is

$$(d1 + a + t - t3 + R/2)(1 + \theta_2) = bd(1 + \theta_1)(1 + \theta_2) + (a + t - lo + R/2)(1 + \theta_2),$$

where $|\theta_2| \leq \epsilon$. Finally, the value of $d3$ is $d2/R$. It follows that

$$d3 = hi + 1/2 + \theta_3,$$

where

$$|\theta_3| \leq R|\theta_1 + \theta_2 + \theta_1\theta_2| + \frac{2.5}{R}|\theta_2|.$$

Therefore, since $R = 2^{30}$ and $|\theta_1|$ and $|\theta_2|$ are both at most $\epsilon = 2^{-53}$, we see that $|\theta_3|$ is at most approximately $2R\epsilon = 2^{-22}$, and so the calculation of $L(d3)$, which drops the fractional part of $d3$, will certainly yield hi, as required.

On 64-bit machines (which eventually replaced 32-bit machines in their ubiquity), the same analysis allows us to prove that *addmulp* works correctly with $R = 2^{50}$. In that case, the value $|\theta_3|$ above is at most approximately $2R\epsilon = 1/4$, which is still small enough to ensure that $L(d3) = hi$.

Algorithm 9.2 Another floating-point implementation of *addmulp*.

```
inline void addmulp(U& a, U b, U d, U& t)
{
    U t1 = b*d + a + t;
    U t2 = D(L(b))*D(L(d))*(1.0/D(L(R)));
    L t3 = L(t1 - (t2 << NBITS)) >> NBITS;
    t = t2 + t3;
    a = t1 & (R-U(1));
}
```

Some Implementation Notes Note that all conversions between integers and floating-point values in Algorithm 9.1 are done from or to signed integers, the reason being that many machines implement such conversions much more efficiently than the corresponding conversions from or to unsigned integers. Also note that the above correctness analysis only depended on the relative error guarantees provided by the IEEE floating-point standard, and not on any 'exact rounding' requirements. Thus, the analysis is still valid, even in the presence of 'double rounding', which can occur on platforms that perform some calculations in extended double precision (such as old x86 machines that use the x87 floating-point instructions), or in the presence of 'contractions', which can occur in implementations that provide 'fused multiply and add' instructions (such as newer x86 machines).

Algorithm 9.2 presents another implementation of *addmulp*, which was introduced in the original version of NTL, and which replaced LIP's implementation of *addmulp*. This implementation works on 32-bit machines (with $R = 2^{30}$) or on 64-bit machines (with $R = 2^{50}$). It also assumes that signed integer arithmetic is two's complement, conversion from unsigned to signed works as expected (i.e., the bit pattern is unchanged), and that right shifts of signed integers are arithmetic shifts.[2]

Under our relative-error assumptions for floating-point arithmetic, we have $\lfloor bd/R \rfloor = t2 + \delta$, where $\delta \in \{0, \pm 1\}$. It follows that $hi = t2 + \delta'$, where $\delta' \in \{-1, \ldots, 3\}$. The calculation of $t3$ uses the high-order bits of $t1$ to compute the value $t3 = \delta'$, which is then added to $t2$ to compute the correct value of hi.

Eventually, NTL was updated to exploit double-word integer arithmetic if possible (and desired). This is done either by using a special type or by using

---

[2] Here, and elsewhere, we make some assumptions about signed arithmetic which technically leads to 'implementation defined' behaviour; however, the code can easily be adapted to use only unsigned arithmetic.

Algorithm 9.3 A double-word integer implementation of *addmulp*.

```
inline void addmulp(U& a, U b, U d, U& t)
{
    UU t1 = UU(b)*UU(d) + UU(a+t);
    a = U(t1) & (R-U(1));
    t = U(t1 >> NBITS);
}
```

Algorithm 9.4 Computing $a \leftarrow a + bd$ using *addmulp*.

```
inline void addmul(U* a, U* b, U d, L n)
{
    U t = 0;
    for (L i = 0; i < n; i++)
    {
        addmulp(a[i], b[i], d, t);
    }
}
```

assembly.[3] Assuming an appropriate double-word unsigned integer type UU, the routine *addmulp* can be implemented as in Algorithm 9.3.

Algorithm 9.4 illustrates how the *addmulp* routine is typically used in LIP and NTL. The routine *addmul* computes $a \leftarrow a + bd$, where a and b point to arrays representing n-digit integers, and d is a single digit.

If GMP is not available, then either the implementation based on Algorithm 9.3 (if a double-word integer type is available) or Algorithm 9.2 (otherwise) is used. The implementation based on Algorithm 9.1 is currently never used. Of course, if GMP is available, that is used instead.

Table 9.1 shows the time (in nanoseconds) needed to multiply two 1000-bit integers, based on several different implementations. Except where otherwise noted, all benchmarks in this chapter were carried out on an Intel Xeon CPU E5-2698 v3 (64-bit Haswell architecture) at 2.30GHz, using GCC version 7.3.1, and GMP version 6.1.0 (note that GMP was compiled with version 4.8.5 of GCC).

The takeaway is that for integers of this size, GMP is three to four times

---

[3] Inline assembly is only implemented on x86 machines. On many 32-bit machines, the type unsigned long long is a 64-bit unsigned integer, and on many 64-bit machines, the type __uint128_t is a 128-bit unsigned integer.

Method	Time (ns)
Algorithm 9.1	1426
Algorithm 9.2	735
Algorithm 9.3	548
GMP	176

Table 9.1 *Time (in nanoseconds) to multiply two 1000-bit integers.*

faster than NTL's native implementations, and eight times faster than LIP's original implementation technique. It should also be noted that to multiply larger integers, NTL's native implementation uses Karatsuba, while GMP employs a number of different algorithms in addition to Karatsuba (including an FFT for very large integers). Thus, GMP's performance relative to NTL gets even better for larger integers.

We also note that LIP (and NTL, when GMP is not used) also exploits floating-point arithmetic in a number of other ways, for operations such as long-integer division and GCDs. However, we shall not discuss these techniques in any detail here.

9.3 Number-Theoretic Transforms

Let p be a prime such that $p - 1$ is divisible by $N = 2^k$. We will assume here that p is a 'single precision' number, i.e., it fits in a single word. In fact, we will usually assume that the bit-length of p is a bit smaller than the width of a word (typically 2 or 4 bits less).

We know that $\mathbb{Z}_p^*$ contains an element ω of multiplicative order N. The N-point Discrete Fourier Transform (DFT) maps the coefficient vector of a polynomial $f \in \mathbb{Z}_p[X]$ to the evaluation vector

$$(f(\omega^0), f(\omega^1), \ldots, f(\omega^{N-1})).$$

The DFT over $\mathbb{Z}_p$ is sometimes called a Number-Theoretic Transform or NTT. As is well known, the inverse transform is also essentially just an NTT. Of course, using the Fast Fourier Transform (FFT) [133], we can compute the NTT and inverse NTT by using $O(N \log N)$ additions, subtractions and multiplications in $\mathbb{Z}_p$.

We shall sometimes call such a prime p an 'FFT prime'. An important application of NTTs is to implement fast polynomial multiplication over $\mathbb{Z}_p$, by using the following well-known technique. Suppose $f, g \in \mathbb{Z}_p[X]$ with $\deg(f) + \deg(g) < N$. Then to compute $h = f \cdot g$, we proceed as follows.

- Compute the evaluation vectors u and v of f and g using the NTT.
- Compute the component-wise product vector $w := u \cdot v$.
- Compute h as the inverse NTT of w.

NTTs play a crucial role in NTL, especially in their use in the multi-modular algorithms for polynomial multiplication, as we will discuss later in Sections 9.4, 9.5, and 9.8. Also, in Section 9.4.5, we shall discuss more recent applications of NTTs to a new type of encryption called fully homomorphic encryption.

The remainder of Section 9.3 will discuss the techniques for efficient implementations of NTTs that are used in NTL.

9.3.1 Single-Precision Modular Arithmetic

Evidently, in order to get a high-performance implementation of a fast NTT algorithm modulo a single-precision prime, one needs high-performance implementations of addition, subtraction and multiplication modulo such primes.

More generally, assume $n > 1$, and a and b are integers in the range $[0, n)$. We can compute $a + b \bmod n$ and $a - b \bmod n$ using the routines *AddMod* and *SubMod* in Algorithm 9.5 (see the `typedefs` in Algorithm 9.1). Correctness is assured provided $n < 2^{BPL-2}$, where *BPL* is the number of bits in the type `long`. In Algorithm 9.5, the routine *CorrectExcess* subtracts n from r if r exceeds n. Similarly, the routine *CorrectDeficit* adds n to r if r is negative. These routines are implemented by using one of two strategies, depending on a compile-time switch. The first strategy is branch free, and uses an appropriate sequence of addition, subtractions, shifts and ANDs; as presented, this code assumes signed integer arithmetic is two's complement, and that right shifts of signed integers are arithmetic shifts. The second strategy uses explicit branching. On modern architectures, which are typically highly pipelined, and which exhibit high branch penalties, the branch-free strategy can be much faster than the branching strategy. However, some modern architectures, such as the x86, provide so-called 'conditional move' instructions, and the branching strategy can be slightly faster than the branch-free strategy on such architectures. NTL installation logic will usually make a good decision on which strategy is better.

To compute $ab \bmod n$, NTL originally employed floating-point arithmetic, essentially as in Algorithm 9.6. This was very similar to the strategy employed by LIP, as well. Correctness is assured provided $n\epsilon \leq 1/8$, where $\epsilon := 2^{-53}$ on a machine that correctly implements the IEEE floating-point standard. This ensures that the computed quotient q is equal to $\lfloor ab/n \rfloor + \delta$, where $\delta \in \{0, \pm 1\}$. The code also assumes that signed integer arithmetic is two's complement,

Algorithm 9.5 Single-precision modular addition and subtraction.

```
const int BPL = 64;
#ifdef AVOID_BRANCHING
  inline long CorrectExcess(L r, L n)
  {
    return (r-n) + (((r-n) >> (BPL-1)) & n);
  }

  inline long CorrectDeficit(L r, L n)
  {
    return r + ((r >> (BPL-1)) & n);
  }
#else
  inline long CorrectExcess(L r, L n)
  {
    return r-n >= 0 ? r-n : r;
  }

  inline long CorrectDeficit(L r, L n)
  {
    return r >= 0 ? r : r+n;
  }
#endif
  inline long AddMod(L a, L b, L n)
  {
    return CorrectExcess(a + b, n);
  }

  inline long SubMod(L a, L b, L n)
  {
    return CorrectDeficit(a - b, n);
  }
```

conversion from unsigned to signed works as expected (i.e., the bit pattern is unchanged), and that right shifts of signed integers are arithmetic shifts.

In a 32-bit machine, the assumption that $n\epsilon \leq 1/8$ is not a real restriction; however, on a 64-bit machine, it essentially restricts the modulus n to be a 50-bit number at most.

Algorithm 9.6 Single-precision modular multiplication based on floating-point arithmetic.

```
inline long MulMod(L a, L b, L n)
{
    D ninv = 1/D(n);
    D bninv = D(b) * ninv;
    L q = L( D(a) * bninv );
    L r = L( U(a)*U(b) - U(q)*U(n) );
    r = CorrectDeficit(r, n);
    r = CorrectExcess(r, n);
    return r;
}
```

Typically, the modulus n will remain fixed for many modular multiplication operations (for example, in NTTs). In this situation, a good optimising compiler may generate code to compute the quantity *ninv* just once. However, an alternative interface is provided in which the value *ninv* can be explicitly pre-computed by the programmer.

Often, in addition to the modulus n, one of the multiplicands, say b, may also remain fixed for many modular multiplication operations (again, for example, in NTTs, where b is a root of unity). Although an optimising compiler may sometimes be able to generate code to compute the quantity *bninv* just once, an alternative interface is provided in which the value *bninv* can be explicitly pre-computed by the programmer. In fact, if the machine provides double-word integer arithmetic, a much faster code sequence is used, which is shown in Algorithm 9.7. The code shown here uses a double-word unsigned integer type UU. Inline assembly code may also be used.

In this code, the value *bninv* should be pre-computed as

$$\lfloor 2^{SPBITS} b/n \rfloor \cdot 2^{BPL-SPBITS}.$$

Here, *SPBITS* is a bound on the bit-length of the modulus n, i.e., $n < 2^{SPBITS}$, and it is assumed that $SPBITS \leq BPL - 2$. The value *bninv* can be computed using floating-point logic similar to that in Algorithm 9.6 (and a function is provided to the programmer to do so).

One can prove that with bninv computed in this way, the quotient q computed in Algorithm 9.7 is equal to either $\lfloor ab/n \rfloor$ or $\lfloor ab/n \rfloor - 1$.

A good optimising compiler targeting the x86 instruction set will generate code that consists of three integer multiplication instructions, two subtractions,

Algorithm 9.7 Single-precision modular multiplication with pre-conditioning.

```
inline long MulModPrecon(L a, L b, L n, U bninv)
{
    U q = U( (UU(U(a)) * UU(bninv)) >> BPL );
    L r = L( U(a)*U(b) - q*U(n) );
    return CorrectExcess(r, n);
}
```

one 'conditional move' instruction, and several register-to-register moves (which nowadays are almost completely cost free). Moreover, two of the multiplications are 'single word' multiplications (i.e., only the low-order word need be computed), and these two multiplications may run concurrently (there are no data dependencies between them).

9.3.2 A Floating-Point-Free Implementation

Although the above floating-point implementations for single-precision modular multiplication are simple and quite effective, they have the disadvantage of restricting the modulus to only 50 bits on a 64-bit machine. In this section, we describe the floating-point-free implementation that NTL currently deploys on machines that provide double-word integer arithmetic. With this technique, one can easily support a 62-bit modulus on a 64-bit machine; however, for a number of reasons, NTL usually restricts the modulus to just 60 bits. Moreover, on modern hardware, this floating-point-free implementation typically outperforms the floating-point implementations.

We assume the modulus n has w bits, so

$$2^{w-1} \le n < 2^w. \tag{9.1}$$

We also assume that $2 \le w \le BPL-2$, although we may also make the stronger assumption that $w \le BPL - 4$.

As a pre-computation step, we compute

$$X := \lfloor (2^v - 1)/n \rfloor. \tag{9.2}$$

Here, $v > w$ is a constant to be discussed below. Note that

$$2^{v-w} \le X < 2^v/n \le 2^{v-w+1}. \tag{9.3}$$

We can also write

$$2^v - 1 = Xn + Y, \quad \text{where } 0 \le Y < n. \tag{9.4}$$

Now suppose we are give an integer C satisfying

$$0 \le C < 2^w n \tag{9.5}$$

that we want to reduce mod n. We split C into high-order and low-order bits:

$$C = A2^s + B, \quad \text{where } 0 \le B < 2^s, \tag{9.6}$$

and then compute the product of A and X, splitting that into high-order and low-order bits:

$$AX = Q2^t + R, \quad \text{where } 0 \le R < 2^t. \tag{9.7}$$

Here, s and t are constants satisfying $s + t = v$, to be discussed below.

We have

$$AX \le \frac{C}{2^s} X < \frac{C}{2^s} \frac{2^v}{n} = \frac{2^t C}{n}.$$

These calculations follow from Eqs. (9.3) and (9.6). In particular, $AX/2^t < C/n$, and by Eq. (9.7), we have

$$Q = \lfloor AX/2^t \rfloor \le \lfloor C/n \rfloor.$$

We will show, with an appropriate choices of parameters, that

$$Q \ge \lfloor C/n \rfloor - 1. \tag{9.8}$$

To this end, we claim that

$$C - Qn < nC/2^v + 2^s + n. \tag{9.9}$$

To see this, observe that

$$\begin{aligned}
Qn &= \frac{AX - R}{2^t} n = \frac{AXn}{2^t} - \frac{R}{2^t} n > \frac{AXn}{2^t} - n \\
&= \frac{A}{2^t}(2^v - 1 - Y) - n \ge \frac{A}{2^t}(2^v - n) - n \\
&= A2^s - A\frac{n}{2^t} - n = C - B - A\frac{n}{2^t} - n > C - 2^s - A\frac{n}{2^t} - n \\
&\ge C - 2^s - C\frac{n}{2^v} - n.
\end{aligned}$$

These calculations follow from Eqs. (9.4), (9.6) and (9.7).

Suppose $BPL \ge w + 4$ (which is the default on 64-bit machines, where $w = 60$). In this case, we set

$$v := 2w + 2, \quad s := w - 2, \quad t := w + 4.$$

Then we have

$$C - Qn < nC/2^v + 2^s + n < n/4 + n/2 + n < 2n.$$

Algorithm 9.8 Single-precision modular multiplication without floating-point (on a 64-bit machine with $w = 60$).

```
inline long NormalizedMulMod(L a, L b, L n, U X)
{
    UU C = UU(U(a)) * UU(U(b));
    U A = U( C >> 58 );
    U Q = (UU(A)*UU(X)) >> 64;
    L r = L( U(C) - Q*U(n) );
    return CorrectExcess(r, n);
}
```

These calculations follow from Eq. (9.9) and the fact that $nC/2^v \le n2^{2w-v} = n/4$ (based on Eqs. (9.5) and (9.1)) and $2^s = 2^{w-2} \le n/2$ (based on Eq. (9.1)). This establishes Eq. (9.8). From Eq. (9.3), we also have

$$X < 2^{v-w+1} = 2^{w+3} < 2^{BPL},$$

so X fits in one machine word. From Eqs. (9.6), (9.5) and (9.1), we also have

$$A \le \frac{C}{2^s} < \frac{2^w n}{2^s} < 2^{2w-s} = 2^{w+2} < 2^{BPL},$$

so A also fits in one machine word.

Suppose $BPL \ge w + 2$ (which is the default on 32-bit machines, where $w = 30$). In this case, we set

$$v := 2w + 1, \quad s := w - 2, \quad t := w + 3.$$

Then we have

$$C - Qn < nC/2^v + 2^s + n < n/2 + n/2 + n \le 2n.$$

This again establishes Eq. (9.8). We also have

$$X < 2^{v-w+1} = 2^{w+2} \le 2^{BPL},$$

so X fits in one machine word. (This is the only place where it is essential that X is computed as in Eq. (9.2), rather than as $\lfloor 2^v/n \rfloor$). We also have

$$A \le \frac{C}{2^s} < \frac{2^w n}{2^s} < 2^{2w-s} = 2^{w+2} \le 2^{BPL},$$

so A also fits in one machine word.

Algorithm 9.8 shows how to use the above strategy to implement modular multiplication on a 64-bit machine with $w = 60$. As usual, L is a synonym

---

Algorithm 9.9 Floating-point-free modular multiplication – the general case.

```
inline long
GeneralMulMod(L a, L b, L n, U X, int shamt)
{
    return
    NormalizedMulMod(a, b << shamt, n << shamt, X)
       >> shamt;
}
```

---

for `long`, `U` is a synonym for `unsigned long`, and `UU` is a synonym for a double-word unsigned integer type. A good optimising compiler targeting the x86 instruction set will generate code that consists of three integer multiplication instructions, one 'shrd' (or 'shld') instruction (which performs the 58-bit double-word right shift), two subtractions, one 'conditional move' instruction, and several register-to-register moves. Some compilers may choose to replace the 'shrd' instruction with two single-word shifts and an addition or logical-or instruction. Note that in contrast to the logic in Algorithm 9.7, only one of the three is a single-word multiplication, and moreover, none of them can run concurrently with the others. It is also possible to adapt the logic to work with $w = 62$, but for a number of reasons, choosing $w = 60$ is advantageous.

Note that the logic in Algorithm 9.8 requires n satisfies Eq. (9.1). For certain applications, this will be true, but in general, one may have to employ the logic in Algorithm 9.9, which only assumes $n < 2^w$. The amount *shamt* is pre-computed so that $2^{w-1} \le n2^{shamt} < 2^w$.

9.3.3 Lazy Butterflies and Truncated Fast Fourier Transforms

Consider again the problem of computing NTTs, that is, an FFT modulo a single-precision prime p. NTL's current NTT implementation incorporates two important optimisations.

Lazy Butterflies

The first optimisation is a 'lazy butterfly' technique. A 'forward FFT' (also known as decimation in frequency) maps naturally ordered inputs to outputs that are in bit-reversed order. The basic operation performed in this algorithm is a 'forward butterfly':

$$\begin{bmatrix} x \\ y \end{bmatrix} \longmapsto \begin{bmatrix} x + y \\ \omega(x - y) \end{bmatrix},$$

where ω is a root of unity. We can assume that the value ω is pre-computed and stored in a table. Thus, we can implement one forward butterfly by using one multiplication, one addition and one subtraction mod p. For the modular multiplication, we can use the pre-conditioned modular multiplication logic discussed above (see Algorithm 9.7). The modular addition and subtraction steps can be implemented as in Algorithm 9.5. This implementation requires a total of three 'correction steps' (specifically, two invocations of *CorrectExcess* and one of *CorrectDeficit*).

David Harvey [261] has shown how to reduce the number of correction steps in the forward butterfly from three to one. This is achieved by keeping intermediate results reduced only mod $2p$ rather than mod p (one might call this a 'lazy reduction' technique). It turns out that this leads to a significant performance improvement in practice. In the same paper, Harvey also gives a similar improvement to the implementation of the 'inverse butterfly' step,

$$\begin{bmatrix} x \\ y \end{bmatrix} \longmapsto \begin{bmatrix} x + \omega y \\ x - \omega y \end{bmatrix},$$

which is used in the 'inverse FFT' (also known as decimation in time) transform that maps bit-reverse-ordered inputs to naturally ordered outputs. This is achieved by keeping intermediate results reduced only mod $4p$ rather than mod p.

The Truncated Fast Fourier Transform

In the application of the FFT to polynomial multiplication, one has to perform N-point FFTs, where N is a power of two that is larger than the degree d of the product polynomial. If d is at or just above a power of two, this leads to an unfortunate inefficiency. Indeed, if one plots the graph of the running time of such a polynomial multiplication algorithm as a function of d, the graph looks like a step function, doubling at each power of two.

Joris van der Hoeven [595] introduced a 'truncated FFT' technique that effectively smooths out these jumps. This 'truncated FFT' technique can be combined with Harvey's 'lazy butterfly' technique, and NTL now incorporates both of these techniques. In fact, NTL's current NTT implementation is derived from code originally developed by Harvey (although it has been extensively rewritten to conform to NTL's internal software conventions).

9.3.4 Comparing Implementation Techniques

Table 9.2 shows the time (in nanoseconds) to perform a modular multiplication using the various techniques outlined above.

Method	Pre-conditioned time (ns)	Normal time (ns)	Non-normal time (ns)
no float / cmov	3.3	4.7	5.4
no float / mask	3.6	5.0	5.7
no float / jump	4.0	5.2	5.9
float / cmov	6.5	7.9	7.9
float / mask	6.7	8.1	8.1
float / jump	4.9	6.4	6.4
hardware div	26.6	26.6	26.6

Table 9.2 *Time (in nanoseconds) for single-precision modular multiplication.*

The first column measures the time for a pre-conditioned modular multiplication (where both one multiplicand and the modulus are fixed). The second column measures the time for a modular multiplication where the modulus is fixed and normalised (i.e., satisfies Eq. (9.1)). The third column measures the time for a modular multiplication where the modulus is fixed but need not be normalized. The first three rows all employ the floating-point-free implementations discussed above (which work with 60-bit primes), where the only difference is how the *CorrectExcess* logic is implemented: using a conditional move instruction, using shifts and masks, and using actual jumps.[4] The next three rows are based on floating-point implementations that do not rely at all on a double-word integer type (and which work with 50-bit primes).

For comparison, the last row in the table is based on assembly code that uses a single hardware instruction for multiplication and a single hardware instruction for division. This method is by far the least efficient method. This is not surprising, as hardware support for integer division is typically very poor: even though it is just a single instruction, it has a very high latency.

The floating-point-free methods are generally the fastest, and have the advantage in that they allow for larger moduli on 64-bit machines (60, or even 62 bits, rather than 50 bits). Notice that among the floating-point-free methods, the implementations based on jumps are the slowest, whereas among the floating-point methods, the implementations based on jumps are the fastest. A reasonable hypothesis to explain this behaviour is that the hardware branch predictor is doing much better in the floating-point methods than in the floating-point-free methods. We tested this hypothesis by empirically measuring the percentage of time the correction logic actually triggered a correction. For the floating-point methods, this was just 1 per cent to 2 per cent, whereas for the floating-point-free methods, this was 15 per cent to 20 per cent of the time.

---

[4] The GCC compiler flags `-fno-if-conversion` `-fno-if-conversion2` `-fno-tree-loop-if-convert` were used to force actual jumps.

Method	Non-lazy time (μs)	Lazy time (μs)
no float / cmov	74	56
no float / mask	80	59
no float / jump	300	130
float / cmov	110	n/a
float / mask	110	n/a
float / jump	290	n/a

Table 9.3 *Time (in microseconds) for multiplying two degree-1023 polynomials modulo a single-precision 'FFT' prime.*

Using the various techniques described above, Table 9.3 shows the time (in microseconds) to multiply two polynomials of degree 1023 modulo a single-precision 'FFT prime', i.e., a prime p such that $p-1$ is divisible by a suitable power of 2. The first three rows all employ the floating-point-free implementations discussed above (which work with 60-bit primes), where, again, the only difference is how the *CorrectExcess* logic is implemented. The last three rows are based on floating-point implementations that do not rely at all on a double-word integer type (and which work with 50-bit primes). The first column measures the time using a 'non-lazy butterfly' implementation, and the second column measures a 'lazy butterfly' implementation. Note that the 'lazy butterfly' implementation is not available in the floating-point-only implementation.

Observe that, in all cases, the implementation of the correction logic based on jumps is always dramatically slower. This is in contrast to what we saw in Table 9.2. The likely explanation for this is that the butterfly steps involve modular additions and/or subtractions, and for these, the branch misprediction rate is likely much higher than for multiplications.

Figure 9.1 shows some timing data comparing the current truncated FFT implementation to an earlier FFT implementation without truncation. The x-axis is the degree bound, y-axis is time (in seconds), shown on a log-log scale. This is the time to multiply two polynomials modulo a single-precision 'FFT' prime (60 bits). This figure nicely illustrates how the truncated FFT smooths out the step-function behaviour exhibited in the non-truncated version.

The integer-only `MulModPrecon` routine (see Algorithm 9.7) was introduced in NTL version 5.4 in 2005. Harvey's 'lazy butterfly' technique was first implemented in NTL version 6.0 in 2013. The integer-only non-preconditioned `MulMod` routines (see Section 9.3.2) were introduced in NTL

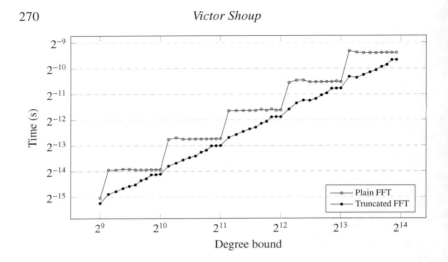

Figure 9.1 Truncated FFT versus plain FFT.

version 9.2 in 2015.[5] Van der Hoeven's 'truncated FFT' was first implemented in NTL version 11.1 in 2018.

9.4 Arithmetic in $\mathbb{Z}_p[X]$ for Multi-Precision p

We next discuss how NTL implements arithmetic in $\mathbb{Z}_p[X]$ for multi-precision p. NTL implements a class ZZ that represents multi-precision integers. In its default configuration, this is a thin wrapper around GMP. Specifically, an object of class ZZ is a pointer to an array of words representing a multi-precision integer, and GMP routines are used to do the arithmetic on such integers. In fact, NTL itself manages the memory allocation and de-allocation (using malloc and free), and (for the most part) relies only on the lower-level mpn routines in GMP that do not themselves manage memory.

NTL also implements a class ZZ_p that represents the ring $\mathbb{Z}_p = \mathbb{Z}/p\mathbb{Z}$ of integers mod p, where p is a multi-precision modulus. Note that, despite the notation, p is not necessarily a prime. Elements of $\mathbb{Z}_p$ are naturally represented as integers in the range $[0, p)$, using the class ZZ. The modulus p is

---

[5] The design of these routines were initially inspired by the logic in [436], but the final form, which is optimised based on assumptions specific to NTL's software conventions, is quite a bit different. Note also that NTL does not implement Montgomery multiplication [438], as it is a bit less convenient to use: it involves a more expensive pre-computation, a non-standard representation, and does not work with even moduli; moreover, it does not offer a significant performance benefit over the algorithms discussed above.

recorded in a global variable containing the value p itself (along with various pre-computed values to make certain computations more efficient). Note that, in multi-threaded applications, this global variable is implemented using 'thread local storage'. Using a global variable in this way surely offends some programming purists' sensibilities, but it also is the only way to allow one to use overload arithmetic operators to express computations in the most natural way.

NTL implements a class ZZ_pX that represents the ring of polynomials $\mathbb{Z}_p[X]$. An object of class ZZ_pX is a vector of ZZ_p objects. The memory for elements of this type of vector are specially managed.[6] The ZZ_p objects in such a vector are represented by ZZ objects, which themselves are represented by word-vectors, each of the same size; moreover, several such word-vectors are packed contiguously into larger blocks of memory. This packing has two benefits: first, when allocating such a vector of ZZ_p objects, the number of calls to `malloc` is significantly reduced; second, when accessing elements of such a vector in order, the cache behaviour is improved, as a result of better locality of reference.

NTL implements several polynomial multiplication algorithms for $\mathbb{Z}_p[X]$: plain, Karatsuba, Schönhage–Strassen, and multi-modular FFT.

For the most part, each of these algorithms reduces the problem of multiplying two polynomials in $\mathbb{Z}_p[X]$ to that of multiplying two polynomials in $\mathbb{Z}[X]$, and then reducing the coefficients of the product polynomial mod p.

The Plain Algorithm If the input polynomials have degree less than n, the plain algorithm performs $O(n^2)$ multiplications of integers of bit-length $\approx \log_2(p)$, as well as $O(n^2)$ additions of integers of bit-length $\approx 2\log_2(p) + \log_2(n)$.

The Karutsuba Algorithm The Karutsuba algorithm [315] performs $O(n^{\log_2 3})$ multiplications of integers of bit-length $\approx \log_2(p) + \log_2(n)$, as well as $O(n^{\log_2 3})$ additions and subtractions of integers of bit-length $\approx 2\log_2(p) + \log_2(n)$.

The Schönhage–Strassen Algorithm The Schönhage–Strassen algorithm [521] is based on the same high-level FFT strategy as outlined at the beginning of Section 9.3: using two FFTs to compute the evaluation vectors of the input

---

[6] For this, and other reasons, NTL implements its own vector template class, and does not rely on the vector class defined in the C++ standard template library (STL). Indeed, NTL was initially implemented long before templates or the STL existed.

polynomials, multiplying these two vectors component-wise to obtain the evaluation vector of the product polynomial, and using one inverse FFT to obtain the product polynomial from its evaluation vector. The difference is that now the FFTs (and inverse FFTs) are performed over the ring $\mathbb{Z}_q$, where $q = 2^{mr} + 1$, m is a power of two, r is odd, and $\log_2(q)$ is at least $\approx \ell := 2\log_2(p) + \log_2(n)$. The element $\omega := 2^r \in \mathbb{Z}_q^*$ is a primitive $2m$th root of unity, which enables the use of the FFT to multiply polynomials of degree less than n, provided $m \geq n$. The requirement that $\log_2(q)$ is at least $\approx \ell$ ensures that computing the coefficients of the product polynomial mod q allows us to recover the coefficients over the integers.

Multiplication in $\mathbb{Z}_q$ by a power of ω can be efficiently implemented in terms of shifts and additions/subtractions. Also, a general multiplication in $\mathbb{Z}_q$ can be implemented by using one integer multiplication (of integers of bit-length $\approx \ell$), as well as some shifts and additions/subtractions. The overall cost of this polynomial multiplication algorithm is $O(n \log n)$ additions and multiplications by powers of ω in $\mathbb{Z}_q$, and $O(n)$ multiplications in $\mathbb{Z}_q$. This polynomial multiplication algorithm works best when $\log_2(p)$ is not too small relative to n, as otherwise we have to work modulo number q whose bit-length (which is at least n) is significantly larger than that of p. One can improve performance for somewhat smaller values of p by making use of an optimisation known as the 'square root of 2 trick'. This optimisation is based on the observation that $\omega_1 := 2^{3mr/4} - 2^{mr/4}$ is a square root of 2 in the ring $\mathbb{Z}_q$, which means that ω_1 is a primitive $4m$th root of unity.

The Multi-Modular Algorithm The multi-modular algorithm reduces the coefficients of the input polynomials modulo several small primes $p_1, \ldots, p_k$. We must have $P := \prod_i p_i > B$, where B is a bound on the magnitude of the coefficients of the product polynomial, so $\sum_i \log_2(p_i)$ is at least $\approx 2\log_2(p) + \log_2(n)$. Each p_i is a single-precision 'FFT prime', so that each $\mathbb{Z}_{p_i}$ contains appropriate roots of unity to enable the use of the FFT to compute the product of the input polynomials mod p_i. After computing the product polynomial mod each p_i, the Chinese remainder algorithm is used to reconstruct the coefficients of the product polynomial over the integers. We have already discussed in Section 9.3 the techniques to multiply polynomials modulo such single-precision 'FFT primes'. To achieve good performance, some care must be taken in implementing the coefficient reduction modulo the small primes, and in the inverse Chinese remaindering step.

Assuming the hardware supports double-word integer arithmetic, the following technique is used for the reduction step. Suppose R is the radix used to represent long-integers. Further, assume we are given a vector $(a_0, \ldots, a_{\ell-1})$

of base-R digits representing a long-integer $a = \sum_j a_j R^j$ that we want to reduce mod each p_i. To do this, we assume we have pre-computed a table of values $r_{ij} := R^j \bmod p_i$. To compute $a \bmod p_i$, we compute the inner product $s_i := \sum_j a_j r_j$, and then reduce $s_i \bmod p_i$. On a typical 64-bit machine, and assuming GMP is used to implement long-integer arithmetic, we have $R = 2^{64}$ and each p_i is a 60-bit integer. Each term $a_j r_j$ in the inner product can be computed by using a single multiplication instruction. We can accumulate up to 16 such terms using a double-word addition (which on an x86 machine can be implemented using one addition and one addition-with-carry instruction). If there are more than 16 terms, then this double-word accumulator needs to be accumulated in a larger triple-word accumulator every 16 terms. After computing s_i, which will be represented as either a two-word or three-word integer, it is reduced mod p_i (using techniques similar to those discussed in Section 9.3.2).

Reducing such an ℓ-digit number a modulo $p_1, \ldots, p_k$, takes time $O(k\ell)$, using a pre-computed table of size $O(k\ell)$. The implied big-O constant in the running time is quite small, and although NTL implements an asymptotically fast algorithm for multi-reduction (relying on GMP's asymptotically fast algorithms), the crossover is extremely high (for $\log_2 p$ in the tens of thousands).

As for the Chinese remaindering step, we are given integers s_i for $i = 1, \ldots, k$, where each s_i is in the range $[0, p_i)$, and we want to compute the integer

$$c = \left(\sum_i P_i s_i \bmod P \right) \bmod p,$$

where $P_i := P/p_i$. A relatively straightforward implementation takes time $O(k\ell)$, but we get an easy speedup by a factor of two with the following observation. Assume that $P > 4B$. Let $A := \sum_i P_i s_i$. Suppose we write $A = PQ + R$, where Q is the integer nearest A/P and $|R| \le P/2$. We want to compute $R \bmod p$. In fact, by the assumption that $P > 4B$, we have $|R| < P/4$, and so $A/P = Q + \delta$, where $|\delta| < 1/4$. We also have $A/P = \sum_i s_i/p_i$. Therefore, under reasonable assumptions, we can compute the quotient Q by computing the sum $\sum_i s_i/p_i$ using double-precision floating-point, and rounding to the nearest integer. Assuming we have pre-computed the values $\bar{P}_i := P_i \bmod p$ (each of which is roughly half the size of P_i), as well as $\bar{P} := P \bmod p$, we can compute c above as

$$c = \left(\sum_i \bar{P}_i s_i - \bar{P} Q \right) \bmod p.$$

The sum $\sum_i \bar{P}_i s_i$ can be computed fairly quickly in the obvious way (GMP provides relatively good support for this); however, when double-word integer arithmetic is available, and especially when $\log_2(p)$ is not too huge, a technique similar to that used above in the reduction step for accumulating double-word products can be somewhat more efficient. Also as above, although NTL implements an asymptotically fast algorithm for Chinese remaindering, the crossover is extremely high.

9.4.1 Comparing Algorithms for Multiplication in $\mathbb{Z}_p[X]$

Figures 9.2–9.4 show the running time (in seconds) for each of the algorithms described above for primes p of bit-length 256, 1024 and 4096, respectively. Each figure shows the time to compute the product of two randomly chosen polynomials over $\mathbb{Z}_p$ of degree $< n$ for values of n ranging between 128 and 8192 (at values of n that are powers of two and midway between powers of two). The graphs are on a log-log scale.

These graphs speak for themselves. For these ranges of p and n, plain and Karatsuba are always slower than the FFT-based Schönhage–Strassen and multi-modular algorithms. One also sees, not surprisingly, that the multi-modular algorithm is significantly faster than Schönhage–Strassen when $\log_2(p)$ is not too huge, but eventually, as $\log_2(p)$ gets large, Schönhage–Strassen is somewhat faster.

For the multi-modular algorithm, we also measured the percentage of the total time spent on coefficient reduction modulo the small primes plus the inverse Chinese remaindering step. For $n = 4096$, this percentage is 35 per cent for 256-bit primes, 51 per cent for 1024-bit primes and 73 per cent for 4096-bit primes.

9.4.2 Multi-Core Implementation of Polynomial Multiplication Algorithms

When available, NTL can exploit multi-core machines to speed up polynomial multiplication, for both the Schönhage–Strassen and multi-modular algorithms.

For Schönhage–Strassen, the FFT and inverse FFT algorithms are recursive divide-and-conquer algorithms, which divide a given problem into two sub-problems of half the size. One thread is assigned to the top-level recursive invocation, two threads to the two second-level recursive invocations, four threads to the four third-level recursive invocations, and so on. The component-wise multiplication of the evaluation vectors is trivially parallelisable.

The multi-modular algorithm is even more straightforward to parallelise. The modular reductions and Chinese remaindering steps are trivially parallelisable across the coefficients, and the FFTs are trivially parallelisable across the

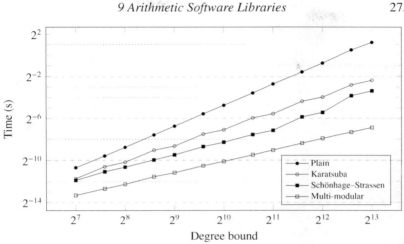

Figure 9.2 Polynomial multiplication modulo a 256-bit prime.

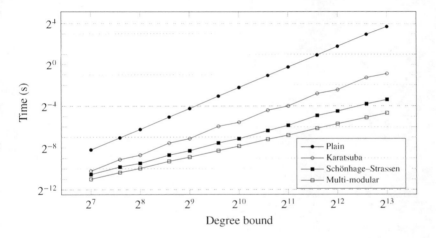

Figure 9.3 Polynomial multiplication modulo a 1024-bit prime.

primes. Figure 9.5 shows the running time (in seconds) to multiply two polynomials of degree less than 4096 modulo a 4096-bit prime, using 1, 2, 4, 8 and 16 threads, with both the Schönhage–Strassen and multi-modular algorithms. This is a log-log plot. A straight line with slope −1 would indicate perfect scaling with respect to the number of threads. As one can see, the scaling is not quite perfect for either algorithm.

Note that NTL implements its own 'thread pool mechanism' on top of

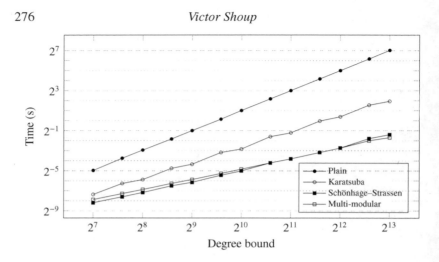

Figure 9.4 Polynomial multiplication modulo a 4096-bit prime.

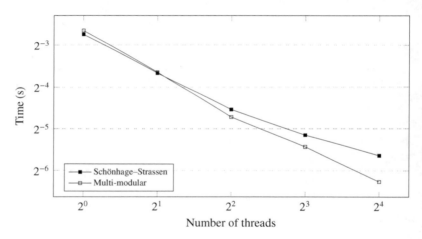

Figure 9.5 Multi-core performance: multiplication of polynomials of degree < 4096 modulo a 4096-bit prime.

standard C++11 threading features, and does not rely on non-standard mechanisms, such as *OpenMP*. This thread pool mechanism is utilised in a number of places throughout NTL to boost performance when multiple cores are available, and may also be used directly by NTL clients.

9.4.3 Other Operations on Polynomials

Of course, NTL implements other operations over $\mathbb{Z}_p[X]$, such as division and GCDs. Asymptotically fast algorithms are used wherever possible. In theory, one can reduce many of these problems to polynomial multiplication, and simply rely on an asymptotically fast algorithm for polynomial multiplication to get asymptotically fast algorithms for these other problems. In practice, one can do better in certain situations. For example, if $f \in \mathbb{Z}_p[X]$ is a polynomial of degree n, and one wants to perform many polynomial multiplications mod f, then one can pre-compute an appropriate polynomial 'inverse' h of f, so that reducing a polynomial of degree less than $2n$ modulo f can be done using just two multiplications of polynomials (of degree less than n). Moreover, if a multi-modular FFT algorithm is used for the polynomial multiplications, appropriate transforms of f and h can also be pre-computed (specifically, the evaluation vectors of f and h modulo each small 'FFT prime'), which speeds things up considerably. In particular, such reduction mod f can be performed at the cost of a single multiplication of polynomials (of degree less than n): a speedup by a factor of two. As another example, a single squaring of polynomials mod f can be computed at the cost of between 1.5 and 1.67 multiplications of polynomials (of degree less than n).

For more details on these and other optimisations, see the paper [541]. That paper also presented an algorithm for factoring polynomials over $\mathbb{Z}_p$, and reported on an experiment, conducted in 1995, that involved factoring a degree 2048 polynomial modulo a 2048-bit prime. Factoring that polynomial took just over 272 h (over 11 days) on a SPARC-10 workstation. With the current version of NTL, factoring the same polynomial by using a single core on our Haswell machine took 188 s, a speedup of over 5000×. Moreover, using 16 cores on our Haswell machine took just under 26 s, a speedup of over 37 000× (which represents a 45 per cent utilisation of these cores). The high-level algorithms have really not changed that much. The SPARC-10 and Haswell release dates differ by 20 years, so Moore's law by itself would predict about a 1000× speedup. In addition to making NTL's polynomial arithmetic exploit multiple cores (when possible), other improvements to the running time include better implementations of:

- single-precision modular arithmetic (most notably, the pre-conditioned modular multiplication code in Algorithm 9.7),
- the NTT itself (most notably, Harvey's lazy butterflies, discussed in Section 9.3.3),
- modular reduction and Chinese remaindering as deployed in the multi-modular polynomial multiplication algorithm (see Section 9.4),

- modular composition, i.e., computing $g(h)$ mod f for $f, g, h \in \mathbb{Z}_p[X]$ – although the high-level algorithm is still the same baby-step/giant-step approach of Brent and Kung [109], much better algorithms for matrix multiplication over $\mathbb{Z}_p$ are used, and these are discussed in Section 9.6.2.[7]

9.4.4 NTL versus FLINT

The Fast Library for Number Theory (FLINT) is a C library started in 2007 by Bill Hart and David Harvey, ostensibly as a 'successor' to NTL, as some people were apparently under the (mistaken) impression that NTL was no longer actively maintained. Since 2007, both NTL and FLINT have evolved considerably, but they still share some common functionality whose performance we can compare.

Multiplication in $\mathbb{Z}_p[X]/(f)$

Table 9.4 compares the relative speed of NTL's ZZ_pX mul routine with the corresponding FLINT routine.[8] The polynomials were generated at random to have degree less than n, and the modulus p was chosen to be a random, odd k-bit number.[9] The unlabelled columns correspond to n-values half-way between the adjacent labelled columns. For example, just to be clear: the entry in the third row and seventh column corresponds to $k = 1024$ and $n = 2048$; the entry in the third row and eighth column corresponds to $k = 1024$ and $n = 2048 + 1024 = 3072$.

The numbers in the table shown are ratios:

$$\frac{\text{FLINT time}}{\text{NTL time}}.$$

So ratios greater than 1 mean NTL is faster, and ratios less than 1 mean FLINT is faster. Ratios outside of the range $(1/1.2, 1.2)$ are in boldface (the others are essentially a tie).

The ratios in the upper right-hand corner of the table essentially compare NTL's multi-modular FFT algorithm with FLINT's Kronecker-substitution algorithm, which reduces polynomial multiplication to integer multiplication,

---

[7] NTL does not implement any of the newer, asymptotically faster algorithms for modular composition by Kedlaya and Umans [318]. Indeed, as far as we are aware, these algorithms remain of only theoretical interest (see [596]).

[8] We compared NTL v11.4.3 with FLINT v2.5.2, which is the most recent public release of FLINT at the time of this writing. These were both built using GMP v6.1.0. The compiler was GCC v4.8.5. All packages were configured using their default configuration flags. All of the programs used for benchmarking can be obtained at www.shoup.net/ntl.

[9] NTL's behaviour is somewhat sensitive to whether p is even or odd, and because odd numbers correspond to the case where p is prime, we stuck with those.

$k/1024$					$n/1024$							
	$1/4$		$1/2$		1		2			4	8		16
$1/4$	1.89	1.98	2.20	2.31	2.49	2.40	2.60	2.51	2.68	2.68	2.80	2.75	2.96
$1/2$	1.50	1.65	1.60	1.84	1.81	2.21	2.18	2.77	3.09	2.90	3.19	3.23	3.37
1	1.11	1.20	1.17	1.30	1.27	1.49	1.46	1.96	1.94	2.93	3.11	2.89	3.07
2	0.87	0.89	0.87	0.92	0.93	1.02	1.01	1.25	1.24	1.81	1.59	2.23	2.35
4	1.00	1.01	0.99	1.03	1.01	0.99	1.00	1.08	1.06	1.25	1.22	1.52	1.40
8	1.05	1.05	1.03	1.01	1.00	0.97	0.96	0.93	0.89	0.98	0.96	0.94	0.90
16	0.96	0.96	0.94	0.95	0.94	0.95	0.93	0.89	0.90	0.87	*0.83*	0.90	0.88

Table 9.4 *Multiplication in $\mathbb{Z}_p[X]$: n = degree bound, k = number of bits in p.*

which is performed by GMP. The ratios in the lower left-hand corner of the table essentially compare NTL's Schönhage–Strassen algorithm with FLINT's Schönhage–Strassen algorithm.

As one can see, NTL's multi-modular-FFT approach can be significantly faster than FLINT's Kronecker-substitution, being more than three times faster in some cases. One can also see that FLINT's Schönhage–Strassen implementation seems slightly better engineered than NTL's, being up to 20 per cent faster than NTL's implementation in some cases.

Squaring in $\mathbb{Z}_p[X]/(f)$

Squaring in $\mathbb{Z}_p[X]/(f)$ is a critical operation that deserves special attention, as it is the bottleneck in many exponentiation algorithms in $\mathbb{Z}_p[X]/(f)$.

Table 9.5 compares the relative performance of NTL's ZZ_pX SqrMod routine with FLINT's corresponding routine. The NTL routine takes as input pre-computations based on f, specifically, a ZZ_pXModulus object. The modulus p was chosen to be a random, odd k-bit number. The polynomial f was a random monic polynomial of degree n, while the polynomial to be squared was a random polynomial of degree less than n.

NTL is using a multi-modular-FFT strategy throughout (combined with the pre-computation techniques briefly discussed in Section 9.4.3), while FLINT is using Kronecker substitution in the upper right region and Schönhage–Strassen in the lower left region. As one can see, NTL's strategy can be over 5 times faster in some cases, while FLINT's can be over 1.5 times faster in others.

Computing GCDs in $\mathbb{Z}_p[X]$

Table 9.6 compares the relative performance of NTL's ZZ_pX GCD routine with FLINT's corresponding routine. The modulus p was chosen to be a random k-bit prime, and the GCD was computed on two random polynomials of degree less than n. Both libraries use a fast 'Half GCD' algorithm.

k/1024	n/1024												
	1/4		1/2		1		2		4		8		16
1/4	3.03	3.10	3.55	3.54	3.95	3.74	4.08	3.85	4.33	4.06	4.50	4.23	4.90
1/2	2.45	2.52	2.60	2.88	2.97	3.43	3.57	4.18	4.68	4.47	5.16	5.06	5.46
1	1.86	1.89	1.96	2.07	2.12	2.40	2.46	3.19	3.26	4.63	5.02	4.42	5.09
2	1.48	1.46	1.49	1.54	1.60	1.72	1.74	2.10	2.12	2.82	2.96	3.51	3.71
4	1.65	1.67	1.67	1.66	1.69	1.64	1.65	1.77	1.73	1.93	2.04	2.31	2.36
8	0.86	0.91	0.87	0.91	0.88	0.93	0.89	0.95	1.04	0.98	0.96	1.11	1.11
16	*0.65*	*0.66*	*0.66*	*0.67*	*0.67*	*0.69*	*0.68*	*0.71*	*0.68*	*0.69*	*0.67*	*0.73*	*0.73*

Table 9.5 *Squaring in* $\mathbb{Z}_p[X]/(f)$: n = *degree bound*, k = *number of bits in* p.

k/1024	n/1024												
	1/4		1/2		1		2		4		8		16
1/4	1.62	1.73	1.81	1.90	2.00	2.00	2.22	2.12	2.41	2.28	2.59	2.41	2.78
1/2	1.43	1.52	1.59	1.63	1.73	1.76	1.89	1.88	2.04	2.05	2.24	2.23	2.46
1	1.25	1.35	1.32	1.41	1.43	1.47	1.52	1.54	1.59	1.65	1.73	1.77	1.89
2	1.22	1.36	1.27	1.34	1.30	1.38	1.36	1.41	1.38	1.43	1.43	1.48	1.50
4	1.24	1.37	1.34	1.43	1.40	1.44	1.46	1.50	1.51	1.51	1.54	1.57	1.58
8	1.04	1.16	1.03	1.15	1.03	1.11	1.03	1.11	1.02	1.07	1.01	1.09	1.02
16	0.98	1.11	0.93	1.04	0.93	1.01	0.89	0.98	0.87	0.95	0.87	0.94	0.87

Table 9.6 *Computing GCDs in* $\mathbb{Z}_p[X]$: n = *degree bound*, k = *number of bits in* p.

k/1024	n/1024								
	1/4		1/2		1		2		4
1/4	7.24	7.32	8.91	9.08	10.87	10.86	13.27	12.75	14.07
1/2	6.15	6.53	7.32	7.87	8.78	9.56	10.38	12.12	14.64
1	5.24	5.50	6.16	6.47	7.26	7.57	8.38	9.51	10.93
2	4.64	4.79	5.35	5.08	6.22	6.52	7.20	7.60	8.37
4	5.00	5.12	5.78	5.94	6.79	6.81	7.72	7.37	8.39

Table 9.7 *Composition modulo a degree-n polynomial in* $\mathbb{Z}_p[X]$, k = *number of bits in* p.

Modular Composition in $\mathbb{Z}_p[X]$

Table 9.7 compares the relative performance of NTL's ZZ_pX CompMod routine and the corresponding FLINT routine. These routines compute $g(h)$ mod f for polynomials $f, g, h \in \mathbb{Z}_p[X]$ using Brent and Kung's [109] modular composition algorithm. The modulus p was chosen to be a random k-bit prime. The polynomial f was chosen to be a random monic polynomial of degree n, and the polynomials g and h were chosen to be random polynomials of degree less than n.

Factoring in $\mathbb{Z}_p[X]$

Table 9.8 compares the relative performance of NTL's ZZ_pX CanZass factoring routine and the corresponding FLINT routine. Both routines implement

k/1024	n/1024								
	1/4		1/2		1		2		4
1/4	3.59	3.62	4.17	4.48	5.73	5.06	6.17	5.28	6.25
1/2	2.79	2.81	14.06	4.22	4.35	4.99	5.08	5.67	8.01
1	2.07	2.19	2.38	2.68	3.29	3.45	3.46	4.39	5.52
2	2.20	1.70	1.76	1.92	2.29	2.56	2.74	3.19	4.04
4	1.72	11.21	7.16	1.92	2.12	2.09	2.50	2.79	2.50

Table 9.8 *Factoring a degree-n polynomial in $\mathbb{Z}_p[X]$, k = number of bits in p.*

the Kaltofen–Shoup algorithm [309], and for the range of parameters that were benchmarked, this is the fastest algorithm that each library has to offer. The modulus p was chosen to be a random k-bit prime. The polynomial to be factored was a random monic polynomial of degree n.

9.4.5 Application to Fully Homomorphic Encryption

Interest in NTTs and corresponding multi-modular techniques for polynomial arithmetic has intensified recently, because of the critical role they play in the design and implementation of several so-called fully homomorphic encryption schemes.

In such a scheme, one party, Alice, can encrypt a value, x, under the public key of a second party, Bob, obtaining a ciphertext c that encrypts x. Next, Alice can send c to a third party Charlie, who can efficiently transform c into another ciphertext c' that encrypts the value $f(x)$, where f is some specific function. Surprisingly, Charlie can do this only knowing c (and an appropriate description of f), without Bob's secret decryption key, and indeed, without learning anything at all about the values x or $f(x)$.

We cannot go into the details of these schemes here. However, for many of these schemes, the bulk of the computation time is spent operating on objects in the ring $R := \mathbb{Z}_q[X]/(\Phi_m(X))$, where q is typically a composite number a few hundred bits in length, and $\Phi_m(X)$ is the mth cyclotomic polynomial, with m typically in the range 16 000–64 000. Multi-modular techniques are especially applicable in this range of parameters. Moreover, there is typically a lot of flexibility in the choice of q.

For example, the library HElib [254], which is built on top of NTL, and which was mainly developed by Shai Halevi and this author, the modulus q can be chosen to be the product of single-precision primes p, where $p - 1$ is divisible by both m and a large power of two. In this setting, in addition to the traditional 'coefficient vector' representation, in which elements of R are represented as the vector of coefficients of a polynomial $f(X) \in \mathbb{Z}_q[X]$ representing a residue class mod $\Phi_m(X)$, one can also advantageously work with a 'double CRT' representation, in which for each prime p dividing q, we store the values $f_p(\omega_p^j)$, for $j \in \mathbb{Z}_m^*$, where ω_p is a primitive mth root of unity in $\mathbb{Z}_p$, and $f_p(X)$ is the image of $f(X)$ in $\mathbb{Z}_p[X]$. In this 'double CRT' representation, elements of R can be added and multiplied in linear time. However, there are still situations where we need to convert back and forth between 'coefficient vector' and 'double CRT' representations. Since each single-precision prime p is also an 'FFT prime', we can use fast NTTs to efficiently implement these conversions. Ignoring the modular reduction or Chinese remaindering, and focusing on the computations mod each small prime p, if $m = 2^r$ is a power of two, so that $\Phi_m(X) = X^{2^{r-1}} + 1$, such a conversion can be done almost directly, using a 2^{r-1}-point NTT (this is sometimes called a nega-cyclic transformation). For general m, HElib uses Bluestein's FFT algorithm [70]. In fact, HElib uses a variation of Bluestein's algorithm that exploits the truncated FFT discussed previously in Section 9.3.3.

9.5 Arithmetic in $\mathbb{Z}_p[X]$ for Single-Precision p

We briefly discuss how NTL implements arithmetic in $\mathbb{Z}_p[X]$ for single-precision p.

NTL implements a class zz_p that represents the ring $\mathbb{Z}_p = \mathbb{Z}/p\mathbb{Z}$ of integers mod p, where p is a single-precision modulus. (Again, despite the notation, p is not necessarily a prime.) Elements of $\mathbb{Z}_p$ are naturally represented as integers in the range $[0, p)$ using the built-in type long.

NTL implements a class zz_pX that represents the ring of polynomials $\mathbb{Z}_p[X]$. An object of class zz_pX is a vector of zz_p objects.

NTL implements several polynomial multiplication algorithms for $\mathbb{Z}_p[X]$: plain, Karatsuba and multi-modular FFT.

For the multi-modular FFT, if p is a prime such that $\mathbb{Z}_p$ contains appropriate roots of unity, then just a single application of the FFT-based polynomial multiplication algorithm (see Section 9.3) is required. Otherwise, just as for the case of multi-precision p, several 'FFT primes' are used together with Chinese remaindering. Depending on the size of p and the degrees of the polynomials, either one, two or three such 'FFT primes' are used.

Just as for arithmetic in $\mathbb{Z}_p[X]$ for multi-precision p, for single-precision p, when the degrees of the polynomials are large enough, many other operations (such as division and GCD) are reduced to fast polynomial multiplication.

k	1		2		4		8		16		32		64
5	0.47	0.55	0.59	0.63	0.68	0.70	0.75	0.72	0.86	0.99	1.17	1.14	1.27
10	0.68	0.72	0.79	0.83	0.90	1.05	1.16	1.22	1.37	1.41	1.60	1.60	1.58
15	0.85	1.00	1.09	1.14	1.24	1.33	1.46	1.55	1.68	2.04	2.14	2.04	2.28
20	0.52	0.55	0.60	0.63	0.67	0.78	0.87	0.95	1.01	1.15	1.13	1.14	1.11
25	0.61	0.71	0.77	0.82	0.89	0.95	1.06	1.11	1.26	1.35	1.54	1.36	1.48
30	0.83	0.86	0.93	0.97	1.08	1.23	1.40	1.40	1.67	1.50	1.66	1.51	1.66
35	0.93	1.01	1.11	1.17	1.25	1.39	1.55	1.59	1.72	1.86	1.77	1.91	1.82
40	1.08	1.14	1.26	1.27	1.44	1.64	1.81	2.02	2.07	1.95	2.13	2.13	2.24
45	1.19	1.31	1.47	1.49	1.72	1.87	2.01	2.19	2.20	2.29	2.31	2.29	2.35
50	0.97	0.96	1.06	1.11	1.25	1.40	1.54	1.57	1.56	1.56	1.56	1.57	1.61
55	1.00	1.09	1.22	1.31	1.40	1.50	1.64	1.82	1.85	1.87	1.99	1.98	2.14
60	1.17	1.23	1.36	1.46	1.62	1.79	2.03	1.96	2.08	1.98	2.06	2.03	2.17

Table 9.9 *Single precision: multiplication in* $\mathbb{Z}_p[X]$*; n = degree bound, k = number of bits in p.*

9.5.1 NTL versus FLINT

Just as we did for the multi-precision case, in Section 9.4.4, we compare the performance of NTL's polynomial arithmetic over $\mathbb{Z}_p[X]$, for single-precision p, to that of FLINT.

Table 9.9 compares the relative speed of NTL's zz_pX mul routine with FLINT's corresponding routine. As in Section 9.4.4, the numbers in the table are the ratio of FLINT time to NTL, so ratios greater than 1 mean NTL is faster. NTL is using a multi-modular FFT throughout, while FLINT is using Kronecker substitution throughout. Table 9.10 compares the relative performance of NTL's zz_pX SqrMod routine with FLINT's corresponding routine. Table 9.11 compares the relative performance of NTL's ZZ_pX GCD routine with FLINT's corresponding routine while the relative performance of NTL's zz_pX CompMod routine for modular composition and the corresponding FLINT routine is compared in Table 9.12. Table 9.13 compares the relative performance of NTL's zz_pX CanZass factoring routine and the corresponding FLINT routine.

9.5.2 Special Case where $p = 2$

An important special case for arithmetic in $\mathbb{Z}_p[X]$ is when $p = 2$. For this, NTL implements a special class GF2X that represents the ring of polynomials $\mathbb{Z}_2[X]$.

In this implementation, each coefficient is a single bit, and these bits are packed into machine words (actually, an unsigned long type).

To multiply two polynomials over $\mathbb{Z}_2[X]$, NTL recursively performs Karatsuba down to the 'base case', which is the multiplication of two 'word-sized'

k	n/1024												
	1		2		4		8		16		32		64
5	0.81	0.83	0.98	1.03	1.19	1.23	1.43	1.52	1.65	1.57	1.94	1.76	2.01
10	1.21	1.24	1.48	1.52	1.78	1.87	2.23	2.15	2.37	2.29	2.50	2.49	2.71
15	1.64	1.76	2.06	2.14	2.51	2.53	2.92	2.79	2.99	3.00	3.13	3.30	3.48
20	0.99	1.03	1.28	1.28	1.47	1.54	1.69	1.67	1.75	1.77	1.93	1.94	1.96
25	1.26	1.32	1.51	1.64	1.84	1.88	2.09	1.95	2.28	2.21	2.58	2.40	2.69
30	1.61	1.67	1.90	1.92	2.33	2.28	2.51	2.31	2.73	2.38	2.64	2.56	2.93
35	1.91	1.92	2.31	2.29	2.66	2.70	2.88	2.82	2.93	3.04	3.14	3.38	3.37
40	2.22	2.28	2.68	2.77	2.91	2.93	3.38	3.24	3.42	3.22	3.96	3.65	3.99
45	2.56	2.61	3.09	3.19	3.52	3.42	3.65	3.40	3.71	3.97	4.03	4.03	4.06
50	1.93	1.96	2.24	2.26	2.53	2.44	2.68	2.47	2.57	2.58	2.72	2.69	2.86
55	2.17	2.21	2.53	2.49	2.81	2.80	2.89	2.88	2.88	2.84	3.32	3.49	3.45
60	2.49	2.49	2.86	2.81	3.15	3.02	3.35	3.20	3.68	3.30	3.51	3.42	4.11

Table 9.10 *Single precision: squaring in $\mathbb{Z}_p[X]/(f)$; n = degree bound, k = number of bits in p.*

k	n/1024												
	1		2		4		8		16		32		64
5	0.89	*0.82*	0.84	*0.78*	*0.81*	*0.76*	0.84	*0.75*	0.86	*0.77*	0.90	*0.80*	0.96
10	0.93	0.89	0.91	0.86	0.90	0.86	0.97	0.91	1.03	0.92	1.08	1.00	1.19
15	0.99	0.97	1.00	0.95	1.04	0.98	1.10	1.03	1.19	1.08	**1.30**	1.18	**1.44**
20	*0.78*	*0.74*	*0.74*	*0.68*	*0.72*	*0.67*	*0.75*	*0.70*	*0.80*	*0.72*	0.86	*0.79*	0.98
25	0.88	*0.80*	*0.82*	*0.76*	*0.83*	*0.76*	0.86	*0.78*	0.95	0.85	1.02	0.92	1.12
30	1.03	0.95	1.00	0.90	1.01	0.92	1.09	0.97	1.17	1.03	**1.29**	1.10	**1.41**
35	**1.28**	**1.20**	1.19	1.14	**1.20**	1.13	**1.28**	1.17	**1.38**	1.27	**1.47**	**1.35**	**1.67**
40	**1.32**	**1.26**	**1.30**	**1.20**	**1.32**	1.17	**1.42**	**1.27**	**1.53**	**1.39**	**1.64**	**1.50**	**1.86**
45	**1.42**	**1.30**	**1.38**	1.27	**1.42**	**1.29**	**1.50**	**1.37**	**1.67**	**1.50**	**1.86**	**1.61**	**2.03**
50	**1.38**	1.16	1.18	1.07	1.18	1.07	**1.21**	1.13	**1.32**	**1.21**	**1.43**	**1.36**	**1.57**
55	**1.43**	**1.22**	**1.24**	1.14	**1.22**	1.15	**1.31**	**1.22**	**1.40**	**1.32**	**1.58**	**1.43**	**1.72**
60	**1.47**	**1.27**	**1.32**	**1.21**	**1.32**	**1.26**	**1.43**	**1.34**	**1.57**	**1.46**	**1.73**	**1.58**	**1.97**

Table 9.11 *Single precision: computing GCDs in $\mathbb{Z}_p[X]$; n = degree bound, k = number of bits in p.*

k	n/1024								
	1		2		4		8		16
5	3.84	3.19	4.44	3.65	4.88	3.23	3.88	3.86	4.56
10	4.30	3.75	5.07	4.46	5.86	4.93	6.08	5.74	6.81
15	4.99	4.53	6.02	5.53	7.22	6.73	8.20	7.66	8.92
30	4.52	4.08	5.34	4.89	6.24	5.85	6.70	6.21	7.28
60	5.21	4.68	6.67	5.93	8.13	7.75	9.23	8.22	9.94

Table 9.12 *Single precision: composition modulo a degree n polynomial in $\mathbb{Z}_p[X]$; k = number of bits in p.*

polynomials. Note that even before reaching this base case, the recursion switches over at some point to hand-coded, branch-free versions of Karatsuba, with an optimisation from Weimerskirch, Stebila and Shantz [620] for multiplying two three-word polynomials.

k	$n/1024$								
	1		2		4		8		16
5	1.22	1.11	1.39	1.46	1.77	1.98	2.03	2.23	2.40
10	1.49	1.82	1.70	1.99	2.71	2.41	2.94	3.11	3.63
15	2.15	2.32	2.45	2.76	3.22	3.70	3.55	3.89	4.31
30	2.49	2.76	2.97	2.70	3.01	3.16	3.53	3.36	3.73
60	3.61	3.51	4.01	3.96	4.50	4.21	5.97	4.66	4.59

Table 9.13 *Single precision: factoring a degree-n polynomial in* $\mathbb{Z}_p[X]$; $k =$ *number of bits in p.*

To multiply two word-sized polynomials, one of two strategies is used, depending on the available hardware. Modern x86 machines have a built-in instruction called PCLMUL for precisely this task, and NTL will use this instruction (by way of compiler 'intrinsics') if this is possible.

Failing this, NTL will use its default strategy, which is the following 'window' method. To multiply two word-sized polynomials a and b, first compute the polynomial $g \cdot b \bmod X^w$ for all polynomials $g \in \mathbb{Z}_2[X]$ of degree less than a small parameter s. Here, w is the number of bits in a word. Typically, $s = 4$ when $w = 64$, and $s = 3$ when $w = 32$. The values $g \cdot b \bmod X^w$ are computed by a series of shifts and XORs (each entry in the table is computed as either the XOR of two previously computed table entries, or a shift of one previously computed table entry). After this, the bits of a are processed in $\lceil w/s \rceil$ blocks of s bits, where for each block, one table entry is fetched, and two shifts and two XORs are performed. Finally, $s - 1$ simple correction steps are executed, to compensate for the fact the values in the table were only computed mod X^w. Each correction step consists of a handful of shifts, ANDs, and XORs. All of the above is carried out using branch-free code that is generated when NTL is configured.

The above strategy was implemented in v2.0 of NTL in 1998 for a window size of $s = 2$. The strategy was later generalised by Brent *et al.* [110] to arbitrary window size, and incorporated into v5.4 of NTL in 2005.

In fact, the paper [110] reports on an implementation of a highly optimised C library called gf2x for multiplying polynomials over $\mathbb{Z}_2[X]$. Moreover, NTL can be configured to call the multiplication routine in the gf2x library instead of its own version. We compared the performance of the current version of NTL to the current version of gf2x (version 1.3, released December 2019). Table 9.14 shows the time to multiply two random polynomials of degree less than n over $\mathbb{Z}_2[X]$, where $n = 10^3$ and $n = 10^6$, for both NTL and gf2x, and both with and without usage of the PCLMUL instruction. As one can see, NTL is fairly competitive with gf2x in these ranges, except for the case where

	$n = 10^3$		$n = 10^6$	
	NTL	gf2x	NTL	gf2x
with PCLMUL	0.21 µs	0.20 µs	13.3 ms	6.4 ms
w/o PCLMUL	1.54 µs	1.24 µs	91.8 ms	15.7 ms

Table 9.14 *Time to multiply two random polynomials of degree $< n$ over* $\mathbb{Z}_2[X]$.

$n = 10^6$ and no PCLMUL instruction is used (gf2x is almost six times faster here).

9.6 Matrix Arithmetic over $\mathbb{Z}_p$

NTL provides support for matrix arithmetic over $\mathbb{Z}_p$, including basic operations such as addition and multiplication, as well as operations such as inversion and solving matrix-vector equations.

Different strategies are used for single-precision and multi-precision modulus p. NTL also implements specialised strategies for $p = 2$, but we do not discuss these here.

9.6.1 Matrix Arithmetic Modulo Single-Precision p

Here are the strategies that NTL uses for computing the product $C = AB$, where A and B are large matrices over $\mathbb{Z}_p$, and where p is a single-precision modulus.

Strassen's Divide-and-Conquer Algorithm For very large matrices, NTL runs a few levels of Strassen's divide-and-conquer matrix multiplication algorithm [577].

Cache-Friendly Memory Access Although Strassen's divide-and-conquer algorithm already yields somewhat cache friendly-code, to obtain even more cache-friendly code, all the matrices are organised into panels, which are matrices with many rows but only 32 columns. We compute the ith panel of C by computing AB_i, where B_i is ith panel of B. If multiple cores are available, we use them to parallelise the computation, as the panels of C can be computed independently.

Next consider the computation of AP, where P is a single panel. We can write $AP = \sum_j A_j P_j$, where each A_j is a panel of A and each P_j is a 32×32 square sub-matrix of P. We have thus reduced the problem to that of computing

$$Q \leftarrow Q + RS, \tag{9.10}$$

where Q and R are panels, and S is a 32×32 square matrix. The matrix S is small and fits into the first-level cache on most machines – that is why we chose a panel width of 32. While the panels Q and R typically do not fit into the first-level cache, the data in each panel is laid out in contiguous memory in row-major order. In the implementation of Eq. (9.10), we process the panels a few rows at a time, so the data in each panel are processed sequentially, and we rely on hardware pre-fetch (which is common on modern high-performance systems) to speed up the memory access. Moreover, these fetches are paired with a CPU-intensive computation (involving S, which is in the first-level cache), so the resulting code is fairly cache friendly.

Fast Modular Arithmetic The basic arithmetic operation in any matrix multiplication algorithm is the computation of the form $x \leftarrow x + yz$, where x and y are scalars. In our case, the scalars lie in $\mathbb{Z}_p$. Although we could use the techniques in Section 9.3.1, this is not the most efficient approach.

If p is small enough, specifically, at most 23 bits in length, we can use the underlying floating-point hardware that is commonly available and typically very fast. Indeed, if we have 23-bit numbers w and x_i and y_i, for $i = 1, \ldots, k$, then we can compute $w + \sum_i x_i y_i$ exactly in floating-point, provided k is not too big: since standard (double-precision) floating-point can exactly represent 53-bit integers, we can take k up to $2^{53-23 \cdot 2} = 2^7$. If k is larger than this, we can still use the fast floating-point hardware, interspersed with occasional 'clean up' operations which convert the accumulated floating-point sum to an integer, reduce it mod p, and then convert it back to floating-point.

By using this floating-point implementation, we can also exploit the fact that modern x86 CPUs come equipped with very fast SIMD instructions for quickly performing several floating-point operations concurrently. Our code is geared to Intel's AVX, AVX2, and AVX512 instruction sets, which allows us to process floating-point operations four (or eight) at a time.

For p larger than 23 bits, the code reverts to using double-word integer arithmetic (if available) to accumulate inner products (rather than the AVX floating-point instructions), but still uses the same 'cache-friendly' panel/square memory organisation, along with Strassen's divide-and-conquer algorithm, and utilising multiple cores, if available.

Other Matrix Operations NTL also implements other matrix operations modulo single-precision p, such as inversion and solving matrix-vector equations. The same techniques for utilising cache-friendly code and for fast modular arithmetic are employed.

Comparison with FFLAS One of the state-of-the-art implementations of matrix operations over finite fields is FFLAS [171]. For small, single-precision p, FFLAS also employs floating-point techniques; however, it reduces all computations to floating-point matrix operations that are then carried out using the well-know BLAS API (see `netlib.org/blas/blast-forum`). We compared the current version of FFLAS (version 2.4.3), paired with the current version of OpenBLAS (version 0.3.7, see `www.openblas.net`). For example, to multiply two 4096×4096 matrices modulo a 23-bit prime, FFLAS took 3.18 s and NTL took 3.69 s, so FFLAS is about 16 per cent faster on this benchmark. As another benchmark, to invert a matrix of the same size, FFLAS took 4.15 s and NTL took 4.74 s, so FFLAS is about 14 per cent faster on this benchmark.

9.6.2 Matrix Arithmetic Modulo Multi-Precision p

For large, multi-precision p, NTL uses a multi-modular technique for matrix multiplication.

(1) The coefficients of the matrix are reduced modulo several single-precision primes $p_1, \ldots, p_k$.
(2) The matrix product is computed modulo each p_i, using the techniques outlined above.
(3) The results are combined using Chinese remaindering.

Steps (1) and (3) are implemented using the same techniques used for the multi-modular polynomial multiplication algorithm, discussed above in Section 9.4. Step (2) is implemented as discussed above in Section 9.6.1 (if AVX instructions are available, then we use 23-bit primes). Just as for the multi-modular polynomial multiplication algorithm, these steps are all trivially parallelisable, which is exploited if multiple cores are available.

At the current time, although NTL does provide implementations of other matrix operations modulo multi-precision p, such as inversion and solving matrix-vector equations, these implementations are very basic, and not as fast as they could be.

Comparison with FFLAS We compared NTL's multi-modular matrix multiplication performance to that of FFLAS (see discussion of FFLAS at the end of Section 9.6.1), which uses a similar multi-modular approach. For example, to multiply two 1024×1024 matrices modulo a 1024-bit prime, FFLAS took 11.5 s and NTL took 11.9 s, essentially a tie. On the same benchmark, but using integer-only scalar arithmetic (no floating-point or AVX, using 60-bit primes), NTL's multi-modular algorithm takes 22.8 s, which is indicative of NTL's performance on other platforms where it would not be able to take advantage of AVX. As another comparison, on this same benchmark, NTL's 'plain' matrix multiplication code (which carries out the computation naively using multi-precision integer arithmetic) takes 217 s (so the multi-modular algorithm is still much faster even without AVX support).

9.7 Polynomial and Matrix Arithmetic over Other Finite Rings

NTL also provides classes for the ring of univariate polynomials $E[X]$, where E is a quotient ring of the form $\mathbb{Z}_p[T]/(m(T))$. These classes are called ZZ_pEX, zz_pEX and GF2EX, respectively, depending on whether p is multi-precision, single-precision or equal to 2, and are implemented in terms of the classes ZZ_pX, zz_pX and GF2X, respectively. For low-degree polynomials over E, naive, quadratic-time algorithms are used. For large-degree polynomials, multiplication is implemented via Kronecker substitution [346], which reduces the problem of multiplication in $E[X]$ first to multiplication in $\mathbb{Z}_p[X, T]$ and then to multiplication in $\mathbb{Z}_p[X]$. Most other operations in $E[X]$ (e.g., division, GCD) are reduced to polynomial multiplication.

Algorithms for factoring univariate polynomials over E, computing minimal polynomials, and generating irreducible polynomials over E are also available (assuming E is a field).

NTL also provides support for basic matrix operations over E. Currently, these implementations are fairly basic, and there is plenty of room for performance improvements in this area.

9.8 Polynomial and Matrix Arithmetic over $\mathbb{Z}$

NTL provides the class ZZX, which represents the ring $\mathbb{Z}[X]$ of univariate polynomials over the integers. Multiplication in $\mathbb{Z}[X]$ is done by using essentially the same techniques as in Section 9.4. In particular, it implements several algorithms: plain, Karatsuba, Schönhage–Strassen, and multi-modular FFT.

For a number of other operations, such as GCD, a multi-modular approach is employed, reducing these operations to corresponding operations in $\mathbb{Z}_p[X]$ for several small primes p.

An algorithm for factoring univariate polynomials over $\mathbb{Z}$ is also provided. The algorithm is based on the well-known Berlekamp–Zassenhaus strategy [632], but with a number of critical improvements. The algorithm works by first ensuring that the given polynomial is square-free (this is essentially just a GCD computation, and is usually very fast). Second, this square-free polynomial is factored modulo several small primes, and one small prime p is selected as 'best', meaning that the number r of irreducible factors mod p is minimal. Third, this factorisation mod p is lifted to a factorisation mod p^k for a suitably large k. This is done by using an asymptotically fast Hensel lifting procedure.

The final step is to discover those subsets of the irreducible factors mod p^k that multiply out to factors of the polynomial over the integers. The naive strategy, which tries all possible subsets, takes time exponential in r. If r is not too big, NTL employs the heuristic pruning strategy introduced in [1], which speeds up this exponential-time strategy significantly. However, for larger r, NTL switches to van Hoeij's algorithm [599], which reduces the problem of discovering appropriate subsets of factors to that of solving a certain type of knapsack problem, which itself is solved by a lattice-basis reduction algorithm.

NTL also supports a number of operations on matrices over $\mathbb{Z}$. In addition to algorithms for basic arithmetic, NTL provides a number of algorithms for lattice-basic reduction. i.e., variations on the famous LLL algorithm [380].

- An exact integer version of LLL is implemented, which is essentially the same as the one presented in [127].
- A number of faster, heuristic floating-point versions of LLL are implemented. These are variations on algorithms presented by Schnorr and Euchner [519], with significant modifications to improve performance and to deal with rounding errors more robustly.
- A number of heuristic floating-point versions of Block-Korkin–Zolotarev (BKZ) reduction (also from [519]), which can produce much higher-quality reduced bases than LLL, are implemented.

Much of the work on the floating-point LLL and BKZ algorithms in NTL arose out of some very pleasant interactions with Phong Nguyen, who was attempting to use NTL's initial implementations of LLL to break certain lattice-based cryptosystems [453, 456]. These attempts really 'stress tested' NTL's implementation, and led to several heuristic improvements over the original floating-point algorithms proposed by Schnorr and Euchner. One such improvement was utilising a 'lazy' size-reduction condition. Indeed, the floating-point LLL algorithms presented in [519] impose the classical size-reduction condition, in which the Gramm–Schmidt coefficients of the lattice-basis must be at most $1/2$ in absolute value. Unfortunately, using this classical size-reduction was found to lead easily to infinite loops, owing to the Gramm–Schmidt coefficients only being computed approximately. NTL's implementation replaces the bound $1/2$ in the size-reduction condition by $1/2 + \epsilon$, where ϵ starts out very small, but then grows as infinite loops are (heuristically) detected. Other improvements in NTL's LLL implementation include:

- implementation of Givens rotations, in place of Gramm–Schmidt orthogonalisation, which yields better numerical stability;
- implementations of LLL algorithms that use other floating-point types to yield (at the expense of performance) greater stability and/or range.

In support of these floating-point LLL implementations, as well as for other applications, NTL provides several specialised floating-point classes.

- `quad_float`, which provides (essentially) twice the precision of ordinary double-precision floating-point. This class was derived from software developed previously by Keith Briggs (see `keithbriggs.info/doubledouble.html`), which itself was derived from software developed earlier by Douglas Priest [484].
- `xdouble`, which provides double-precision floating-point with an extended exponent range.
- RR, which provides arbitrary precision floating-point (with correct rounding for basic arithmetic and square root) and extended exponent range (note that RR provides functionality similar to the MPFR library [191], which was developed some time later than NTL's RR).

In recent years, better floating-point LLL algorithms have been developed [455] and implemented [587]. However, these algorithms have not yet been incorporated into NTL. For an excellent survey on more recent developments on the LLL algorithm, see [457].

9.9 The Future of NTL

Currently, the author has no plans to add significant new functionality to NTL. Rather, the plan is to continue to improve the performance of NTL, either by implementing new algorithms or by exploiting new hardware features. In addition, the author hopes to make it easier for others to contribute to NTL by using a code-hosting facility. At that stage, other contributors may perhaps wish to add new functionality to NTL.

10

XTR and Tori

Martijn Stam

Abstract

At the turn of the century, 80-bit security was the standard. When considering discrete-log-based cryptosystems, it could be achieved by using either subgroups of 1024-bit finite fields or using (hyper)elliptic curves. The latter would allow more compact and efficient arithmetic, until Lenstra and Verheul invented XTR. Here XTR stands for 'ECSTR', itself an abbreviation for Efficient and Compact Subgroup Trace Representation. XTR exploits algebraic properties of the cyclotomic subgroup of sixth-degree extension fields, allowing representation only a third of their regular size, making finite-field DLP-based systems competitive with elliptic-curve ones.

Subsequent developments, such as the move to 128-bit security and improvements in finite-field DLP, rendered the original XTR and closely related torus-based cryptosystems no longer competitive with elliptic curves. Yet, some of the techniques related to XTR are still relevant for certain pairing-based cryptosystems. This chapter describes the past and the present of XTR and other methods for efficient and compact subgroup arithmetic.

10.1 The Birth of XTR

10.1.1 The Rise of Subgroup Cryptography

When Diffie and Hellman introduced the concept of public-key cryptography [163], they gave the world the core key-agreement mechanism still in use today. Suppose Anna and Bob have already agreed on a cyclic group of known order with generator g. For Anna and Bob to agree on a secret key, they both select a random exponent modulo the group order, say x for Anna and y

for Bob, and send each other $X = g^x$ and $Y = g^y$, respectively. Upon receiving the other party's X, respectively Y value, Anna and Bob can raise it to their own private exponent to derive the shared key $k = X^y = Y^x = g^{xy}$.

Diffie and Hellman originally suggested use of the multiplicative group $\mathbb{Z}_p^*$ of integers modulo a large prime p as their cyclic group. This group has known order $p - 1$ and two problems relevant to Diffie–Hellman key agreement are believed to be hard in it: the discrete logarithm problem (given g^x, find x) and what later became known as the computational Diffie–Hellman problem (CDH; given g^x and g^y, find g^{xy}).

At the time, in 1976, not much was known about the discrete logarithm problem in $\mathbb{Z}_p^*$; from a mathematical perspective the group isomorphism from $\mathbb{Z}_{p-1}$ to $\mathbb{Z}_p^*$ by exponentiation of a known, fixed generator g was well understood and, prior to Diffie and Hellman's breakthrough paper, there appeared no urgent need to invert the isomorphism efficiently for what seemed like rather large primes p. Shanks's baby-step–giant-step method achieving 'birthday' complexity $O\left(p^{1/2}\right)$ had been published several years prior [534] but that was about it. Naturally, the concept of key agreement put forward by Diffie and Hellman changed this perception, and soon the discrete logarithm problem in $\mathbb{Z}_p^*$ was studied in more detail, but for the moment a 200-bit modulus p seemed acceptable.

However, at the time, Hellman had already submitted a paper together with Pohlig to speed up the discrete logarithm problem in groups of composite order with known factorisation [472]. This Pohlig–Hellman algorithm would essentially solve the discrete logarithm problem in the prime-power subgroups first and then use the Chinese remainder theorem to retrieve the discrete logarithm modulo the group order. The Pohlig–Hellman algorithm necessitates that $p - 1$ has at least one large prime factor N, but working in $\mathbb{Z}_p^*$ still appeared fine. Indeed, when ElGamal turned the Diffie–Hellman key-agreement protocol into a public-key encryption scheme, he stuck to the $\mathbb{Z}_p^*$ setting.

It was not until 1989, when Schnorr introduced his eponymous signature scheme [513], that working in a prime-order subgroup of $\mathbb{Z}_p^*$ became popular. At the time, Schnorr suggested the use of primes p and N with $N|p - 1$, N around 140 bits and p around 512 bits (ostensibly targeting 70-bit security). The advantage of using these 'Schnorr subgroups' was primarily computational: advances in solving the discrete logarithm problem had pushed up the size of p. Staying fully in $\mathbb{Z}_p^*$ would require working with exponents the size of p. Schnorr's innovation allowed the size of the exponent to be trimmed back to the minimum (i.e., twice the security level to protect against Shanks's generic baby-step–giant-step attack, or by that time, Pollard's rho [476]), which also led to more compact signatures. Thus for efficiency reasons working in subgroups is beneficial.

For ElGamal encryption [182], using Schnorr subgroups is mildly annoying as the message needs to be embedded in the subgroup; thus working in $\mathbb{Z}_p^*$ remained popular. However, in 1993 Brands introduced the decisional Diffie–Hellman (DDH) problem [76, 106], which is potentially much easier for an adversary. For this DDH problem, an adversary is still given g^x and g^y, but this time instead of having to compute g^{xy}, it is given a candidate value g^z and only needs to decide whether $g^z = g^{xy}$ or not. Five years later, Tsiounis and Yung [592] showed that the semantic security of ElGamal encryption is essentially equivalent to the DDH problem in the group being used. Moreover, where the Pohlig–Hellman algorithm implied the hardness of the CDH problem was linked to the hardest subgroup, for the DDH problem it actually links to the weakest subgroup. For $\mathbb{Z}_p^*$ we are guaranteed a subgroup of order 2 and thus DDH is easy: ElGamal encryption as originally proposed is not semantically secure. Consequently, for security reasons, working in subgroups became beneficial.

10.1.2 The Search for Compactness and Efficiency

Although the role and need of Schnorr subgroups became ever clearer when building cryptosystems loosely based on the discrete logarithm problem in $\mathbb{Z}_p^*$, two other developments took place. On the one hand, improved subexponential algorithms for finding those logarithms were developed, culminating in the number field sieve. At the turn of the millenium, to achieve 80-bit security one would need a 1024-bit prime p but only a 160-bit prime N. On the other hand, a new competitor arrived by using elliptic curve groups instead.

In the mid 1980s, elliptic-curve cryptography had been introduced by Miller [433] and Koblitz [340] and it was becoming more and more attractive: several families of 'weak' curves had been identified and efficient point counting had been solved [522], resulting in a large number of curves for which the best attacks were believed to be generic 'birthday bound' ones such as Van Oorschot and Wiener's version of Pollard rho [600]. For a prime-order subgroup G_N of an elliptic curve over a prime field $E(\mathbb{F}_p)$, this meant $N \approx p$ and believed discrete log complexity $\Theta(p^{1/2}) = \Theta(N^{1/2})$. Representing a point on the elliptic curve would naively takes two elements X and Y (not to be confused with the X and Y used previously for Diffie–Hellman key agreement), both in $\mathbb{F}_p$, satisfying the curve equation, e.g., $Y^2 = X^3 + aX + b$ for shortened Weierstrass when $\mathbb{F}_p$ is a large prime field. Those two elements take $2 \lg p$ bits, but it is straightforward to compress by representing only X and a bit indicating which root to take for Y, resulting in only $\lg p \approx \lg N$ bit representations. Thus, for 80-bit security, only 160 bits are needed rather than the 1024 bits when using $\mathbb{Z}_p^*$'s Schnorr subgroups.

The problem with Schnorr subgroups of $\mathbb{Z}_p^*$ was that there was (and is) no known way to exploit being in a prime-order subgroup, other than using smaller exponents: all operations and representations still rely directly on the supergroup $\mathbb{Z}_p^*$, resulting in wastefully large representations and, notwithstanding the smaller exponent, inefficient exponentiation. To counter these problems, Lenstra suggested working in finite fields with small extension degrees instead [366]. Moreover, to avoid ending up in a smaller subfield, he suggested working in a prime-order subgroup of the cyclotomic subgroup (Definition 10.1).

Definition 10.1 (Cyclotomic Subgroup)**.** Let p be a prime power and let $n \in \mathbb{Z}_{\geq 1}$ be an extension degree. Then $|\mathbb{F}_{p^n}^*| = p^n - 1 = \prod_{d|n} \Phi_d(p)$ where Φ_d is the dth cyclotomic polynomial; the unique $\mathbb{F}_{p^n}^*$-subgroup of order $\Phi_n(p)$ is called the cyclotomic subgroup of $\mathbb{F}_{p^n}^*$.

If a prime N divides $\Phi_d(p)$ for $d|n, d \neq n$, then the subgroup G_N of order N can be embedded in a proper subfield $\mathbb{F}_{p^d}$ of the larger $\mathbb{F}_{p^n}$ and the discrete logarithm problem in G_N can be solved in that smaller subfield instead. The true hardness of the discrete logarithm problem in $\mathbb{F}_{p^n}^*$ ought to reside in those prime-order subgroups that cannot be embedded into proper subfields. To support this claim, Lenstra also showed a bound on the greatest common prime divisor of $\Phi_d(p)$ and $\Phi_n(p)$.

Lemma 10.2 (Lenstra)**.** *Let $N > n$ be a prime factor of $\Phi_n(p)$. Then N does not divide any $\Phi_d(p)$ for divisors d of n with $d < n$.*

Lenstra demonstrated how exponentiation can be speeded up moderately in cyclotomic subgroups. The cyclotomic subgroup of $\mathbb{F}_{p^n}$ has size $\Phi_n(p) \approx p^{\varphi(n)}$, which raises the prospect of utilising its structure to represent elements of that subgroup more compactly as well, using $\varphi(n) \lg p$ bits instead of the naive $n \lg p$. If we assume that the discrete logarithm problem (and related Diffie–Hellman problems) are roughly as hard in a finite field $\mathbb{F}_{p^n}$ as they are in a prime field $\mathbb{Z}_{p'}$ with $p' \approx p^n$, such a compact representation would yield a compression by a factor $n/\varphi(n)$.

10.2 The Magic of XTR

The XTR cryptosystem was the first cryptosystem based on the finite-field discrete logarithm problem that combined good compression, namely by a factor 3, with efficient exponentiation in the compressed form. XTR is shorthand for ECSTR, which itself stands for Efficient and Compact Subgroup Trace Representation and was developed by Arjen Lenstra and Eric Verheul [377]. Sometimes XTR is also considered a homophone in Dutch of 'ekster', or magpie.

The XTR cryptosystem works in the cyclotomic subgroup of $\mathbb{F}_{p^6}^*$, for prime p, although the generalisation to extensions fields $\mathbb{F}_{q^6}^*$ with $q = p^m$ is mostly straightforward. XTR operates in a prime-order-N subgroup $G_N \subset G_{p^2-p+1}$, where N divides $\Phi_6(p) = p^2 - p + 1$. Elements in G_{p^2-p+1} can be compactly represented by their trace over $\mathbb{F}_{p^2}$, which is defined by

$$\mathrm{Tr} : x \to x + x^{p^2} + x^{p^4} .$$

Lenstra and Verheul showed that, if $g \in G_{p^2-p+1}$ and $c = \mathrm{Tr}(g)$, then g is a root of the polynomial

$$X^3 - cX^2 + c^p X - 1 , \tag{10.1}$$

thus given c it is possible to recover g up to conjugacy, as g^{p^2} and g^{p^4} will also be roots of the equation above. Moreover, if $c_x = \mathrm{Tr}(g^x)$, they derived the recurrence relation

$$c_{x+y} = c_x c_y - c_x^p c_{x-y} + c_{x-2y} . \tag{10.2}$$

This recurrence relation allows fast 'exponentiation' of c_x given a compressed base c and an exponent x without first having to decompress to obtain g (or one of its conjugates). We will discuss efficiency of XTR in Section 10.2.1.

Notwithstanding the title of the original paper 'The XTR Public Key System', XTR is not really a cryptosystem, but rather a method for working efficiently and compactly in the 'DLP-hard' part of $\mathbb{F}_{p^6}$. The main limitation of the XTR method is the kind of exponentiations that can be performed easily. Whereas a single exponentiation $\mathrm{Tr}(g^x)$ is easy, a double exponentiation $\mathrm{Tr}(g^x h^y)$ is already more challenging, whereas triple exponentiations and beyond are not really feasible in compressed format. As a consequence, not all discrete-log cryptosystems can be ported to the XTR setting.

One interpretation is to consider the exponent group $\mathbb{Z}_N$ acting on the set of trace-representations (cf. hard homogeneous spaces for isogeny-based cryptography [141]). Cryptosystems that can be phrased by using this abstraction should be suitable for direct application of XTR. They notably include Diffie–Hellman key exchange and a variety of ElGamal-based KEMs (key encapsulation mechanisms). However, the ability to perform double exponentiations widens the scope to include, for instance, the Nyberg–Roeppel signature scheme. We give some concrete examples of XTRified schemes in Section 10.2.2.

The security of an XTR-based cryptosystem is tightly linked to the relevant underlying hard problem in $G_N \subseteq G_{p^2-p+1} \subseteq \mathbb{F}^*_{p^6}$, be it the discrete logarithm problem, the computational Diffie–Hellman problem, or its decisional version. Note that XTR predates the Diffie–Hellman alphabet soup assumption explosion triggered by the constructive use of elliptic-curve pairings [103]; although some of these newer assumptions do make sense in pairing-free groups such as $G_N \subseteq \mathbb{F}^*_{p^6}$, the cryptosystems based on those assumptions typically involve operations beyond the XTR-friendly single and double exponentiations.

Precursors. Some early cryptosystems pre-dating XTR realised partial benefits from working in the cyclotomic subgroup of finite fields, without necessarily realising the structure being exploited. LUC works in the cyclotomic subgroup of $\mathbb{F}_{p^2}$, achieving a compression factor 2, although historically it was not presented thus. In 1981, Müller and Nöbauer [446] suggested replacing the exponentiation in RSA by the evaluation of Dickson polynomials $g_x(1, h)$ modulo an RSA modulus, where x takes the place of the exponent and h that of the base. Twelve year laters, Smith and Lennon [554] suggested to use the Lucas function $V_x(h, 1)$ instead as an alternative to RSA, again still modulo an RSA modulus. They called this new cryptosystem LUC, yet, as $g_x(1, h) = V_x(h, 1)$, this Lucas cryptosystem was, and is, equivalent to the Dickson scheme.

Then, in 1994, Smith and Skinner [553] suggested the used of the Lucas function modulo a prime to be used for discrete-log-based cryptosystems such as Diffie–Hellman key exchange or ElGamal encryption. They hoped that this 'prime' LUC cryptosystem would not allow any sub-exponential attack, oblivious of the mathematically much cleaner interpretation of LUC as being based on the cyclotomic subgroup of $\mathbb{F}_{p^2}$ using traces to compress and speed up calculations. The observation that LUC was, in fact, just $\mathbb{F}_{p^2}$ in disguise was first made by Lenstra, together with Bleichenbacher and Bosma [67].

The first scheme achieving compression by a factor 3, using properties of the cyclotomic subgroup of $\mathbb{F}_{p^6}$, was proposed by Brouwer, Pellikaan and Verheul [112] in 1999. It formed the inspiration for XTR, but unlike XTR it did not offer computation in the compressed domain. A slightly different compression method was developed around the same time by Gong and Harn [228, 229], though it offered only a factor 1.5 compression.

For a slightly more expanded discussion of the history, see also [558, Sections 4.4.3 and 4.6.4].

10.2.1 Efficient Implementation

When implementing XTR, essentially three efficient routines are needed: fast parameter generation of p, N and a base element c on the one hand, efficient membership tests, and efficient double exponentaion. It turns out that efficient single exponentiation is best regarded as a special, speeded-up version of double exponentiation.

Parameter Generation and Membership Tests Generating the primes p and N, where N divides $p^2 - p + 1$, is relatively easy provided that N is considerably smaller than p. Select N first, find a root of r of $X^2 - X + 1$ modulo N and try $p = r + N\ell$ for integer ℓ until a suitable prime p is found.

To find c, the naive approach would be to sample g at random from $\mathbb{F}_{p^6}^*$, raise it to $(p^6 - 1)/(p^2 - p + 1) = (p^3 - 1)(p + 1)$ and verify whether the result has the desired order $p^2 - p + 1$. Subsequently take the trace and raise to the power $(p^2 - p + 1)/N$. Finding a generator of G_{p^2-p+1} this way is quite expensive and can, in fact, be done far more efficiently by exploiting fast irreducibility testing of the polynomial (10.1), as explored by Lenstra and Verheul [375, 378].

For a membership test, the goal is to test whether a purported compressed element $c \neq 3$ is indeed the trace of an element in G_N. One could check this by evaluating c_N using a single exponentiaton routine (that does not first reduce the exponent modulo N) and checking whether the result equals $3 = \mathrm{Tr}(1)$. The techniques to generate c faster can also be used to speed up membership tests [378].

Single Exponentiation Normally, a single exponentiation refers to the problem of calculating g^x in the group G_N for a given generator g of G_N and exponent $x \in \mathbb{Z}_N$. For XTR, the problem translates to calculating $c_x = \mathrm{Tr}(g^x)$ given compressed generator $c = \mathrm{Tr}(g)$ and exponent x, based on the recurrence relation for c_{x+y} given in Eq. (10.2).

In general, evaluating $c_{x+y} = c_x c_y - c_y^p c_{x-y} + c_{x-2y}$ given all required elements on the right-hand side would take four $\mathbb{F}_p$-multiplications. However, if $x = y$, then the relation simplifies to $c_{2x} = c_x^2 - 2c_x^p$, which effectively costs only half as much, namely two $\mathbb{F}_p$-multiplications.

Armed with this knowledge, Lenstra and Verheul devised an elegant single exponentiation routine based on a left-to-right binary expansion of the exponent x by keeping track of the triplet $(c_{2k}, c_{2k+1}, c_{2k+2})$, where k is the exponent processed so far (corresponding to the most significant bits of x). A useful property of the triplet is that it always contains two even 'exponents', namely $2k$ and $2k+2$. As a result, processing the next bit of x can always be done using two invocations of the c_{2x} rule and one invocation of the general recursion rule. Overall, a single exponentiation would cost around $8 \lg N$ $\mathbb{F}_p$-multiplications. At the time, they believed this to be roughly three times faster than a direct exponentiation in G_N.

Double Exponentiation In the context of XTR, a double exponentiation consists of the problem of, given basis c_κ, c_λ, their 'quotients' $c_{\kappa-\lambda}$ and $c_{\kappa-2\lambda}$ and two exponents x and y, compute $c_{\kappa x+\lambda y}$. The original double exponentiation by Lenstra and Verheul [377] was somewhat cumbersome and slow. A far more efficient routine was developed by Stam and Lenstra [559] based on Montgomery's PRAC algorithm [437]. The original PRAC algorithm was developed by Montgomery for the efficient calculation of Lucas sequences, that is, recurrence relations of the form $L_{x+y} = f(L_x, L_y, L_{x-y})$. These occur for instance when implementing scalar multiplication on elliptic curves using Montgomery representation.

At the core of the PRAC algorithm, including its XTR variant, is the extended Euclidean algorithm. Recall that, to calculate the greatest common divisor of two positive integers x and y, the extended Euclidean algorithm eventually outputs not just the gcd d, but also the Bézout coefficients a and b such that $ax+by = d$. To create a Euclidean exponentiation routine for exponents x and y given bases g^κ and g^λ, introduce random variables $\tilde{x}, \tilde{y}$ as well as α, β and keep as invariant $\tilde{x}\alpha + \tilde{y}\beta = x\kappa + y\lambda$ and $\gcd(\tilde{x}, \tilde{y}) = \gcd(x, y)$. It is easy to initialise by setting $\tilde{x} \leftarrow x$ and $\tilde{y} \leftarrow y$. The Greek lettered variables are not typically known or efficiently computable, but we can keep track of g^α and g^β instead. Just as in the extended Euclidean algorithm, in each step $(\tilde{x}, \tilde{y})$ can be reduced while maintaining the invariant, e.g., by setting $(\tilde{x}, \tilde{y}) \leftarrow (\tilde{y}, \tilde{x} - \tilde{y})$ and updating the other variables accordingly. Eventually $y = 0$, at which point $g^{x\kappa+y\lambda} = (g^\alpha)^d$, where $d = \gcd(x, y)$. Adapting the algorithm to Lucas sequences, one also needs to keep track of $g^{\alpha-\beta}$; for XTR one additionally needs $g^{\alpha-2\beta}$.

There are many different steps possible to reduce $(\tilde{x}, \tilde{y})$. We already mentioned $(\tilde{y}, \tilde{x}-\tilde{y})$, but if $\tilde{y}$ is even, say, one can also try $(\tilde{x}, \tilde{y}/2)$. Which steps to use when to reduce $(\tilde{x}, \tilde{y})$ are governed by rules that have been determined heuristically. We refer to [558, Tables 3.4 and 4.2] for possible collections of rules (and their precedence). These rules result in an XTR double exponentiation on average costing roughly $6 \lg N$ $\mathbb{F}_p$-multiplications [558, Corollary 4.13.ii].

Single Exponentiation, Revisited The PRAC-based double exponentiation appears 25 per cent faster than the binary single exponentiation. Unsurprisingly, it is possible to leverage this speed up of the double exponentiation for single exponentiation as well, by casting the latter as a case of the former by writing $g^x = g^r g^{x-r}$ for some arbitrary r. The choice $r = 1$ and a slight trimming of PRAC's Euclidean reduction rules will lead back to a (costly) binary method, but Montgomery already suggested the use of the golden ratio ϕ by setting r to $\lfloor r/\phi \rfloor$ to ensure that the PRAC algorithm will initially use its most advantageous reduction rule (i.e., delivering the largest reduction in the size of the exponent per $\mathbb{F}_p$-multiplication). With this speed up, calculating c_x costs an average $5.2 \lg N$ $\mathbb{F}_p$-multiplications [558, Corollary 4.13.i].

Further speed ups are possible when allowing pre-computation, as it allows to split the exponent evenly: if the triplet $(c_{\tau-1}, c_\tau, c_{\tau+1})$ has been pre-computed with $\tau = \lfloor \sqrt{N} \rfloor$, then one can write $x = x_1 + x_2\tau$ with both x_1 and x_2 of length roughly $\frac{1}{2} \lg N$. The resulting double exponentiation brings the costs of a single exponentiation with pre-computation down to an average of $3 \lg N$ $\mathbb{F}_p$-multiplications [558, Corollary 4.13.v].

Stam and Lenstra [559] introduced another neat trick to speed up and compress the pre-computation by exploiting the Frobenius endomorphism. The idea is to rewrite the exponent as $x = x_1 + x_2 p \bmod N$ with x_1 and x_2 both as short as possible, so ideally roughly $\frac{1}{2} \lg N$ bits (as above). The difference between using Frobenius and $\tau = \lfloor N \rfloor$ is that for Frobenius, the pre-computation is a lot simpler as $c_p = c^p$ and $c_{p-1} = c$ can be computed for free and, re-using an earlier observation by Lenstra and Verheul [378, Proposition 5.7], c_{p-2} can be computed at the cost of a square root computation (the only caveat here is that an additional bit is needed to resolve which root to use). Another difference is that, in the case of Frobenius, it is not a priori clear whether x_1 and x_2 can computed that small. It turns out that the extended Euclidean algorithm suffices.

Coincidentally, concurrently and independently of the work on XTR, Gallant, Lambert and Vanstone (GLV) suggested a very similar method to use efficient non-trivial automorphisms to speed up elliptic-curve scalar point multiplication [204]. The GLV method is more general than the method for XTR just described and, depending on the automorphism, the exponent might be split in more than two parts. For the special XTR setting, Stam and Lenstra exploited that $N | p^2 - p + 1$ to prove the split was guaranteed to result in short exponents x_1 and x_2 [558, Lemma 2.29] (essentially by framing the problem of finding a short vector in a two-dimensional lattice). For the GLV method, Sica, Ciet and Quisquater [545] provided a more general analysis.

One potential disadvantage of the PRAC-based single exponentiation compared with the Lenstra–Verheul binary single exponentiation is the increased variation in not just the runtime of the algorithm, but also the sequence of underlying operations. From a side-channel perspective, such exponent-dependent variation should be considered leakage that can likely be exploited.

Page and Stam [466] analysed idealised single power analysis (SPA) against XTR in more detail and although their attack was still computationally expensive, the binary routine appears intrinsically safer. In fact, a simpler version of the binary routine proposed by Montgomery for the efficient calculation of Lucas sequences [437] has been adapted for ordinary scalar multiplication on elliptic curves with the express purpose of boosting side-channel resistance [305] and Han, Lim and Sakurai showed that Lenstra and Verheul's binary XTR exponentiation routine is SPA secure, although it is still susceptible to differential power attacks [256].

Twofold Exponentiation Knuth and Papadimitriou [337] established a beautiful result linking the complexity of a double exponentiation $g^x h^y$ with that of a twofold exponentiation (g^x, g^y), and vice versa. The result is that any improvement for double exponentiation can also be used for a twofold exponentiation, meaning that computing g^x and g^y jointly is a lot cheaper than computing both separately. For recurrence relations as used by XTR, the duality concept is not quite as clean, but jointly calculating c_x and c_y can still be done quite efficiently with a modification of the PRAC double-exponentiation routine, leading to an overall cost of $6 \lg N$ $\mathbb{F}_p$-multiplications on average [558, Corollary 4.13.iii].

10.2.2 Examples of XTRified Primitives

Diffie–Hellman Key Agreement At the beginning of this chapter, we mentioned Diffie–Hellman key agreement. In a cyclic group G_N of order N and with generator g, Anna selects ephemeral exponent $x \in \mathbb{Z}_N$, calculates $X \leftarrow g^x$ and sends X to Bob. Bob had selected exponent $y \in \mathbb{Z}_N$ and sent $Y = g^y$ to Anna. They both want to compute the shared key g^{xy}.

Moving to XTR, we take traces of all the G_N group elements, so for instance the new shared key will be $\mathrm{Tr}(g^{xy})$. Moreover, the communication will be compressed as well: Anna selects ephemeral exponent $x \in \mathbb{Z}_N$ but this time calculates $c_x = \mathrm{Tr}(g^x)$. Upon receipt, Bob can then use c_x as its base XTR exponent 'd' and calculate $K \leftarrow d_y = \mathrm{Tr}(g^{xy})$ as desired.

The benefits of using XTR here over direct calculation in G_N are immediate: the public representation of the group parameters is compressed, the communication overhead is reduced by a factor of three and the trace-exponentiations are faster.

ElGamal-Style Key Encapsulation Mechanisms Bare-bones ElGamal key encapsulation is almost the same as Diffie–Hellman key agreement, with the only notable difference that Bob's ephemeral value $c_y = \mathrm{Tr}(g^y)$ is now declared to be his public key, with the exponent y promoted to his long-term private key. If Anna wants to send a message to Bob, select ephemeral exponent $x \in \mathbb{Z}_q$, create ciphertext $c_x = \mathrm{Tr}(g^x)$ and calculate $K \leftarrow \mathrm{Tr}(g^{xy})$ as the encapsulated key.

It is immediately clear that Hashed ElGamal, where K is calculated as $\mathcal{H}(\mathrm{Tr}(g^{xy}))$ with $\mathcal{H}$ a hash function, works fine with XTR as well, giving rise to an efficient IND-CCA2 secure public-key cryptosytem in the random oracle model. With a little more effort, Damgård's ElGamal with explicit ciphertext rejection can be seen to work as well, yet a closely related variant with implicit ciphertext rejection is more troublesome (cf. [320]).

Signature Schemes Lenstra and Verheul [377] described how to create XTR-based Nyberg–Rueppel signatures [459], supporting message recovery. Below we look at Schnorr signatures instead; these look similar to Nyberg–Rueppel signatures but without offering message recovery. They are created by applying the Fiat–Shamir transform to Schnorr's sigma protocol for proving knowledge of a discrete logarithm.

We assume that public parameters contain the group description, consisting of p, N and $c_1 = \mathrm{Tr}(g)$. Then a user's private key is an exponent $x \in \mathbb{Z}_q$ and the corresponding public key is the element $c_x = \mathrm{Tr}(X)$, where $X = g^x$, plus the auxiliary elements $c_{x-1} = \mathrm{Tr}(X/g)$ and $c_{x+1} = \mathrm{Tr}(Xg)$.

The auxiliary elements enable the calculation of double exponentations of the type $\mathrm{Tr}(g^y X^z)$. If the public key is represented as the triplet $(c_{x-1}, c_x, c_{x+1}) \in (\mathbb{F}_{p^2})^3$ no compression is taking place and one might as well send $y \in \mathbb{F}_{p^6}$ instead. However, the public parameters are still being compressed (namely the generator of the group); moreover, Lenstra and Verheul also showed that a few additional bits allow unique and efficient recovery of c_{x-1} and c_{x+1} based on c_x and those public parameters.

To sign a message $\tilde{m}$, the signer generates a random exponent $w \in \mathbb{Z}_N$ and evaluates $a \leftarrow c_w = \mathrm{Tr}(g^w)$. Then calculate $s \leftarrow \mathcal{H}(c_x, a, \tilde{m})$ and $r \leftarrow w + x \cdot s \bmod N$, where $\mathcal{H} : \mathbb{F}_{p^2} \times \mathbb{F}_{p^2} \times \{0,1\}^* \to \mathbb{Z}_N$ is a hash function. The signature consists of the pair $(r, s) \in (\mathbb{Z}_N)^2$. To verify a signature, re-calculate $a \leftarrow \mathrm{Tr}(g^r \cdot X^{-s})$ by using XTR double exponentiation and accept the signature if and only if $\mathcal{H}(c_x, a, \tilde{m})$ equals s.

As the original Schnorr signature only ever sent across elements of the exponent group $\mathbb{Z}_N$ to begin with, there is no compression gain to be had here; however, both the public parameters and the public key can be compressed and both signature generation and verification are considerably speeded up compared with naive operations directly in G_N.

10.3 The Conservative Use of Tori

10.3.1 Direct Compression using Tori

One downside of using traces, as XTR does, is that the compression is not lossless: conjugates are mapped to the same compressed element. Another downside is that the exponentiation in compressed form relies on a third-order recurrence relation and does not support arbitrary multiplications. As a consequence, some more complicated discrete-logarithm-based cryptosystems cannot easily be implemented using XTR. Although arguably one could compute directly in G_N and only use the trace-plus-a-trit to compress and decompress elements (where the trit is used to indicate which conjugate has been compressed), a much neater lossless compression method is based on algebraic tori, as proposed by Rubin and Silverberg in 2003 [503].

They presented two new systems: $\mathbb{T}_2$ as an alternative to LUC (based on quadratic extension fields) and CEILIDH, as an alternative to XTR. CEILIDH, pronounced Cayley, was presented as an acronym for 'Compact, Efficient, Improves on LUC, Improves on Diffie–Hellman', but was really named after Silverberg's deceased cat [504]. It is unclear whether naming CEILIDH after an undoubtedly adorable cat was at all inspired by XTR being a homophone for a bird.

In any case, like XTR, CEILIDH allows us to represent elements of $G_N \subset G_{p^2-p+1} \subset \mathbb{F}_{q^6}^*$ by using only two elements of $\mathbb{F}_q$. Note that, whereas XTR is usually presented as defined over a prime field, for CEILIDH it is customary to allow any underlying finite field $\mathbb{F}_q$, including characteristic-p fields $q = p^m$ with extension degree $m > 1$. Lenstra and Verheul already observed the same generalisation works in principle for XTR, the real advantage of CEILIDH is that the compression is injective so, given a compressed element, it is always possible to recover uniquely the original element.

We refer to the original papers by Rubin and Silverberg [503, 504] for a precise mathematical definition of algebraic tori (see also [199]) but, given a finite field $\mathbb{F}_q$ and its n-degree extension $\mathbb{F}_{q^n}$, one way of characterising the algebraic torus $\mathbb{T}_n(\mathbb{F}_q)$ is as the intersection of the kernels of the norm maps from $\mathbb{F}_{q^n}$ to $\mathbb{F}_{q^d}$ for $d|n$ (Definition 5.2) [504, Theorem 5.7.(ii)]. Moreover, the algebraic tori neatly coincide with the cyclotomic subgroups [504, Proposition 5.8], so for instance $\mathbb{T}_6(\mathbb{F}_p) = G_{p^2-p+1}$.

Loosely speaking, an algebraic torus $\mathbb{T}_n$ is rational over $\mathbb{F}_q$ if there exist a map from $\mathbb{T}_n(\mathbb{F}_q)$ to $(\mathbb{F}_q)^{\varphi(n)}$ that is defined almost everywhere as quotients of polynomials and the same holds true for its inverse. The rationality of an algebraic torus can be exploited to compress its elements with compression factor

$n/\varphi(n)$; moreover $\mathbb{T}_2$ and $\mathbb{T}_6$ are known to be rational, leading to compression factors 2 and 3, respectively. Moreover, Rubin and Silverberg provided concrete rational maps (in both directions) for $\mathbb{T}_2$ and $\mathbb{T}_6$, coining the latter system CEILIDH.

As mentioned already, CEILIDH stands for 'Compact, Efficient, Improves on LUC, Improves on Diffie–Hellman'. In a way, this acronym can be misleading as Diffie–Hellman is generally understood to be a key-agreement protocol that can be phrased independent of the underlying group; LUC turned out to be an trace-based factor-2 compression method for 'hard' subgroups of quadratic extension fields that additionally allowed efficient arithmetic on the compressed form (just like XTR). In contrast, CEILIDH is really just a factor-3 compression/decompression method for 'hard' subgroups of sextic extension fields without any efficient method for exponentiations. However, the perspective offered by algebraic tori is useful and, as we will discuss in Section 10.3.2, one can complement CEILIDH to arrive at efficient torus-based cryptography.

Amortised Compression If an efficient compression mechanism exists for $\mathbb{T}_n(\mathbb{F}_q)$ with compression factor $n/\varphi(n)$ whenever the torus $\mathbb{T}_n$ is rational, a natural question is for which n the tori are rational and which compression factors can be achieved. From a practical perspective, it is advantageous to look at the smallest n achieving a certain compression factor, which boils down to looking at the products of the successive smallest primes. So first 2, then $6 = 2 \cdot 3$, and next up would be $30 = 2 \cdot 3 \cdot 5$ providing slightly better compression than XTR.

Pre-CEILIDH, Brouwer, Pellikaan and Verheul [112] conjectured that extension degree 30 would allow for improved compression, yet Verheul later seemingly changed his mind and, with Bosma and Hutton, argued that such a system is unlikely to exist [99]. The introduction of algebraic tori by Rubin and Silverberg brought to bear a rich field of mathematics to draw upon to settle whether better compression for degree 30 is possible or not.

Rationality of algebraic tori has been well studied and Voskresenskii had conjectured that $\mathbb{T}_n$ is rational for all n (actually, that is a consequence of the conjecture; the original statement is more general and does not restrict to finite fields). For n the product of at most two primes, the conjecture has been proven, thus the torus $\mathbb{T}_n$ is known to be rational (enabling CEILIDH). On the other hand, for n the product of three primes the conjecture is still open; in particular the case $n = 30$ is still open.

Yet it is still possible to obtain almost optimal compression, but with a small caveat. Van Dijk and Woodruff [597] point out that the tori $\mathbb{T}_n$ are known to be stably rational for all n, meaning that there exist rational maps from $\mathbb{T}_n(\mathbb{F}_q) \times$

$\mathbb{F}_q^d \to \mathbb{F}_q^{\varphi(n)+d}$ for some $d \geq 0$. In many scenarios, this stable rationality allows compression for $\mathbb{T}_{30}(\mathbb{F}_q)$ that beats the factor 3 provided by XTR or CEILIDH.

Consider a hybrid encryption scheme where the ephemeral key is encapsulated by using an element of $\mathbb{T}_{30}(\mathbb{F}_q)$ and the data is encrypted symmetrically leading to some ciphertext bitstring. One could peel off an appropriate amount of bits of said ciphertext bitstring, embed them into $\mathbb{F}_q^d$, and then compress that part of the ciphertext together with the $\mathbb{T}_{30}(\mathbb{F}_q)$ element. Thus the overhead of the key encapsulation is reduced by a factor $30/\varphi(30)$.

In the original paper, van Dijk and Woodruff showed a stably rational map for $\mathbb{T}_{30}(\mathbb{F}_q)$ with $d = 32$, but less than a year later the idea was considerably refined and d was reduced from 32 to 2 with, as a side benefit, much faster rational maps [598].

10.3.2 Efficient Arithmetic

When Lenstra and Verheul introduced XTR, they compared its efficiency relative to direct operation in uncompressed form, i.e., in $\mathbb{F}_{p^6}$. As a rough efficiency measure, one can count the number of $\mathbb{F}_p$ squarings and multiplications, as these tend to be the most costly (see also Chapter 8). Cohen and Lenstra [126] had previously worked out that a $\mathbb{F}_{p^6}$-multiplication could be done in 18 $\mathbb{F}_p$ multiplications, whereas a $\mathbb{F}_{p^6}$-squaring required only 12 $\mathbb{F}_p$-multiplications. Using a standard square-and-multiply exponentiation routine, a single, respectively double, exponentiation directly in $\mathbb{F}_{p^6}$ would then cost $21 \lg N$, respectively $25.5 \lg N$, $\mathbb{F}_p$-multiplications. Note this is slightly lower than the cost of $23.4 \lg N$, respectively $27.9 \lg N$, $\mathbb{F}_p$-multiplications when using 18 $\mathbb{F}_p$-squarings for a single $\mathbb{F}_{p^6}$-squaring [377, Lemma 2.12].

Speeding Up the Cyclotomic Subgroup G_{p^2-p+1} However, it turned out that, also in non-compressed form, working in $\mathsf{G}_N \subset \mathsf{G}_{p^2-p+1}$ can be used to speed up calculations considerably. Stam and Lenstra [560] introduced a number of useful techniques in the case that $p \equiv 2 \bmod 9$ or $p \equiv 5 \bmod 9$. The congruence ensures that p generates $\mathbb{Z}_9^*$, facilitating the use of a normal base, which means that the Frobenius endomorphims can be evaluated essentially for free.

The first observation is that $p^2 - p + 1$ divides $p^3 + 1$, thus for an element $g \in \mathsf{G}_{p^2-p+1}$ it holds that $g^{p^3+1} = 1$, or equivalently $g^{-1} = g^{p^3}$. In other words, the Frobenius endomorphism can be used to invert essentially for free in the group G_{p^2-p+1}. As for elliptic curves, free inversions can be exploited for faster exponentiation based on signed representations of the exponent(s), such as the non-adjacent form (NAF) for single exponentation or the joint sparse form (JSF) for double exponentiation. The resulting speed up leads to an average cost of $18 \lg N$, respectively $21 \lg N$, $\mathbb{F}_p$-multiplications for a single, respectively double, exponentiation.

The second observation is that, as for improved XTR pre-computation, the Frobenius endomorphism can be used to split the exponent, similar to the GLV method for elliptic curves. Specifically, an exponent $x \in \mathbb{Z}_N$ can be re-written as $x_1 p + x_0$ with $\lg x_1 \approx \lg x_0 \approx \frac{1}{2} \lg N$, thus replacing a single $\lg N$-bit exponentiation with a much faster double $\frac{1}{2} \lg N$-bit exponentiation. Combining this observation with Solinas's joint sparse form, the cost of a single exponentiation goes down to $10.5 \lg N$ $\mathbb{F}_p$-multiplications, i.e., already twice as fast as the benchmark originally used by Lenstra and Verheul.

Yet, the most surprising observation made by Stam and Lenstra is that, for elements in $\mathsf{G}_{p^2 - p + 1}$, the squaring operation can be simplified and speeded up considerably, so it only takes 6 $\mathbb{F}_p$-multiplications. Combining with the previous two observations, the average cost of a single, respectively double, exponentiation in G_N can be brought down to $6 \lg N$, respectively $9 \lg N$, $\mathbb{F}_p$-multiplications [560, Theorem 4.31], which is slightly faster than the original XTR exponentiation routines, yet slightly slower than the improved XTR routines.

Speeding Up the Algebraic Torus $\mathbb{T}_6(\mathbb{F}_q)$ As CEILIDH is only a compression/decompression method for $\mathbb{T}_6(\mathbb{F}_q)$, a natural question is how the techniques to speed up calculations in $\mathsf{G}_{p^2 - p + 1}$ can be best combined with techniques to compress elements in $\mathbb{T}_6(\mathbb{F}_q)$. A straightforward approach would be to restrict CEILIDH to the case $q = p \equiv 2, 5 \bmod 9$, compress and decompress using CEILIDH, and perform all operations directly in $\mathsf{G}_{p^2 - p + 1}$, possibly incurring some additional overhead when changing basis representations.

Granger, Page and Stam [240] provide a more detailed analysis. For instance, by treating $\mathbb{T}_6(\mathbb{F}_p)$ as part of $\mathbb{T}_2(\mathbb{F}_{p^3})$, it is possible to look at partial compression and decompression and for some operations to work in $\mathbb{T}_2(\mathbb{F}_{p^3})$ on partially compressed elements (based on the work by Rubin and Silverberg on $\mathbb{T}_2$). Essentially, for exponentiations, working in $\mathsf{G}_{p^2 - p + 1}$ is alwasy optimal, but there is considerable scope in optimising intermediate operations by changing representation as required. In that sense, these optimisations are reminiscent of mixed coordinates as used for fast scalar multiplication over elliptic curves.

One obvious downside of using the Stam–Lenstra techniques for $\mathsf{G}_{p^2 - p + 1}$ is the requirement that $q = p \equiv 2, 5 \bmod 9$ to achieve fast squarings in $\mathbb{T}_6(\mathbb{F}_q)$. Granger and Scott [236] partly resolved this restriction by showing a more general method that allowed squaring in $\mathbb{T}_6(\mathbb{F}_q)$ for only 6 $\mathbb{F}_q$-multiplications for any $q \equiv 1 \bmod 6$. Finally, Karabina [313] suggested a slightly different partially compressed format using 4 $\mathbb{F}_q$ elements for $\mathbb{T}_6(\mathbb{F}_q)$ with, again, $q \equiv 1 \bmod 6$. Squarings with that representation cost as little as 4 $\mathbb{F}_q$-multiplications, but the representation does not allow for direct multiplication, necessitating decompression at an amortised cost of 3 $\mathbb{F}_q$-multiplications per decompression using Montgomery's simultaneous inversion trick. Thus, if for an exponentiation in $\mathbb{T}_6(\mathbb{F}_q)$ the number of $\mathbb{T}_6(\mathbb{F}_q)$-multiplications is less than two-thirds of the number of $\mathbb{T}_6(\mathbb{F}_q)$-squarings, the Karabina representation might be fastest of all. However, optimal use of Montgomery's simultaneous inversion trick does require that all $\mathbb{T}_6(\mathbb{F}_q)$-multiplications are done in parallel. This restriction limits the kind of exponentation routines possible, plus it requires considerable storage.

10.4 Pairings with Elliptic Curves

10.4.1 An (In)Equivalence

XTR was first presented at Crypto 2000; moreover, XTR was presented first at Crypto 2000. This prime slot in the programme allowed other cryptographers to respond with their thoughts on XTR during the very same conference at the rump session. Menezes and Vanstone [417] cheekily rebranded ECSTR as 'Elliptic Curve Singular Trace Representation' suggesting that XTR can be considered as an elliptic curve in disguise. Of course, recurrence-based 'prime' LUC turned out to be $\mathbb{F}_{p^2}$ exponentiation in disguise and elsewhere polynomial-based NTRU turned out to be a lattice cryptosystem in disguise, so it was a perspective worth entertaining. Let us investigate a little further the reasons behind Menezes and Vanstone's observation.

First, recall that an elliptic curve $E_{a,b}$ over $\mathbb{F}_q$ consists of the points $(X, Y) \in \mathbb{F}_q{}^2$ satisfying the curve equation $Y^2 = X^3 + aX + b$ for shortened Weierstrass. Together with the point at infinity, these $\mathbb{F}_q$-rational points form an additive group. If $q = p^2$, then the order of this group lies in the Hasse interval $p^2 - 2p+1, p^2+2p+1]$, indeed one often writes the order as p^2-t+1, where t is the Frobenius trace number. If $t = p$, then we have that $p^2 - t + 1 = \Phi_6(p)$, so the order of the elliptic curve matches the order of cyclotomic subgroup G_{p^2-p+1}.

Menezes [416] characterised those curves as 'Class Three' supersingular elliptic curves over $\mathbb{F}_{p^2}$ with positive parameter t (as $t = p$ rather than $-p$). For those curves, the Menezes–Okamoto–Vanstone (MOV) embedding [421] provides an efficient group isomorphism from the curves to the cyclotomic subgroup G_{p^2-p+1}. Thus any problem like DLP, CDH or DDH that is easy in G_{p^2-p+1} will also be easy for the elliptic curve. During the Crypto 2000 rump session, Menezes and Vanstone suggested that the MOV embedding would be efficiently invertible, which would immediately imply that the DLP, CDH and DDH problems are equally hard in G_{p^2-p+1} and for those supersingular curves. That would be bad news for XTR and CEILIDH, as the MOV embedding also implies that the DDH problem is in fact easy for those curves.

Verheul [605, 606] promptly took up the challenge and he showed a rather remarkable result, namely that if the MOV embedding is efficiently invertible, then the CDH problem is easy in both G_{p^2-p+1} and for those supersingular curves. Slightly more precisely, those supersingular curves are known to be of the form $E_a(\mathbb{F}_{p^2}) : Y^2 = X^3 + a$, where $a \in \mathbb{F}_{p^2}$ is a square but not a cube in $\mathbb{F}_{p^2}$. Let us call such an a suitable and, as before, let $N|p^2 - p + 1$. Furthermore, denote with $\mathbb{G}_a[N]$ the subgroup of order N of the curve $E_a(\mathbb{F}_{p^2})$ (including the point at infinity). Then Verheul showed that if for some a there exists an efficiently computable injective homomorphism from $G_N \subset G_{p^2-p+1}$ to $\mathbb{G}_a[N]$, then the CDH problem in both G_N and $\mathbb{G}_a[N]$ is easy. As the CDH problem in both is still believed to be hard, the MOV embedding is unlikely to be efficiently invertible and thus the cyclotomic subgroup G_{p^2-p+1} is not an elliptic curve in disguise.

The original MOV embedding was based on the Weil pairing, which is an efficiently computable bilinear map $e : \mathbb{G} \times \mathbb{G} \to G_{p^2-p+1}$, where $\mathbb{G}$ denotes the elliptic curve group and Verheul's result could be rephrased in terms of the search of an efficient isomorphism $\psi : G_{p^2-p+1} \to \mathbb{G}$. In modern parlance, the Weil pairing is symmetric, or of type 1 [197]. Asymmetric pairings have become more popular in practice: in those cases $\mathbb{G}_1$ and $\mathbb{G}_2$ are two distinct groups related to the elliptic curve (defined over $\mathbb{F}_q$) and the bilinear map is defined as $e : \mathbb{G}_1 \times \mathbb{G}_2 \to G_{\Phi_n(q)} \subset \mathbb{F}_{q^n}^*$. Here n is known as the embedding degree, which is the smallest n such that N divides $\Phi_n(q)$. Note that, in the wider literature, the embedding degree is more commonly referred to as k, we use n instead to highlight the connection with the degree of the algebraic torus. Verheul's results were subsequently generalised and refined to this setting as well [203, 314].

10.4.2 Improved Pairings

Once one realises that elliptic-curve pairings map into cyclotomic subgroups, or algebraic tori, one immediately obtains that the compression and efficiency techniques developed for those groups are applicable to the outputs of a pairing, leading to compressed pairings [524]. However, often the interaction between techniques for cyclotomic subgroups and pairings is a little bit more intricate [242].

A typical pairing evaluation consists of two phases: a Miller loop that returns a finite-field element and a final exponentiation to the power $(q^n - 1)/N$ that ensures membership of the group $G_N \subset \mathbb{T}_n(q)$. For the final exponentiation it is often possible to deploy ideas similar to those described in this chapter. The improved squaring routines by Granger and Scott and the compressed squaring by Karabina are especially beneficial as they work for $p \equiv 1 \bmod 6$, as commonly used for pairing-friendly curves. See also the overview by Beuchat *et al.* [56] for details.

Similarly, exponentiations in pairing groups can benefit. When the embedding degree is some multiple of 6, both XTR and CEILIDH, as well as fast squarings in the cyclotomic subgroup, apply. Which method is fastest depends very much on the situation at hand. Although XTR allows the fastest squarings, it is hard to exploit the Frobenius endomorphism beyond a split of the exponent in two; moreover the output will be compressed and can be hard to process further. Trace-less computations can benefit from a higher-dimensional 'GLV' split of the exponent, but the squarings are more expensive. See also the overview by Bos *et al.* [97, Section 6.5] for some concrete comparisons.

Characteristic 3 Galbraith [198] advocated the use of supersingular curves in characteristic 2 or 3 for pairing-based cryptography. Here the binary case has only embedding degree $n = 4$, whereas the ternary case gives embedding degree $n = 6$. The higher embedding degree offsets working with the slightly awkward characteristic [241]. Moreover, Duursma and Lee [176] proposed an efficient algorithm to compute pairings for supersingular curves over $\mathbb{F}_{3^m}$, adding to their popularity. As the embedding degree is $n = 6$, the target group of the pairing lies in $\mathbb{T}_6(\mathbb{F}_{3^m})$, which makes both XTR and CEILIDH relevant for the computations involved. Granger *et al.* [234] worked out the details, including how to slightly improve the Duursma–Lee algorithm for this case by incorporating trace- or torus-based speed ups as part of the pairing.

Surprisingly, Shirase *et al.* [536, 537] showed that, in some cases, even better compression is possible than the $\varphi(n)/n$ that was targeted by LUC, XTR, CEILIDH and van Dijk *et al.* The key observation for $\mathbb{T}_6(\mathbb{F}_q)$ is that, if $q = 3^{2k-1}$, then $\Phi_6(q) = q^2 - q + 1$ factors as $(q + 3^k + 1)(q - 3^k + 1)$. Elements in the $\mathbb{F}_{q^6}^*$ subgroup of order $(q - 3^k + 1)$ can then be represented by their trace over $\mathbb{F}_q$, rather than by their trace over $\mathbb{F}_{q^2}$ as would be the case for XTR. Thus a factor 6 compression is achieved. Karabina [313] later extended the factor 6 compression using a torus-based approach, while also showing factor 4 for the binary case.

Unfortunately, we now know that the discrete logarithm problem in finite fields of small characteristic, such as 3, is essentially broken (see Section 5.5.3), rendering these improvements moot.

Parameter Generation XTR, as originally proposed, focussed on using a prime field $\mathbb{F}_p$, but Lenstra and Verheul already mentioned the use of an extension field $\mathbb{F}_q$ instead. One challenge in that case is parameter selection: if $q = p^m$, then one requires prime p and N such that $N|\Phi_{6m}(p)$. First selecting N and then ensuring that p is a root of Φ_{6m} modulo N works only if p is at least a few bits longer than N. However, larger m are especially interesting if they allow smaller p, potentially quite a bit smaller than N.

An alternative approach then is to simply select prime p first and factor $\Phi_{6m}(p)$ in the hope to find a prime divisor N of the desired. This approach was suggested by Lenstra [366] and adopted by van Dijk and Woodruff [597] for $\mathbb{T}_{30}$.

There is, however, a third approach, as pointed out by Galbraith and Scott [201, section 9]. Pairing-friendly curves give suitable parameters for XTR (or CEILIDH), so the polynomial families of parameters used to generate those curves can also be used to generate XTR parameters. Such an approach is comparable to the fast generation suggested by Lenstra and Verheul [377, Algorithm 3.1.1], namely searching for r such that $N = r^2 - r + 1$ and $p = r^2 + 1$ are simultaneously prime (and $p \equiv 2 \bmod 3$). However, as Lenstra and Verheul already warn 'such "nice" p may be undesirable from a security point of view because they may make application of the Discrete Logarithm variant of the Number Field Sieve easier'. As explained in Chapter 5 and also Section 11.5.1, the NFS has improved for parameters for pairing-friendly curves, necessitating larger parameters for a given security level. Although the precise recommendations are still somewhat in flux [30, 173, 251, 507], for XTR and CEILIDH it seems prudent to restrict to the prime variants G_{p^2-p+1} and $\mathbb{T}_6(\mathbb{F}_p)$ and ignore 'pairing-friendly' parameter sets.

10.5 Over the Edge: Cyclotomic Subgroups Recycled

When XTR was proposed in 2000, the standard security level was still 80 bits, elliptic curves were not all that widely deployed, and people hardly worried about post-quantum security. RSA moduli were 1024 bits long and, similarly, 160-bit Schnorr subgroups were based on 1024-bit prime fields. In that historical context, for XTR over a finite field $\mathbb{F}_{p^6}$ with prime p, this meant that $6 \lg p \approx 1024$ and $\lg q \approx 160$. Thus the '160-bit' group elements in G_N could be represented by using the trace with two elements in $\mathbb{F}_{p^2}$, for only about 340 bits in total.

Compared with compressed elliptic-curve cryptography of the same strength, that is just over a doubling in the number of bits for a single group element; moreover, at the time, an XTR exponentiation was faster than an elliptic-curve scalar multiplication. For suitably chosen elliptic curves, the best-known algorithms to solve CDH/DDH still entails solving an elliptic-curve discrete logarithm problem (ECDLP); moreover, solving the ECDLP appears as hard as it is generically, thus its assumed hardness is governed by the birthday bound and, to achieve k-bit security, it suffices to work with a curve defined over a $2k$-bit field. In other words, barring the emergence of new cryptanalytical techniques or the advent of quantum computers, elliptic curves scale extremely well in the targeted security level. These days, the standard security level is 128 bits for the near term and 256 bits for long term. A natural question is therefore how the comparison between XTR, or working in cyclotomic subgroups/algebraic tori in general, holds up against using elliptic curves.

For the finite-field discrete logarithm problem, much better algorithms are known, as described in detail in Chapter 5. This immediately implies that qualitatively XTR and related systems scale considerably less well, but how does the comparison evolve concretely? To answer this question, we will consider both the perspective from the turn of the century, as well as a modern one based on current understanding of the hardness of the various discrete logarithm problems.

In 2001, Lenstra provided estimates for different security levels based on understood performance of index calculus methods at the time, as well as extrapolating the observed rate of algorithmic and computational improvements for future predictions [368]. For prime-field systems, he essentially used estimates for the NFS for factoring integers and divided by the extension degree as appropriate. Although he did not include $\mathbb{T}_{30}$ in his comparison, we used that same methodology below. For small characteristic fields, he used slightly different formulae to take into account known speed ups for that case [368, Table 6].

For instance, in 2001, 128-bit security would require an RSA modulus between 2644 and 3224 bits [368, Table 1], which, after division by six, gives us $441 \leq \lg p \leq 538$ for an XTR prime field [368, Table 5]. Similarly, his then prediction for 2020 security levels was that we would need $550 \leq \lg p \leq 660$ for an XTR prime field. Thus already by 2001 standards, the comparison skews considerably in favour of elliptic curves. Representing a single group element is roughly four times as costly (in bits) for XTR than for ECC. Lenstra also

	Recommended 80-bit in 2001	Estimated 128-bit in 2001	Predicted 128-bit for 2020	Recommended 128-bit in 2020	Recommended 256-bit in 2020
p	1024	2644–3224	3296–3956	3072	15 360
p^2	512	1322–1612	1648–1978	1536	7680
p^6	340	882–1076	1100–1320	1024	5120
p^{30}	256	712–864	880–1056	*discouraged*	*discouraged*
3^{6m}	n.a.	1260–1565	1603–1954	**broken**	**broken**

Table 10.1   *Compact representations (in bits) for cyclotomic subgroups based on various security levels and estimates.*

provided run-time estimates for various algorithms [368, Table 10] supporting the conclusion that, for higher security levels, ECC would handily beat XTR.

In Table 10.1, we summarise Lenstra's findings, but we added an initial column for 80-bit security, where for p^{30} we use the recommendation by van Dijk and Woodruff [597] (which was a bit lower, so an $\mathbb{F}_p$ element narrowly would fit in a single 32-bit word) and, for simplicity, pretend as if a single group element could be compressed all on its own. Additionally, we appended two columns with current NIST recommendations targeting either 128-bit or 256-bit security [35, page 55] (these recommendations match the ones suggested by ECRYPT [178, Table 4.6] that have been reproduced as Table 11.2). Moreover, for $\mathbb{T}_{30}$ the Granger–Vercauteren attack [238] (see Section 5.4.3) and later developments need to be taken into account and the limited potential for higher compression is not worth the risk. In general, for $\mathbb{T}_6(q)$ having non-prime q seems to offer no discernible benefit over prime $\mathbb{T}_6$ or $\mathbb{T}_2$, unless the group arises in the context of an elliptic-curve pairing, in which case $q = p^m$ for $m \in 2, 3, 4$ may be observed. Finally, the small characteristic case has been completely broken, as explained in Section 5.5.3.

Acknowledgement I am grateful to Eric Verheul and Robert Granger for their inspiration and assistance while writing this chapter.

11

History of Cryptographic Key Sizes

Nigel P. Smart and Emmanuel Thomé

11.1 Introduction

Picking a cryptographic algorithm to use is often the simplest part in designing a system that uses cryptography. The difficult part comes in how one combines different algorithms together to meet some security objective, and how the related parameters to the underlying cryptographic algorithms are selected. Probably the most important parameter is the key size, and one needs to select this for each algorithm separately, so that the combined system meets the desired security level.

Different algorithms have different metrics for measuring key strength; thus breaking a 128-bit key length for the Advanced Encryption Standard (AES) does not have the same difficulty as breaking a 128-bit key length for the Rivest, Shamir and Adleman cryptosystem RSA (the former is infeasible even with a quantum computer, whilst the latter is an undergraduate student project). To ease this issue, in 2000 Lenstra and Verheul published an influential paper titled 'Selecting Cryptographic Key Sizes' [376], with a longer version in [379]. This was the first systematic attempt to measure the relative difficulty of different problems and provide some form of scientific guidance.

Determining key sizes is not an exact science, but it is an combination of four related factors.

(1) A stable understanding of the best algorithm(s) to solve the computational problem.
(2) Good asymptotic estimates of running time and storage requirements of such algorithm(s).
(3) Some data points from large-scale computations, e.g., of cryptographic challenges.

(4) The 'tightness' of any security proof that the cryptographic scheme relies upon.

In this chapter we will ignore the latter point as, in practice, security proofs are often used to validate designs rather than used to set key sizes. Thus 'tightness' in security proofs, i.e., the gap between the difficulty of breaking the scheme versus the difficulty of breaking the underlying hard mathematical problem, is often ignored in practice.

At the time of publication of [376] in 2000 the default key size for many symmetric algorithms was 80 bits of security, although no standardised algorithm had this level of security or higher (AES did not become a standard until 2002), whereas RSA keys were suggested to be around 1024 bits in length, with elliptic-curve cryptography (ECC) keys of 160 bits in length. In practice 512-bit RSA keys were still widespread, despite a 512-bit RSA modulus being factored in the previous year [118].

Following Lenstra's paper a number of official and community recommendations followed. The most important of the official recommendations were those of NIST [447], ANSSI [20] and BSI [113]. The most influential of the community-led recommendations was the series of reports published by the ECRYPT projects in various years spanning from 2004 to 2018, with ENISA sponsoring the reports in 2013 and 2014.[1] The ECRYPT reports covered not only key size, but also other aspects of how a cryptographic primitive should be used, such as recommendations for modes of operation, etc.

To derive recommendations for cryptographic key size one needs to take into account all the possible algorithms, and ways of implementing them, that can be used to recover keys. In doing so, one needs to also take into account that one's attacker might not be just a 'script kiddy' working in their bedroom, but could be a well-funded nation-state actor with access to large software and hardware resources.

11.2 Attacking Symmetric Algorithms with Software and Hardware

Attacking symmetric algorithms (block ciphers and hash functions) by using hardware (and in some cases software) has a long history. Some of this work has been motivated by the desire to show that existing algorithms are insecure and need replacing (such as in the case of the Data Encryption Standard (DES) and MD5), some has been motivated to show just how inefficient such

---

[1] The various ECRYPT reports can be found at https://www.ecrypt.eu.org/

approaches will be (such as in the case of AES), whereas some has been to enable new applications (such as in the case of Bitcoin mining of SHA-256).

11.2.1 Attacking DES

The most famous, and impactful, research in this area has been the long history of breaking the DES algorithm. Recall that DES is one of the oldest crypto-graphic primitives to be standardised for public use and, when proposed, the key length was set to be 56 bits. Even at the time, Martin Hellman and Whit Diffie estimated that DES would be insecure against a determined attacker, by estimating that one could build a parallel computer to evaluate one million DES operations per second at a cost of $20 million [164]. Accounting for inflation this would amount to $86 million in today's money.

This idea was left, in the open community at least, as an intellectual curiosity for 20 years until, in 1998, the Electronic Frontier Foundation (EFF) built the Deep Crack machine [221]. Deep Crack could find a single DES key in 22 hours, with a cost of building the machine of $200 000. The reduction in cost, and the time needed to perform the attack, came from the relentless march of improvements in computer hardware.

With the advent of field-programmable gate arrays (FPGAs) the design and build costs for such machines came within the range of university researchers. Fast forward another decade to 2006, and the COPACOBANA FPGA-based cracker [348] was built, which cost about $10 000 and could break a single DES key in 6.4 days. At roughly the same time it was estimated [488] that, after a pre-computation of a week on a $12 000 device, one could break any DES key in half an hour by using just a $12 FPGA.

Although with special-purpose hardware one has a high upfront cost and then the cost of attacking each key is minimal, and only really costs in terms of time, pure software attacks on DES keys are possible with standard computers. Lenstra and his co-authors [328] estimated that the dollar cost of finding a key in as short a time as possible on the Amazon EC2 cloud service was $14, 700 in 2012. This reduced to $4500 by 2018, purely owing to the reduction in prices by Amazon for its cloud service. This hardness measure, in terms of dollar cost per key, is a more telling cost for weak algorithms such as DES. For a secure algorithm one expects the dollar cost per key to be so high that it dwarfs the size of the world economy.

11.2.2 Attacking AES

The attacks mentioned above on the venerable DES cipher led NIST to look for a replacement. In 1997, a competition was announced to find a replacement and in 2000, the winner, a cipher called Rijndael from Belgium, was selected. The Rijndael algorithm was then selected as the Advanced Encryption Standard (AES). AES comes in three variants, each with block size 128 bits; the variants are a cipher with key size 128 bits, one with 192 bits and one with 256 bits. Although in theory the same approaches that were applied to DES can also be applied to AES, the effects are noticeably different. For example, [328] estimates that in 2018 it would cost around 10^{19} million to break AES-128 using Amazon's EC2 cloud service.

11.2.3 Attacking Hash Functions

Not all breaking of mainstream symmetric algorithms has been done via raw computer power; some have been broken by advances in cryptanalysis, combined with raw computing power. A classic example of this is the MD5 hash algorithm. This was proposed in 1991 by Ron Rivest and it saw widespread deployment in many applications. Soon after deployment (in 1996), Dobbertin found a collision in the compression function used within the algorithm, and cryptographers began recommending that MD5 should not be used. However, this did not stop the widespread deployment of MD5 in the subsequent decades.

However, in 2004, a collision for the full MD5 was announced by a group of cryptographers from China led by Wang, effectively breaking the hash function at the CRYPTO Rump Session (see [264] for a contempory discussion and [616] for the paper). However, this was a random collision and thus was not a direct threat to computer systems. Soon after, Lenstra and co-authors constructed two X.509 certificates with the same hash value [570].

Rapid progress followed; using off-the-shelf computers one can now find collisions in specific situations in a matter of hours. That MD5 was broken led to other attacks on computer systems; for example, the FLAME malware from 2012 attacked Microsoft operating systems by exploiting problems with MD5.

When flaws in MD5 were discovered, the recommendation was to switch to the SHA-1 algorithm. This was a very similar algorithm, but cryptographers thought it would provide longer-term security compared with MD5. However, although it was stronger, it was not that much stronger. A year after finding a collision in MD5, Wang's team made another announcement at the CRYPTO Rump Session, this time giving a theoretical attack against SHA-1. Further improvements to this theoretical attack were announced over the following decade. Then, in 2015, using GPU time on Amazon's EC2 cloud service, Stevens *et al.* announced the first collision in SHA-1 requiring a $2000 cost per new collision on EC2[2]. In 2020, Leurent and Peyrin announced the first practical chosen prefix collision on SHA-1. This method requires an expenditure of less than $100 000 per collision, and builds upon the work in [390].

The flaws in SHA-1 announced in 2005 led NIST to release a modified, more secure version, called SHA-2 (which comes in various output sizes, the most popular being SHA-256). In addition, NIST announced a competition for a new hash algorithm, to be called SHA-3. The competition started in 2006, and in 2012 was won by a Belgium team with their Keccak algorithm. Lenstra *et al.* estimated the cost of breaking SHA-256 via running EC2 instances at 10^{58} million dollars. Currently, no estimates exist for the cost of finding a collision in SHA-3.

The algorithm SHA-256 is used in the Bitcoin system as the hard problem on which miners operate to authenticate a block. Each miner needs to evaluate two evaluations of SHA-256 and then see if the resulting output has a special form before they can claim their reward. The profitability of this mining operation led first to FPGA and then to ASIC mining hardware. The total Bitcoin system (as of 2020) is hashing at a rate of roughly 2^{66} hashes per second. To find a collision in SHA-256 is expected to take 2^{128} such hashes; thus even with the entire Bitcoin network we can expect to find a hash only in 2^{36} years!

11.3 Software Attacks on Factoring and Discrete Logarithms

The integer factorisation problem and the discrete logarithm problem in finite fields were the two hard problems underpinning the security of most of public-key cryptography in its first decades of existence. Although a variety of primitives that rely on the assumed hardness of other mathematical problems have appeared since, the prevalence of the primitives that rely on integer factorisation and discrete logarithm, such as RSA and the Diffie–Hellman key exchange using multiplicative groups of finite fields, is clear.

One, at times, may encounter the argument that the hardness of factoring is a well-studied problem because it has been actively studied for a long time (decades, centuries, or millenia for the most bodacious statements of this kind). In fact, the first important algorithmic improvements over naive methods date

---

[2] https://sites.google.com/site/itstheshappening/

back to the early 1970s, with work by Shanks [534], Pollard [475] and Morrison and Brillhart [444]. The motivations for these early endeavours towards smarter and smarter factoring methods were, for example, the factorisation of discriminants of number fields, or the factorisation of numbers of the form $b^n \pm 1$, which are potentially useful in a variety of settings ranging from combinatorics to group theory.

The advent of the RSA cryptosystem in the late 1970s, and its dependence on the hardness of factoring, led to a considerable surge in academic interest (and certainly non-academic interest as well). A good measure of the perceived (in)feasibility limit by 1978 is given in the RSA paper itself [501]: factoring a 100-decimal digit integer was estimated to require about a century of computation time. We will see shortly that this key size withstood factoring attempts for barely a decade. This provides two immediate lessons. First, in hindsight, four decades of algorithmic progress have evidently gone a long way. Second, estimating hardness is a tricky matter.

The Morrison–Brillhart CFRAC algorithm can be regarded as the ancestor of a series of algorithms that includes in particular the quadratic sieve and the Number Field Sieve, both of which have numerous variants. For four decades, it has been a cryptographically important task to provide, on a regular basis, good estimates of how hard factoring actually is in practice, using either commodity hardware or various kinds of more special-purpose equipment, ranging from academic supercomputers to specially designed hardware accelerators. Such estimates include the development of algorithmic ideas that are very effective in improving the computation time, often based on the constraints of the chosen computation platform.

11.3.1 Factoring by Email

The early 1980s saw several improvements in the state of the art of integer factorisation. In their 1988 paper 'Factoring by Electronic Mail' [371], Lenstra and Manasse addressed the variety of existing factoring records at the time, and provided a pragmatic answer to the question: how big are the integers we can factor within one month of elapsed time, if we only want to use computing time that we can get for free? This article was influential in defining the line of work of academic records for the next few decades. It firmly installs, however, a distinction between academic records, done with this strategy, and what can be done by governments, or possibly criminal organisations, who are ready to spend money to meet their goals.

Lenstra and Manasse used the MPQS algorithm, which is the multiple polynomial variant of the quadratic sieve. This algorithm is described in [547]. In

MPQS (and also in CFRAC, which MPQS superseded), one can separate the computation into two main steps. First, the relation-collection step searches for multiplicative relations involving integers modulo N. In a later matrix step, these relations are combined to form a congruence of squares. Finding this combination is equivalent to finding an element in the nullspace of a sparse matrix over the binary field GF(2). As a matter of fact, these two steps have also been the two main steps of all large factoring and discrete logarithm computations since.

The relation-collection step in MPQS can be distributed massively. By 1988, standard hardware that was capable of contributing significantly to the relation collection towards the factorisation of numbers of, say, 100 decimal digits, was relatively common in universities and research laboratories. Furthermore, the amount of output that was produced by each machine participating in the computation was relatively low. This made it possible to gather the results using any commodity means, and electronic mail was a perfect fit for that goal, usefully leveraging existing infrastructure.

The largest computation done with MPQS, following this approach of gathering contributions from hundreds of enthusiasts, was the factorisation of RSA-129. This computation solves a challenge that was concocted by the RSA authors and presented in Martin Gardner's column in the August 1977 issue of *Scientific American* [207]. Atkins, Graff, Lenstra and Leyland factored the public modulus in 1994 and were able to unveil the secret RSA-encrypted message that was 'The Magic Words are Squeamish Ossifrage' [24]. A striking fact about this computation is the number of contributors (more than 600) and the variety of machines used, ranging from 16MB 80386sx PCs to Cray C90s (the code was even ported to fax machines).

The RSA-129 challenge was the most famous of a number of challenge factorisation problems. The RSA company itself published a list of factorisation challenges at various key lengths to encourage people to look into the security of cryptography based on factoring. This became known as the RSA Challenge list.

11.3.2 The Development of the Number Field Sieve

The RSA-129 paper concluded, quite naturally, with extrapolations on the hardness of reaching 'most wanted' targets such as the factorisation of 512-bit RSA moduli. Such keys were in widespread use at the time. With MPQS, it was estimated this goal was almost feasible by using massive power, but also qualified as 'the last gasp of an elderly workhorse'. The more recent Number

Field Sieve, with better asymptotical behaviour, appeared probably more fit for that goal.

The Number Field Sieve originated in an astute way, designed by Pollard in 1988, to factor the seventh Fermat number $2^{128} + 1$. Pollard used cubic integers. More mathematics were entering the scene. Pollard's method was understood by many as a curiosity. Its reliance on an almost coincidental match between cubic integers and the special form of $2^{128} + 1$ raised doubts as to the possibility of extending the approach to other numbers.

Generalisations did come, however. By the end of 1989, Lenstra, Lenstra, Manasse and Pollard had extended Pollard's original approach to further numbers, and, in 1990, were able to factor the ninth Fermat number $2^{512} + 1$, which was something of a holy grail to factoring enthusiasts. Several challenges still had to be overcome in order to factor general numbers, not the least of which was the fact that the matrix step, it seemed, had to be solved with coefficient ring $\mathbb{Z}$ – it later appeared that solving over $\mathbb{Z}/2\mathbb{Z}$ was enough. Several people contributed to loosening the constraints. The 1993 book 'The Development of the Number Field Sieve' [360] includes further work by Bernstein, Buhler, Couveignes and Pomerance, leading to a General Number Field Sieve (GNFS) algorithm, which was convincingly able to factor large numbers with good performance.

While early successes with GNFS were obtained while the algorithm was still being developed, the RSA-129 MPQS mark was improved in 1996, with the factorisation of RSA-130 by GNFS. Because the Number Field Sieve had the peculiarity of being able to take advantage of the special form of integers to be factored in some cases, a good benchmark could no longer be the good old way of picking challenge numbers from the Cunningham tables, or otherwise defined numbers that had 'some mathematical interest'. This good benchmark was defined in 1991 by the RSA Challenge list, formed of good RSA moduli, with two secretly chosen prime factors of similar size.

Following RSA-130, several records followed, most often picked from the RSA Challenge list. All were factored with GNFS. This provided a nice view of the progress of computational power on the one hand, and also on the progress on implementation and algorithmic refinements on the other. The landmark factorisation of a 512-bit RSA modulus was reached in 1999, the culmination of a significant effort from many contributors, coordinated by CWI in Amsterdam. The list of factored RSA-challenge numbers at the time of writing is given in Table 11.1.[3]

---

[3] Source `https://en.wikipedia.org/wiki/RSA_Factoring_Challenge`.

Challenge name	Digits	Bits	Date factored	Factored by
RSA-100	100	330	1 Apr. 1991	A. K. Lenstra
RSA-110	110	364	14 Apr. 1992	A. K. Lenstra and M. S. Manasse
RSA-120	120	397	9 Jul. 1993	T. Denny *et al.*
RSA-130	130	430	10 Apr. 1996	A. K. Lenstra *et al.*
RSA-140	140	463	2 Feb. 1999	H. te Riele *et al.*
RSA-150	150	496	16 Apr. 2004	K. Aoki *et al.*
RSA-155	155	512	22 Aug. 1999	H. te Riele *et al.*
RSA-160	160	530	1 Apr. 2003	J. Franke *et al.*
RSA-170	170	563	29 Dec. 2009	D. Bonenberger and M. Krone
RSA-576	174	576	3 Dec. 2003	J. Franke *et al.*
RSA-180	180	596	8 May 2010	S. A. Danilov and I. A. Popovyan
RSA-190	190	629	8 Nov. 2010	A. Timofeev and I. A. Popovyan
RSA-640	193	640	2 Nov. 2005	J. Franke *et al.*
RSA-200	200	663	9 May 2005	J. Franke *et al.*
RSA-210	210	696	26 Sep. 2013	R. Propper
RSA-704	212	704	2 Jul. 2012	S. Bai, E. Thomé and P. Zimmermann
RSA-220	220	729	13 May 2016	S. Bai, P. Gaudry, A. Kruppa, E. Thomé and P. Zimmermann
RSA-230	230	762	15 Aug. 2018	S. S. Gross
RSA-768	232	768	12 Dec. 2009	T. Kleinjung *et al.*
RSA-240	240	795	24 Nov. 2019	F. Boudot, P. Gaudry, A. Guillevic, N. Heninger, E. Thomé and P. Zimmermann
RSA-250	250	829	28 Feb. 2020	F. Boudot, P. Gaudry, A. Guillevic, N. Heninger, E. Thomé and P. Zimmermann

Table 11.1 *The solved RSA challenges.*

11.3.3 Discrete Logarithms in Finite Fields

The development of algorithms to solve discrete logarithms in large-prime characteristic finite fields $\mathbb{F}_p$ follows roughly the same development of the case of factoring.[4] The basic algorithm is to perform a sieving stage, followed by a large matrix step. The key difference is in the matrix step. For factoring the matrix equation one is trying to solve is modulo 2, whereas for discrete logarithms it is modulo q, where q is the large prime factor of $p - 1$.

The methods used to collect relations are also almost identical. Indeed, in 2000 Schirokauer [510] showed how to use the Number Field Sieve to solve discrete logarithms in the field $\mathbb{F}_p$, and that is now the method of choice. As with factoring, there is the basic Number Field Sieve, and a more-efficient

---

[4] We leave the case of non-prime fields to later.

variant called the Special Number Field Sieve, which can use special properties of the representation of the number p if they exist.

At the time of writing, a record for discrete logarithms has just been announced for a general 240-decimal digit, 795-bit prime field [101]. The record for primes of special form are slightly larger, with a discrete logarithm computation in 2016 for a 308-decimal digit, or 1024-bit, prime [195]. Note that discrete logarithms are slightly harder than factoring, judging by the publication time of records for general 768-bit factoring versus prime discrete logarithms (2009 and 2016, respectively) or of the records for special 1024-bit factoring versus prime discrete logarithms (2007 [22] and 2016). However, the largest known records were done at the same time, and indicate that the difference gap between the two problems is not as large as usually considered [101].

11.4 Hardware for Factoring

At the heart of modern factoring algorithms, and to some extent almost all algorithms for solving discrete logarithms, is the so-called sieving step. This sieving step is essentially trying to solve a more-complex version of the classical problem of finding simultaneous solutions to modular diophantine equations

$$x = a_i \quad (\text{mod } m_i) \quad \text{for} \quad i = 1, \dots, t.$$

Classically 'machines' to solve this problem date back to as far back as 1896, when Lawrence presented a mechanical design for such a machine. In 1912, the first actual design was implemented by four French mathematicians Gérardi, Kraitchik, and Pierre and Eugène Carissan. The most famous of the mechanical devices were those built by the Lehmers in the 1930s, some of which can now be found in the Computer History Museum in Silicon Valley. For a discussion of the history of such early machines see [530] and [356, 357, 358].

The advent of modern factoring algorithms led to a renewed interest in such 'sieving machines'. In the early 1980s, a special-purpose computer called the 'Georgia Cracker' was designed to implement the CFRAC factoring algorithm by Smith and Wagstaff [482]. This was accompanied by another design by Rudd, Buell and Chiarull who designed a machine in 1984 that could factor numbers ten times faster than existing general-purpose computers [505].

In the early 2000s there was renewed interest in special-purpose hardware for the sieving step. This was initiated by a Eurocrypt 1999 Rump Session talk in which Adi Shamir introduced his idea for the TWINKLE machine. In classical sieving we sieve a region by assigning each region some space in a computer memory, then we use time to go through all the primes in the factorbase to encode the contribution of each prime to each location in the space. TWINKLE switches this around, and uses light-based computation (much like one of Lehmer's devices [357]). The idea now is that space is used up by lights that flash according to a timing signal dependent on the prime, and thus each light corresponds to a prime, although we use time to represent the sieving region. At a time signal corresponding to when more lights flash we have an element in the sieving region that needs to be investigated further. The initial TWINKLE design was presented in [532] and then analysed in [373] by Lenstra and Shamir. It was deemed too costly to build in order to factor numbers of interest.

TWINKLE was soon followed in 2003 by another idea from Shamir and Tromer called TWIRL [533]. This dispensed with the optical computing methodology, but kept the time/space reversal of TWINKLE. The design was based on novel routing networks to represent the sieving operation. The design was never built, but was considered feasible for factoring 1024-bit integers in a paper by Lenstra and co-authors [384]; one estimate gives a cost of $1.1 million dollars to factor a 1024-bit integer in one year. A related design, based on an idea of Bernstein, was presented by Geiselmann and Steinwandt [215, 216]. It, too, was considered practical for 1024-bit moduli, but slightly more expensive to build than TWIRL. Related to TWIRL is a design from 2005 by Franke and others called SHARK [192]. This used a different form of routing network, but could utilise more off-the-shelf components. The authors estimated that one could factor 1024-bit numbers in a year at a cost of $200 million dollars. This is higher than TWIRL, but was considered easier to build with current technology; thus the design risk was less.

It is not only the sieving step that has been the subject of hardware design proposals. Modern factoring also involves a large matrix step. In 2001, Bernstein presented a proposal for the matrix step, which in fact influenced the methods of Geiselmann and Steinwandt mentioned above. Bernstein's idea was to reduce the amount of idle storage in the standard matrix algorithm by using a form of mesh sorting. The design was analysed, again by Lenstra and co-authors, in [383], where they concluded that the method was not cost-effective in factoring 1024-bit integers.

Special-purpose but commodity hardware, has also been used to factor integers. The graphical processing units (GPUs) in modern gaming consoles provide a cheap form of computing. Often the hardware is sold at a discount as the manufacturer aims to get their investment back in sales of games. Combined with the amazing parallel processing capability of modern GPUs, games consoles form an attractive way of obtaining super-computer performance at a fraction of the price. In 2011 it was shown how to utilise such hardware to factor integers, by using a variant of the elliptic-curve factoring method (ECM) [87, 90].

11.5 Attacking Cryptosystems Based on Elliptic Curves

For an elliptic curve $E(\mathbb{F}_q)$ over the finite field $\mathbb{F}_q$ to be used in cryptography, one usually selects a curve that has a subgroup of prime order p, with $p \approx q$. By Hasse's Theorem we cannot have $p \gg q$, and for efficiency we want to keep q as small as possible, but the complexity of the best algorithm for attacking an elliptic curve has complexity $O(\sqrt{p})$. Thus, to obtain 80-bit security one selects a 160-bit curve and to obtain 128-bit security one selects a 256-bit curve, and so on.

Inspired by the RSA Challenges in 1997, Certicom (a Canadian company that specialised in ECC-based cryptography) launched the Certicom ECC challenge. This comprised a set of toy challenges (for 79-bit, 89-bit and 97-bit curves), Level I challenges (for 109-bit and 131-bit curves), plus Level II challenges (for 163-bit, 191-bit, 239-bit and 359-bit curves). The challenges were also sub-divided into those over finite fields of characteristic two and those over large-prime fields.

By the end of 1999 all of the toy challenges had been solved in both categories of fields. These attacks were performed via the method of van Oorschot and Wiener [602], which are themselves extensions of the ideas of Pollard [476, 477]. Clients were distributed across the Internet on people's idle machines, with communication back to the central control point being performed either by email or via socket connections. The 109-bit challenges were then all solved by 2004. Some teams have attempted to solve the 131-bit challenge, but so far no reports of success are available.

The application of gaming consoles mentioned earlier has also been applied to attacking elliptic-curve-based systems. For example, breaking a 112-bit elliptic-curve instance by using gaming consoles is discussed in [91].

In [328] Lenstra and his co-authors consider the cost of breaking ECC keys using Amazon EC2 instances alone. They estimate that determining a 256-bit ECC key will require one to spend $\$10^{20}$ million on EC2. With the estimated costs for the Level II Certicom challenges being $\$10^6$, $\$10^{11}$, $\$10^{18}$ and $\$10^{36}$ million respectively. Current key sizes are recommended to be 256-bits for ECC, to correspond to the 128-bit security of AES-128, or 512-bits, to correspond to the 256-bit security of AES-256. Thus, barring the advent of a quantum computer, current recommended key sizes for ECC can be considered secure.

However, one needs to avoid 'special curves', which have attacks that perform better than $O\left(\sqrt{p}\right)$. One such class of curves is those defined over finite fields with $q = \ell^n$. An example of such attacks is the Weil descent attack when $q = 2^n$ and n is composite, as explained in [212], the so-called GHS attack. This attack exploits special properties of elliptic curves over such fields for composite n. The attack works much like factoring methods; in the first stage one collects relations, and then in the second stage one uses linear algebra to find specific discrete logarithms. Thus although this attack has no effect on key sizes it does have an effect on the choice of specific key types.

In recent years, the method of Weil descent has been extended to elliptic curves over general fields of the form $\mathbb{F}_{\ell^n}$. This study was initiated by Semaev [529], who introduced the Semaev summation polynomials into the Weil restriction methodology. This idea was then developed by Gaudry [210] and Diem [158]. A nice succinct statement of these results is that, if ℓ is a prime power and n is such that

$$a \cdot \sqrt{\log(\ell)} \le n \le b \cdot \sqrt{\log(\ell)}$$

for some fixed positive real numbers $a < b$, then asymptotically one can solve the discrete logarithm problem on an elliptic curve $E(\mathbb{F}_{\ell^n})$ in time

$$e^{O\left(\log(\ell^n)^{2/3}\right)}.$$

Another class of weak curves is the so-called anomalous curves, which are the curves for which $p = q$. These were shown to be very weak in three papers which appeared at roughly the same time [509, 528, 550].

11.5.1 Pairing-Based Cryptography

The most interesting class of curves for which there is a possible weakness is those that possess a pairing to a finite field. In particular there are classes of elliptic curves $E(\mathbb{F}_q)$ for which there are numbers $k \le n$ for which there is an efficiently computable map

$$\hat{t} : E(\mathbb{F}_q) \times E(\mathbb{F}_{q^k}) \longmapsto \mathbb{F}_{q^n}.$$

In fact, such pairs (k, n) exist for all elliptic curves, but curves for which (k, n) are small are very rare indeed.

In the early days of ECC it was common to select curves, for efficiency reasons, which had $k = 1$ and $n \le 6$. However, this means one can map the discrete logarithm problem in $E(\mathbb{F}_q)$ into the discrete logarithm problem in

$\mathbb{F}_{q^n}$ [421]. Suppose $E(\mathbb{F}_q)$ is chosen to have 80-bit symmetric security, i.e., 160-bit ECC security, then this means the finite field $\mathbb{F}_{q^n}$ is of size less than 960. In the early 1990s this killed off such curves in the case when $n \neq 6$.

When $n = 6$ one seemed to have a nice balance in the security, as in those days 1000-bit finite-field discrete logarithms were considered as secure as 160-bit elliptic-curve discrete logarithms. This 'magic' property of the number 6, i.e., $6 \cdot 160 \approx 1000$, was used by Lenstra and Verheul to construct the XTR cryptosystem in 2000 [377].

However, the move to 128-bit symmetric security, and the equivalent 256-bit elliptic-curve security and the supposed equivalence of 3072-bit finite-field discrete logarithm security, led to the number six losing its magic properties and being replaced by the magic number 12, because $12 \cdot 256 = 3072$. We shall return to the number 12 below. Thus the existence of such pairings seemed to imply that such curves should be avoided at all costs in cryptographic applications.

However, the interest in pairings on elliptic curves was ignited when Joux [292] found a constructive application of such mappings. This was quickly followed by other applications, such as identity-based encryption [80, 506]. Soon a whole zoo of applications of pairings was created. But such a zoo needed secure curves to work on.

The main security requirement of pairing-based systems is that the curve $E(\mathbb{F}_q)$ needs a large subgroup of prime order p with $\log_2 p$ being twice the desired symmetric security level, e.g., $p = 2^{256}$. We also would like q to be small to enable good efficiency. But we need a pairing to exist for which n is such that the finite field $\mathbb{F}_{q^n}$ gives hard discrete logarithms.

There turned out to be three popular choices.

(1) Pick $q^n = \ell^m$, for $\ell = 2$ or $\ell = 3$. The advantage here is one obtains very good implementation efficiency. The problem is the discrete logarithm in low characteristic fields turned out to be not as hard as originally thought (see below).

(2) Pick $p \approx q$ and select n to make the discrete logarithm hard. Thus we return to picking n to be (say) the magic number 12. Such curves are usually called 'pairing-friendly'. A typical choice here was the so-called Barreto–Naehrig (BN) curves [40]. The problem here is that such curves are subject to a potential 'medium-prime' attack (again see below).

(3) Pick $n = 2$, and make q huge, e.g., $q \approx 2^{1500}$, but keep p of size 2^{256} (say). Here one loses efficiency, but we are fairly certain of security.

We now discuss these three cases in turn.

When $q^n = \ell^m$ for Small ℓ Algorithms to solve discrete logarithms in small characteristic field have had a very interesting history. For $\ell = 2$, the first algorithms can be dated back to 1982 [268]. But the real advance came in 1984 when Coppersmith [135], building upon earlier ideas of Blake *et al.* [63], gave an algorithm with heuristic complexity $L_{q^n}(1/3, c + o(1))$. This was the first algorithm to have such a complexity, and this was a decade before the NFS algorithm was known for the factoring problem with a similar complexity.

The idea behind Coppersmith's algorithm was shown, after the invention of the Number Field Sieve, to be closely related to another algorithm called the Function Field Sieve, which was invented by Adleman in 1994 [6]. This algorithm was improved slightly by Joux and Lercier in 2002 [295], but there seemed to be little further improvements that could be made.

Then, in 2013, a series of papers and email announcements started coming out. In looking at fields of the form $q = 2^m$, an announcement by Joux in February 2013 gave a $L_{q^n}(1/4, c + o(1))$ complexity algorithm and a new record of $m = 1778$, which was way above existing records [294]. Then in the same month, Granger *et al.* solved discrete logarithms in a field with $m = 1971$, using a variant of the Function Field Sieve [226]. In March, Joux responded with another record of $m = 4080$, with Granger *et al.* responding with $m = 6120$ in April. The back-and-forth continued, with Joux posting a record of $m = 6168$ in May. In June 2013, Barbulescu *et al.* presented a quasi-polynomial time algorithm for fields of the form $q^n = 2^m$ [32]. Finally, in January 2014, Granger *et al.* posted the solution to a problem with $m = 9234$. Thus in one year the entire area of pairing-based cryptography over finite fields of characteristic two had been destroyed. The current record is a huge value of $m = 30750$ set by Granger *et al.* in 2019 [246].

The case of characteristic three finite fields, which held a lot of promise for pairing based cryptography, faired not much better. In the period from 2010 to 2016 the record advanced from a 676-bit field, up to 4841 [4].

Medium-Prime Finite-Field Discrete Logarithms The 'medium-prime' case of the finite-field discrete logarithm problem is for the case $q^n = \ell^m$, where ℓ is not small and m is not one. It was not really until the advent of pairing-based cryptography and XTR in the early 2000s that this problem became of interest to cryptographers. The first major work on this case of the problem was by Granger and Vercauteren [238], and Joux and Lercier [296], which showed that a variant of the Function Field Sieve could be applied in this case to obtain a complexity of $L_{q^n}(1/3, c + o(1))$. In [302] the analysis was then carried out for the Number Field Sieve algorithm as well. This enabled a

$L_{q^n}(1/3, c + o(1))$ algorithm to be applied in all finite fields; with the precise algorithm to be used, depending on the sizes of ℓ and m.

In 2015, Barbulescu *et al.* [34] initiated the renewed study of an earlier algorithm of Schirokauer called the Tower Number Field Sieve (TNFS), which had appeared in [510]. There followed a rapid sequence of papers looking again at the TNFS as a means to attack the third type of pairing parameters mentioned above, e.g., [33, 321, 322, 508].

No efficient implementation of the TNFS algorithm exists as of 2019. Yet, hardness estimates can be made by using rough extrapolations. This was done, for example, by Barbulescu and Duquesne in [30] or Guillevic in [252], who provided comprehensive analyses of key sizes for such pairing-based systems. In particular, their recommendation is that, for 128-bits of symmetric security, a BN curve needs to be selected with at least $q \approx 2^{446}$, which makes the implementation advantages of BN curves disappear, because q is no longer of the order of 2^{256}. Thus curves would need to be selected to obtain the same balance between security and performance, with a higher embedding degree n. This is a particular problem, as such BN curves have recently been standardised for use in systems such as the Trusted Computing Module.

When $n = 2$ In this case the theoretical algorithm of choice is that of Joux *et al.* [302]. Using an adaption of this method, in 2014 Barbulescu *et al.* [33] presented a discrete logarithm record in a finite field of the form $\mathbb{F}_{q^2}$, for a prime q of 298 bits, where the interesting subgroup had order $p \approx 2^{294}$. To date this is the largest problem solved, and hence it appears that picking q of approximately 512 bits to 1500 bits should still result in a secure pairing system.

11.6 Post-Quantum Cryptography

The problem with the above analysis of the difficulty of the factoring and discrete logarithm problems, is that it becomes totally redundant if a quantum computer is ever built. In 1994, Shor [538] came up with a polynomial time algorithm for both the factoring problem and the discrete logarithm problem in any finite abelian group. The only problem with Shor's algorithm is that one requires a quantum computer to run it.

A quantum computer uses quantum bits (so-called qubits) as opposed to classical bits. Currently, very small numbers of quantum bits can be processed for a short period of time without falling prey to noise. Thus, the issue is whether one can build a quantum computer, with enough qubits, that can run noise-free for long enough to factor an interesting number, or break an interesting discrete logarithm.

The current quantum-factoring records are much less than can be done with classical computers. In 2001, a team from IBM factored the number 15 using a quantum computer with seven qubits [603]. At the time of writing the current record is to factor $4\,088\,459 = 2017 \times 2027$, which was done on a 5-qubit processor [150]. However, despite these numbers being small, the threat of a quantum computer means that people have turned to examining so-called post-quantum systems.

These post-quantum public-key cryptographic systems are necessarily based on different hard problems (lattice-based problems, coding-theory-based problems, problems in isogeny theory, and so on). However, these problems need to go through the same analysis of difficulty to determine key sizes as has previously happened for the pre-quantum algorithms. At the time of writing, NIST is running a 'competition' to determine what type of post-quantum cryptosystem will be most suitable in the coming decades.

Lattice-Based PQC

One of the most promising areas of post-quantum cryptography (PQC) is those systems based on lattices. The most basic lattice problem is the shortest-vector problem, which is the problem of determining the smallest non-zero vector in the set

$$\{\, \mathbf{x} = A \cdot \mathbf{k} : \mathbf{k} \in \mathbb{Z}^n \,\}$$

when A is a random square integer matrix of dimension n. The most-famous classical algorithm to solve this problem is the LLL algorithm [380], and its modern improvements such as the BKZ algorithm, which was introduced in [512] and then improved in other works such as [122]. However, none of the classical algorithms to solve this problem can be significantly improved by using a quantum computer. Thus, the closest-vector problem seems a very good candidate to use to build post-quantum cryptographic systems. As with factoring and elliptic-curve-based systems before them, lattice-based cryptanalysis has been spurred on by a series of challenge problems. The most famous of these is the Darmstadt lattice challenge (https://www.latticechallenge.org). This presents a number of challenges for different dimensions n, and for different variants of lattice problems.

The most-important variant of the challenges is that given by the Learning-With-Errors (LWE) problem. This is the problem to recover the vector $\mathbf{s} \in \mathbb{Z}^n$ given only a random matrix A and the vector $\mathbf{b}$, which satisfies

$$\mathbf{b} = A \cdot \mathbf{s} + \mathbf{e} \pmod{q}$$

for some vector **e** of small Euclidean norm. The LWE problem is related to the classical closest-vector problem in lattices, but it can be used more easily to construct cryptographic systems.

The complication with lattice problems, and LWE problems in particular, is that the potential attack algorithms have many different parameters that can be tweaked, and many different sub-procedures may have (small) potential quantum speed-ups. Thus determining a precise 'key-size' estimate is hard. To help with this task for the LWE problem, there is software which performs the calculation, called the lwe-estimator (`https://bitbucket.org/malb/lwe-estimator/src/master`).

Coding-Based PQC

Basing public-key cryptography on the difficult problem of decoding random linear codes is a long-standing construction. The earliest designs go back to McEliece [413] in 1978. What is quite remarkable about code-based cryptography is that the state of the art in terms of cryptanalysis has remained relatively constant over the past 40 years. This is, in part, due to the problem of decoding random linear codes being known to be NP-hard.

Again to spur the community into studying these problems, especially due to the NIST 'competition', a set of challenge problems have been created (`http://decodingchallenge.org`). The problems, much like the Darmstadt lattice challenge problems are divided into generic decoding problems, and problems related more closely to proposed cryptographic systems.

MQ-Based PQC

It is well known that the boolean satisfiability (SAT) problem is NP-complete, thus it is clearly an attractive proposition to try to base cryptographic schemes on problems related to SAT. One such attempt is to use so-called MQ-based problems. In these problems one is given a large set of multivariate quadratic equations over a finite field (usually one of characteristic two) and one is asked to find a solution. The first such important proposal in this space was by Matsumoto and Imai in 1988 [403], who presented a proposal for public-key signatures and public-key encryption. This was, however, broken in 1995 by Patarin [470].

The basic idea remains attractive. Whereas almost all work on public-key encryption based on MQ systems has not led to a good system, the situation for public-key signatures is very different. Patarin and his co-authors introduced the 'Balance Oil and Vinegar' signature scheme in 1997 [471] and then the 'Unbalanced Oil and Vinegar' scheme in 1999 [323]. These ideas are still being worked on, and a number of submissions to the NIST 'competition' are based on the Oil and Vinegar construction.

Again, to help research in cryptanalysis of MQ-based cryptography, a website with a set of challenges has been set up (`https://www.mqchallenge.org`).

Symmetric-Key-Based PQC

Another way to construct post-quantum signatures is to dispense with complex constructions based on hard mathematical problems, and simply base public-key cryptography upon symmetric-key cryptography. After all, quantum computers are not expected to pose much of a threat to standard symmetric-key cryptography. Two prominent post-quantum signature schemes do exactly this.

The first construction based on hash functions goes back to an old idea from 1979, namely the one-time signature scheme of Lamport [351]. This can be combined with Merkle trees to form a stateful many-time signature scheme [422]. To remove the stateful nature of the signature scheme, an algorithm called SPHINCS was proposed in 2015 [54].

The second construction makes use of the difficulty of inverting one-way functions, and is based on a technique from theoretical cryptography called MPC-in-the-Head [285]. This methodology was extended in a series of works [12, 219]. When instantiated with low-complexity block ciphers such signature schemes are rather efficient.

11.7 Key-Size Recommendation

The asymptotic complexity of GNFS to factor an integer N is of the following form, as $N \to \infty$:

$$L(N)^{1+o(1)}, \text{ with } L(N) = \exp\left((64/9)^{1/3}(\log N)^{1/3}(\log \log N)^{2/3}\right).$$

Almost 30 years ago, the complexity of MPQS was written with a similarly shaped equation, with a larger exponent $1/2$ instead of $1/3$. But the $(1 + o(1))$ part was present there too. In that MPQS context, the RSA-129 paper contained the following cautious words, to guard against the temptation to ignore the $o(1)$ and use simply the analogue of $L(N)$ as an operation count: since $o(1)$ is neither 0 nor constant, this practice hardly makes sense. Even knowing the computation time from one number N_1 gives only shallow ground for an extrapolation to another size N_2 based on the ratio $L(N_2)/L(N_1)$: this cannot account for the growth of what is hidden by $o(1)$. Yet, with due care, extrapolations based on such ratios were proposed over the years, based on known

	Parameter	Legacy	Future system use	
			Near term	Long term
Symmetric-key size	k	80	128	256
Hash-function output size	m	160	256	512
MAC output size	m	80	128	256
RSA problem	$\ell(n) \geq$	1024	3072	15 360
Finite-field DLP	$\ell(p^n) \geq$	1024	3072	15 360
	$\ell(p), \ell(q) \geq$	160	256	512
ECDLP	$\ell(q) \geq$	160	256	512
Pairing	$\ell(q^n) \geq$	1024	3072	15 360
	$\ell(p), \ell(q) \geq$	160	256	512

Table 11.2 *Key-size analysis, where $\ell(\cdot)$ refers to the bit length of the parameter.*

GNFS records, in order to answer the much-needed question: what key size will match the appropriate security level by year X?

Table 11.2 is given in [178] as a summary of recommendations from various sources. It summarises the key-size recommendations for the 'standard' algorithms currently in use, i.e., it ignores the current zoo of post-quantum algorithm proposals. The column 'Legacy' refers to key sizes that are currently in use and that should be retired as soon as possible (but not urgently); whereas 'Near term' refers to a time period until 2026 (10 years ahead given the 2016 publication) and 'Long term' refers to projecting security beyond that horizon.

Another way of measuring security was also introduced by Lenstra and co-authors in [387]. In this paper, the authors suggested measuring the energy required to break a cryptosystem. This idea stemmed from a remark made about the factorisation of the RSA-768 challenge; in that the authors estimated that it used as much energy as would have sufficed to bring two Olympic-size swimming pools to the boil from a starting temperature of 20 °C.

The traditional 80-bit security level is equivalent to boiling the average daily rain fall of the Netherlands (which is apparently according to [387] the healthy equivalent to 2^{15} swimming pools per day). To obtain 128-bit security one needs to expend the energy to boil off all the water on the planet. Thus with this metric one can rest assured that ciphers that require 2^{128} operations to break them will remain secure for a very long time indeed.

References

[1] Abbott, J., Shoup, V., and Zimmermann, P. 2000. Factorization in Z[x]: the Searching Phase. Pages 1–7 of: Traverso, C. (ed.), *International Symposium on Symbolic and Algebraic Computation – ISSAC*. ACM. (Cited on page 290.)

[2] Acar, T. 1997. *High-Speed Algorithms & Architectures for Number Theoretic Cryptosystems*. PhD thesis, Department of Electrical & Computer Engineering, Oregon State University. (Cited on page 232.)

[3] Acar, T., and Shumow, D. 2010. *Modular Reduction without Pre-Computation for Special Moduli*. Technical Report. Microsoft Research. (Cited on page 241.)

[4] Adj, G., Canales-Martínez, I., Cruz-Cortés, N., Menezes, A., Oliveira, T., Rivera-Zamarripa, L., and Rodríguez-Henríquez, F. 2018. Computing Discrete Logarithms in Cryptographically-Interesting Characteristic-Three Finite Fields. *Advances in Mathematics of Communications*, **12**(4), 741–759. (Cited on page 328.)

[5] Adkins, H. 2011. *An update on attempted man-in-the-middle attacks.* https://security.googleblog.com/2011/08/update-on-attempted-man-in-middle.html. (Cited on page 154.)

[6] Adleman, L. M. 1994. The Function Field Sieve. Pages 108–121 of: Adleman, L. M., and Huang, M. A. (eds.), *Algorithmic Number Theory, First International Symposium, ANTS-I, Ithaca, NY, USA, May 6-9, 1994, Proceedings*. LNCS, vol. 877. Springer, Heidelberg, Germany. (Cited on page 328.)

[7] Adrian, D., Bhargavan, K., Durumeric, Z., Gaudry, P., Green, M., Halderman, J. A., Heninger, N., Springall, D., Thomé, E., Valenta, L., VanderSloot, B., Wustrow, E., Zanella-Béguelin, S., and Zimmermann, P. 2015. Imperfect Forward Secrecy: How Diffie-Hellman Fails in Practice. Pages 5–17 of: Ray, I., Li, N., and Kruegel, C. (eds.), *ACM CCS 2015*. Denver, CO, USA: ACM Press. (Cited on pages 157, 158, 159, 160, 162, 163, 164, and 167.)

[8] Aggarwal, D., and Maurer, U. 2009. Breaking RSA Generically Is Equivalent to Factoring. Pages 36–53 of: Joux, A. (ed.), *EUROCRYPT 2009*. LNCS, vol. 5479. Cologne, Germany: Springer, Heidelberg, Germany. (Cited on page 141.)

[9] Agrawal, M., Kayal, N., and Saxena, N. 2004. PRIMES is in P. *Annals of Mathematics*, **160**(2), 781–793. (Cited on pages 47 and 77.)

[10] Ajtai, M. 1998. The Shortest Vector Problem in L_2 is NP-hard for Randomized Reductions (Extended Abstract). Pages 10–19 of: *30th ACM STOC*. Dallas, TX, USA: ACM Press. (Cited on page 79.)

[11] Albrecht, M. R., Fitzpatrick, R., and Göpfert, F. 2014. On the Efficacy of Solving LWE by Reduction to Unique-SVP. Pages 293–310 of: Lee, H.-S., and Han, D.-G. (eds.), *ICISC 13*. LNCS, vol. 8565. Seoul, Korea: Springer, Heidelberg, Germany. (Cited on page 31.)

[12] Albrecht, M. R., Rechberger, C., Schneider, T., Tiessen, T., and Zohner, M. 2015a. Ciphers for MPC and FHE. Pages 430–454 of: Oswald, E., and Fischlin, M. (eds.), *EUROCRYPT 2015, Part I*. LNCS, vol. 9056. Sofia, Bulgaria: Springer, Heidelberg, Germany. (Cited on page 332.)

[13] Albrecht, M. R., Papini, D., Paterson, K. G., and Villanueva-Polanco, R. 2015b. *Factoring 512-Bit RSA Moduli for Fun (and a Profit of $ 9,000)*. https://martinralbrecht.files.wordpress.com/2015/03/freak-scan1.pdf. (Cited on page 144.)

[14] Albrecht, M. R., Bai, S., and Ducas, L. 2016. A Subfield Lattice Attack on Overstretched NTRU Assumptions - Cryptanalysis of Some FHE and Graded Encoding Schemes. Pages 153–178 of: Robshaw, M., and Katz, J. (eds.), *CRYPTO 2016, Part I*. LNCS, vol. 9814. Santa Barbara, CA, USA: Springer, Heidelberg, Germany. (Cited on pages 35, 36, and 37.)

[15] Albrecht, M. R., Göpfert, F., Virdia, F., and Wunderer, T. 2017. Revisiting the Expected Cost of Solving uSVP and Applications to LWE. Pages 297–322 of: Takagi, T., and Peyrin, T. (eds.), *ASIACRYPT 2017, Part I*. LNCS, vol. 10624. Hong Kong, China: Springer, Heidelberg, Germany. (Cited on pages 23, 31, 32, and 33.)

[16] Albrecht, M. R., Massimo, J., Paterson, K. G., and Somorovsky, J. 2018. Prime and Prejudice: Primality Testing Under Adversarial Conditions. Pages 281–298 of: Lie, D., Mannan, M., Backes, M., and Wang, X. (eds.), *ACM CCS 2018*. Toronto, ON, Canada: ACM Press. (Cited on page 142.)

[17] Albrecht, M. R., Ducas, L., Herold, G., Kirshanova, E., Postlethwaite, E. W., and Stevens, M. 2019. The General Sieve Kernel and New Records in Lattice Reduction. Pages 717–746 of: Ishai, Y., and Rijmen, V. (eds.), *EUROCRYPT 2019, Part II*. LNCS, vol. 11477. Darmstadt, Germany: Springer, Heidelberg, Germany. (Cited on page 33.)

[18] Alford, W. R., and Pomerance, C. 1993. Implementing the Self-Initializing Quadratic Sieve on a Distributed Network. Pages 163–174 of: van der Poorten, A., Shparlinski, I., and Zimmer, H. G. (eds.), *Number Theoretic and Algebraic Methods in Computer Science*. Moscow: World Scientific, Singapore, Singapore. (Cited on page 57.)

[19] Alkim, E., Ducas, L., Pöppelmann, T., and Schwabe, P. 2016. Post-quantum Key Exchange - A New Hope. Pages 327–343 of: Holz, T., and Savage, S. (eds.), *USENIX Security 2016*. Austin, TX, USA: USENIX Association. (Cited on pages 22, 23, 31, and 34.)

[20] ANSSI. 2010. *Référentiel Général de Sécurité, Annexe B1 Mécanismes cryptographiques : Régles et recommandations concernant le choix et le dimensionnement des mécanismes cryptographiques*. Version 1.20. (Cited on page 315.)

[21] Antipa, A., Brown, D. R. L., Menezes, A., Struik, R., and Vanstone, S. A. 2003. Validation of Elliptic Curve Public Keys. Pages 211–223 of: Desmedt, Y. (ed.), *PKC 2003*. LNCS, vol. 2567. Miami, FL, USA: Springer, Heidelberg, Germany. (Cited on page 172.)

[22] Aoki, K., Franke, J., Kleinjung, T., Lenstra, A. K., and Osvik, D. A. 2007. A Kilobit Special Number Field Sieve Factorization. Pages 1–12 of: Kurosawa, K. (ed.), *ASIACRYPT 2007*. LNCS, vol. 4833. Kuching, Malaysia: Springer, Heidelberg, Germany. (Cited on pages 3, 169, and 323.)

[23] Applebaum, B., Cash, D., Peikert, C., and Sahai, A. 2009. Fast Cryptographic Primitives and Circular-Secure Encryption Based on Hard Learning Problems. Pages 595–618 of: Halevi, S. (ed.), *CRYPTO 2009*. LNCS, vol. 5677. Santa Barbara, CA, USA: Springer, Heidelberg, Germany. (Cited on page 16.)

[24] Atkins, D., Graff, M., Lenstra, A. K., and Leyland, P. C. 1995. The Magic Words are Squeamish Ossifrage. Pages 263–277 of: Pieprzyk, J., and Safavi-Naini, R. (eds.), *ASIACRYPT'94*. LNCS, vol. 917. Wollongong, Australia: Springer, Heidelberg, Germany. (Cited on pages 3, 49, 57, and 320.)

[25] Aviram, N., Schinzel, S., Somorovsky, J., Heninger, N., Dankel, M., Steube, J., Valenta, L., Adrian, D., Halderman, J. A., Dukhovni, V., Käsper, E., Cohney, S., Engels, S., Paar, C., and Shavitt, Y. 2016. DROWN: Breaking TLS Using SSLv2. Pages 689–706 of: Holz, T., and Savage, S. (eds.), *USENIX Security 2016*. Austin, TX, USA: USENIX Association. (Cited on pages 150 and 151.)

[26] Bai, S., Stehlé, D., and Wen, W. 2016. Improved Reduction from the Bounded Distance Decoding Problem to the Unique Shortest Vector Problem in Lattices. Pages 76:1–76:12 of: Chatzigiannakis, I., Mitzenmacher, M., Rabani, Y., and Sangiorgi, D. (eds.), *ICALP 2016*. LIPIcs, vol. 55. Rome, Italy: Schloss Dagstuhl - Leibniz-Zentrum fuer Informatik. (Cited on page 29.)

[27] Bai, S., Stehlé, D., and Wen, W. 2018. Measuring, Simulating and Exploiting the Head Concavity Phenomenon in BKZ. Pages 369–404 of: Peyrin, T., and Galbraith, S. (eds.), *ASIACRYPT 2018, Part I*. LNCS, vol. 11272. Brisbane, Queensland, Australia: Springer, Heidelberg, Germany. (Cited on pages 24 and 25.)

[28] Baillie, R., and Wagstaff, Jr, S. S. 1980. Lucas Pseudoprimes. *Mathematics of Computation*, **35**, 1391–1417. (Cited on page 46.)

[29] Barbulescu, R. 2013. *Algorithmes de logarithmes discrets dans les corps finis*. PhD thesis, Université de Lorraine, France. (Cited on page 159.)

[30] Barbulescu, R., and Duquesne, S. 2019. Updating Key Size Estimations for Pairings. *Journal of Cryptology*, **32**(4), 1298–1336. (Cited on pages 123, 311, and 329.)

[31] Barbulescu, R., and Pierrot, C. 2014. The Multiple Number Field Sieve for Medium- and High-Characteristic Finite Fields. *LMS Journal of Computation and Mathematics*, **17**(A), 230–246. (Cited on page 123.)

[32] Barbulescu, R., Gaudry, P., Joux, A., and Thomé, E. 2014. A Heuristic Quasi-Polynomial Algorithm for Discrete Logarithm in Finite Fields of Small Characteristic. Pages 1–16 of: Nguyen, P. Q., and Oswald, E. (eds.), *EUROCRYPT 2014*. LNCS, vol. 8441. Copenhagen, Denmark: Springer, Heidelberg, Germany. (Cited on pages 130, 132, 135, 137, and 328.)

[33] Barbulescu, R., Gaudry, P., Guillevic, A., and Morain, F. 2015a. Improving NFS for the Discrete Logarithm Problem in Non-prime Finite Fields. Pages 129–155 of: Oswald, E., and Fischlin, M. (eds.), *EUROCRYPT 2015, Part I*. LNCS, vol. 9056. Sofia, Bulgaria: Springer, Heidelberg, Germany. (Cited on page 329.)

[34] Barbulescu, R., Gaudry, P., and Kleinjung, T. 2015b. The Tower Number Field Sieve. Pages 31–55 of: Iwata, T., and Cheon, J. H. (eds.), *ASIACRYPT 2015, Part II*. LNCS, vol. 9453. Auckland, New Zealand: Springer, Heidelberg, Germany. (Cited on pages 123 and 329.)

[35] Barker, E. 2020. *NIST Special Publication 800-57 Part 1, Revision 5, Recommendation for Key Management: Part 1 – General*. Technical Report. National Institute of Standards and Technology. `https://doi.org/10.6028/NIST.SP.800-57pt1r5`. (Cited on page 313.)

[36] Barker, E., and Dang, Q. 2011. *NIST SP 800-57 Part 3 Revision 1: Recommendation for Key Management–Application-Specific Key Management Guidances*. Technical Report. Gaithersburg, MD, USA. `https://nvlpubs.nist.gov/nistpubs/SpecialPublications/NIST.SP.800-57Pt3r1.pdf`. (Cited on page 144.)

[37] Barker, E., Chen, L., Roginsky, A., Vassilev, A., and Davis, R. 2018. *Recommendation for Pair-Wise Key Establishment Schemes Using Discrete Logarithm Cryptography*. `https://nvlpubs.nist.gov/nistpubs/SpecialPublications/NIST.SP.800-56Ar3.pdf`. (Cited on pages 162 and 167.)

[38] Barker, E., Chen, L., Roginsky, A., Vassilev, A., Davis, R., and Simon, S. 2019. *Recommendation for Pair-Wise Key Establishment Using Integer Factorization Cryptography*. `https://nvlpubs.nist.gov/nistpubs/SpecialPublications/NIST.SP.800-56Br2.pdf`. (Cited on page 145.)

[39] Barker, E. B., and Roginsky, A. L. 2011. *NIST SP 800-131A. Transitions: Recommendation for Transitioning the Use of Cryptographic Algorithms and Key Lengths*. Technical Report. Gaithersburg, MD, USA. `https://doi.org/10.6028/NIST.SP.800-131A`. (Cited on pages 143 and 157.)

[40] Barreto, P. S. L. M., and Naehrig, M. 2006. Pairing-Friendly Elliptic Curves of Prime Order. Pages 319–331 of: Preneel, B., and Tavares, S. (eds.), *SAC 2005*. LNCS, vol. 3897. Kingston, Ontario, Canada: Springer, Heidelberg, Germany. (Cited on page 327.)

[41] Barrett, P. 1987. Implementing the Rivest Shamir and Adleman Public Key Encryption Algorithm on a Standard Digital Signal Processor. Pages 311–323 of: Odlyzko, A. M. (ed.), *CRYPTO'86*. LNCS, vol. 263. Santa Barbara, CA, USA: Springer, Heidelberg, Germany. (Cited on page 237.)

[42] Bauer, A., and Joux, A. 2007. Toward a Rigorous Variation of Coppersmith's Algorithm on Three Variables. Pages 361–378 of: Naor, M. (ed.), *EUROCRYPT 2007*. LNCS, vol. 4515. Barcelona, Spain: Springer, Heidelberg, Germany. (Cited on pages 98 and 103.)

[43] Bellare, M. 1998. PSS: Provably Secure Encoding Method for Digital Signatures. *Submission to the IEEE P1363a: Provably Secure Signatures Working Group*. (Cited on page 149.)

[44] Bellare, M., and Rogaway, P. 1995. Optimal Asymmetric Encryption. Pages 92–111 of: Santis, A. D. (ed.), *EUROCRYPT'94*. LNCS, vol. 950. Perugia, Italy: Springer, Heidelberg, Germany. (Cited on pages 86, 149, and 225.)

[45] Bellare, M., and Rogaway, P. 1996. The Exact Security of Digital Signatures: How to Sign with RSA and Rabin. Pages 399–416 of: Maurer, U. M. (ed.), *EUROCRYPT'96*. LNCS, vol. 1070. Saragossa, Spain: Springer, Heidelberg, Germany. (Cited on page 154.)

[46] Bentahar, K., and Smart, N. P. 2007. Efficient 15, 360-bit RSA Using Woop-Optimised Montgomery Arithmetic. Pages 346–363 of: Galbraith, S. D. (ed.), *11th IMA International Conference on Cryptography and Coding*. LNCS, vol. 4887. Cirencester, UK: Springer, Heidelberg, Germany. (Cited on pages 236 and 237.)

[47] Bernstein, D. J. 2006a (September). *Elliptic vs. Hyperelliptic, part I*. Talk at ECC (slides at http://cr.yp.to/talks/2006.09.20/slides.pdf). (Cited on page 238.)

[48] Bernstein, D. J. 2015. *Error-Prone Cryptographic Designs*. Real World Cryptography (RWC) invited talk. https://cr.yp.to/talks/2015.01.07/slides-djb-20150107-a4.pdf. (Cited on page 226.)

[49] Bernstein, D. J. 2006b. Curve25519: New Diffie-Hellman Speed Records. Pages 207–228 of: Yung, M., Dodis, Y., Kiayias, A., and Malkin, T. (eds.), *PKC 2006*. LNCS, vol. 3958. New York, NY, USA: Springer, Heidelberg, Germany. (Cited on pages 172 and 242.)

[50] Bernstein, D. J., Duif, N., Lange, T., Schwabe, P., and Yang, B.-Y. 2011. High-Speed High-Security Signatures. Pages 124–142 of: Preneel, B., and Takagi, T. (eds.), *CHES 2011*. LNCS, vol. 6917. Nara, Japan: Springer, Heidelberg, Germany. (Cited on page 242.)

[51] Bernstein, D. J., Duif, N., Lange, T., Schwabe, P., and Yang, B.-Y. 2012. High-Speed High-Security Signatures. *Journal of Cryptographic Engineering*, **2**(2), 77–89. (Cited on page 176.)

[52] Bernstein, D. J., Chang, Y.-A., Cheng, C.-M., Chou, L.-P., Heninger, N., Lange, T., and van Someren, N. 2013. Factoring RSA Keys from Certified Smart Cards: Coppersmith in the Wild. Pages 341–360 of: Sako, K., and Sarkar, P. (eds.), *ASIACRYPT 2013, Part II*. LNCS, vol. 8270. Bengalore, India: Springer, Heidelberg, Germany. (Cited on pages 80, 148, and 149.)

[53] Bernstein, D. J., Chou, T., Chuengsatiansup, C., Hülsing, A., Lambooij, E., Lange, T., Niederhagen, R., and van Vredendaal, C. 2015a. How to Manipulate Curve Standards: A White Paper for the Black Hat http: //bada55.cr.yp.to. Pages 109–139 of: Chen, L., and Matsuo, S. (eds.), *Security Standardisation Research - Second International Conference, SSR 2015, Tokyo, Japan, December 15-16, 2015, Proceedings*. LNCS, vol. 9497. Springer. (Cited on page 172.)

[54] Bernstein, D. J., Hopwood, D., Hülsing, A., Lange, T., Niederhagen, R., Papachristodoulou, L., Schneider, M., Schwabe, P., and Wilcox-O'Hearn, Z. 2015b. SPHINCS: Practical Stateless Hash-Based Signatures. Pages 368–397 of: Oswald, E., and Fischlin, M. (eds.), *EUROCRYPT 2015, Part I*. LNCS, vol. 9056. Sofia, Bulgaria: Springer, Heidelberg, Germany. (Cited on page 332.)

[55] Bernstein, D. J., Chuengsatiansup, C., Lange, T., and van Vredendaal, C. 2019.

NTRU Prime. Technical Report. National Institute of Standards and Technology. Available at https://csrc.nist.gov/projects/post-quantum-cryptography/round-2-submissions. (Cited on pages 37 and 38.)

[56] Beuchat, J.-L., Perez, L. J. D., Fuentes-Castaneda, L., and Rodriguez-Henriquez, F. 2017. Final Exponentiation. Chap. 7, pages 7–1–7–28 of: El Mrabet, N., and Joye, M. (eds.), *Guide to Pairing-Based Cryptography*. CRC Press. (Cited on page 310.)

[57] Beurdouche, B., Bhargavan, K., Delignat-Lavaud, A., Fournet, C., Kohlweiss, M., Pironti, A., Strub, P.-Y., and Zinzindohoue, J. K. 2015. A Messy State of the Union: Taming the Composite State Machines of TLS. Pages 535–552 of: *2015 IEEE Symposium on Security and Privacy*. San Jose, CA, USA: IEEE Computer Society Press. (Cited on page 157.)

[58] Bhargavan, K., and Leurent, G. 2016. Transcript Collision Attacks: Breaking Authentication in TLS, IKE and SSH. In: *NDSS 2016*. San Diego, CA, USA: The Internet Society. (Cited on page 202.)

[59] Biasse, J.-F., and Song, F. 2016. Efficient quantum algorithms for computing class groups and solving the principal ideal problem in arbitrary degree number fields. Pages 893–902 of: Krauthgamer, R. (ed.), *27th SODA*. Arlington, VA, USA: ACM-SIAM. (Cited on page 38.)

[60] Biasse, J.-F., Espitau, T., Fouque, P.-A., Gélin, A., and Kirchner, P. 2017. Computing Generator in Cyclotomic Integer Rings - A Subfield Algorithm for the Principal Ideal Problem in $L_{|\Delta_{\mathbb{K}}|}(\frac{1}{2})$ and Application to the Cryptanalysis of a FHE Scheme. Pages 60–88 of: Coron, J.-S., and Nielsen, J. B. (eds.), *EURO-CRYPT 2017, Part I*. LNCS, vol. 10210. Paris, France: Springer, Heidelberg, Germany. (Cited on page 38.)

[61] Biham, E. 1997. A Fast New DES Implementation in Software. Pages 260–272 of: Biham, E. (ed.), *FSE'97*. LNCS, vol. 1267. Haifa, Israel: Springer, Heidelberg, Germany. (Cited on page 232.)

[62] Biham, E., and Chen, R. 2004. Near-Collisions of SHA-0. Pages 290–305 of: Franklin, M. (ed.), *CRYPTO 2004*. LNCS, vol. 3152. Santa Barbara, CA, USA: Springer, Heidelberg, Germany. (Cited on page 189.)

[63] Blake, I. F., Mullin, R. C., and Vanstone, S. A. 1984. Computing Logarithms in $GF(2^n)$. Pages 73–82 of: Blakley, G. R., and Chaum, D. (eds.), *CRYPTO'84*. LNCS, vol. 196. Santa Barbara, CA, USA: Springer, Heidelberg, Germany. (Cited on page 328.)

[64] Bleichenbacher, D. 1998. Chosen Ciphertext Attacks Against Protocols Based on the RSA Encryption Standard PKCS #1. Pages 1–12 of: Krawczyk, H. (ed.), *CRYPTO'98*. LNCS, vol. 1462. Santa Barbara, CA, USA: Springer, Heidelberg, Germany. (Cited on pages 149 and 151.)

[65] Bleichenbacher, D. 2000. *On the Generation of One-Time Keys in DL Signature Schemes*. Presentation at IEEE P1363 working group meeting. (Cited on page 180.)

[66] Bleichenbacher, D., and May, A. 2006. New Attacks on RSA with Small Secret CRT-Exponents. Pages 1–13 of: Yung, M., Dodis, Y., Kiayias, A., and Malkin, T. (eds.), *PKC 2006*. LNCS, vol. 3958. New York, NY, USA: Springer, Heidelberg, Germany. (Cited on pages 80 and 99.)

[67] Bleichenbacher, D., Bosma, W., and Lenstra, A. K. 1995. Some Remarks on Lucas-Based Cryptosystems. Pages 386–396 of: Coppersmith, D. (ed.), *CRYPTO '95*. LNCS, vol. 963. Santa Barbara, CA, USA: Springer, Heidelberg, Germany. (Cited on page 298.)

[68] blockchain. 2015. *The Most Repeated R Value on the Blockchain.* https://bitcointalk.org/index.php?topic=1118704.0. (Cited on page 177.)

[69] Blömer, J., and May, A. 2005. A Tool Kit for Finding Small Roots of Bivariate Polynomials over the Integers. Pages 251–267 of: Cramer, R. (ed.), *EURO-CRYPT 2005*. LNCS, vol. 3494. Aarhus, Denmark: Springer, Heidelberg, Germany. (Cited on pages 101 and 103.)

[70] Bluestein, L. 1970. A Linear Filtering Approach to the Computation of Discrete Fourier Transform. *IEEE Transactions on Audio and Electroacoustics*, **18**(4), 451–455. (Cited on page 282.)

[71] Bluher, A. W. 2004. On $x^{q+1} + ax + b$. *Finite Fields and Their Applications*, **10**(3), 285–305. (Cited on page 133.)

[72] Blum, L., Blum, M., and Shub, M. 1986. A Simple Unpredictable Pseudo-Random Number Generator. *SIAM Journal on Computing*, **15**(2), 364–383. (Cited on page 104.)

[73] Böck, H., Somorovsky, J., and Young, C. 2018. Return Of Bleichenbacher's Oracle Threat (ROBOT). Pages 817–849 of: Enck, W., and Felt, A. P. (eds.), *USENIX Security 2018*. Baltimore, MD, USA: USENIX Association. (Cited on page 151.)

[74] Boneh, D. 1999. Twenty Years of Attacks on the RSA Cryptosystem. *Notices of the American Mathematical Society*, **46**(2), 203–213. (Cited on pages 70, 71, and 227.)

[75] Boneh, D., and Durfee, G. 2000. Cryptanalysis of RSA with Private Key d less than $N^{0.292}$. *IEEE Transactions on Information Theory*, **46**(4), 1339–1349. (Cited on page 74.)

[76] Boneh, D. 1998. The Decision Diffie-Hellman Problem. In: *Third Algorithmic Number Theory Symposium (ANTS)*. LNCS, vol. 1423. Springer, Heidelberg, Germany. Invited paper. (Cited on pages 162 and 295.)

[77] Boneh, D. 2000. Finding Smooth Integers in Short Intervals using CRT Decoding. Pages 265–272 of: *32nd ACM STOC*. Portland, OR, USA: ACM Press. (Cited on page 78.)

[78] Boneh, D. 2001. Simplified OAEP for the RSA and Rabin Functions. Pages 275–291 of: Kilian, J. (ed.), *CRYPTO 2001*. LNCS, vol. 2139. Santa Barbara, CA, USA: Springer, Heidelberg, Germany. (Cited on pages 79 and 86.)

[79] Boneh, D., and Durfee, G. 1999. Cryptanalysis of RSA with Private Key d Less than $N^{0.292}$. Pages 1–11 of: Stern, J. (ed.), *EUROCRYPT '99*. LNCS, vol. 1592. Prague, Czech Republic: Springer, Heidelberg, Germany. (Cited on pages 80, 98, 102, and 104.)

[80] Boneh, D., and Franklin, M. K. 2001. Identity-Based Encryption from the Weil Pairing. Pages 213–229 of: Kilian, J. (ed.), *CRYPTO 2001*. LNCS, vol. 2139. Santa Barbara, CA, USA: Springer, Heidelberg, Germany. (Cited on pages 67 and 327.)

[81] Boneh, D., and Venkatesan, R. 1996. Hardness of Computing the Most Significant Bits of Secret Keys in Diffie-Hellman and Related Schemes. Pages

129–142 of: Koblitz, N. (ed.), *CRYPTO'96*. LNCS, vol. 1109. Santa Barbara, CA, USA: Springer, Heidelberg, Germany. (Cited on page 179.)

[82] Boneh, D., and Venkatesan, R. 1998. Breaking RSA May Not Be Equivalent to Factoring. Pages 59–71 of: Nyberg, K. (ed.), *EUROCRYPT'98*. LNCS, vol. 1403. Espoo, Finland: Springer, Heidelberg, Germany. (Cited on page 141.)

[83] Boneh, D., Halevi, S., and Howgrave-Graham, N. 2001. The Modular Inversion Hidden Number Problem. Pages 36–51 of: Boyd, C. (ed.), *ASIACRYPT 2001*. LNCS, vol. 2248. Gold Coast, Australia: Springer, Heidelberg, Germany. (Cited on pages 80, 104, and 105.)

[84] Bos, J. N. E. 1992. *Practical Privacy*. PhD thesis, Technische Universiteit Eindhoven. (Cited on page 236.)

[85] Bos, J. W. 2010. High-Performance Modular Multiplication on the Cell Processor. Pages 7–24 of: Hasan, M. A., and Helleseth, T. (eds.), *Workshop on the Arithmetic of Finite Fields – WAIFI 2010*. LNCS, vol. 6087. Springer, Heidelberg, Germany. (Cited on pages 6 and 243.)

[86] Bos, J. W., and Kaihara, M. E. 2010. Montgomery Multiplication on the Cell. Pages 477–485 of: Wyrzykowski, R., Dongarra, J., Karczewski, K., and Wasniewski, J. (eds.), *Parallel Processing and Applied Mathematics – PPAM 2009*. LNCS, vol. 6067. Springer, Heidelberg, Germany. (Cited on page 6.)

[87] Bos, J. W., and Kleinjung, T. 2012. ECM at Work. Pages 467–484 of: Wang, X., and Sako, K. (eds.), *ASIACRYPT 2012*. LNCS, vol. 7658. Beijing, China: Springer, Heidelberg, Germany. (Cited on page 325.)

[88] Bos, J. W., Kaihara, M. E., Kleinjung, T., Lenstra, A. K., and Montgomery, P. L. 2009. *On the Security of 1024-bit RSA and 160-bit Elliptic Curve Cryptography*. Cryptology ePrint Archive, Report 2009/389. http://eprint.iacr.org/2009/389. (Cited on page 143.)

[89] Bos, J. W., Kleinjung, T., and Lenstra, A. K. 2010. On the Use of the Negation Map in the Pollard Rho Method. Pages 66–82 of: Hanrot, G., Morain, F., and Thomé, E. (eds.), *Algorithmic Number Theory Symposium – ANTS-IX*. LNCS, vol. 6197. Springer. (Cited on page 6.)

[90] Bos, J. W., Kleinjung, T., Lenstra, A. K., and Montgomery, P. L. 2011. Efficient SIMD Arithmetic Modulo a Mersenne Number. Pages 213–221 of: Antelo, E., Hough, D., and Ienne, P. (eds.), *IEEE Symposium on Computer Arithmetic – ARITH-20*. IEEE Computer Society. (Cited on pages 3, 6, 243, 247, and 325.)

[91] Bos, J. W., Kaihara, M. E., Kleinjung, T., Lenstra, A. K., and Montgomery, P. L. 2012. Solving a 112-bit Prime Elliptic Curve Discrete Logarithm Problem on Game Consoles Using Sloppy Reduction. *International Journal of Applied Cryptography*, 2(3), 212–228. (Cited on pages 6, 243, and 325.)

[92] Bos, J. W., Costello, C., Hisil, H., and Lauter, K. 2013a. Fast Cryptography in Genus 2. Pages 194–210 of: Johansson, T., and Nguyen, P. Q. (eds.), *EUROCRYPT 2013*. LNCS, vol. 7881. Athens, Greece: Springer, Heidelberg, Germany. (Cited on pages 238 and 241.)

[93] Bos, J. W., Costello, C., Hisil, H., and Lauter, K. 2013b. High-Performance Scalar Multiplication Using 8-Dimensional GLV/GLS Decomposition. Pages 331–348 of: Bertoni, G., and Coron, J.-S. (eds.), *CHES 2013*. LNCS, vol. 8086. Santa Barbara, CA, USA: Springer, Heidelberg, Germany. (Cited on page 241.)

[94] Bos, J. W., Lauter, K., Loftus, J., and Naehrig, M. 2013c. Improved Security for a Ring-Based Fully Homomorphic Encryption Scheme. Pages 45–64 of: Stam, M. (ed.), *14th IMA International Conference on Cryptography and Coding.* LNCS, vol. 8308. Oxford, UK: Springer, Heidelberg, Germany. (Cited on page 37.)

[95] Bos, J. W., Halderman, J. A., Heninger, N., Moore, J., Naehrig, M., and Wustrow, E. 2014a. Elliptic Curve Cryptography in Practice. Pages 157–175 of: Christin, N., and Safavi-Naini, R. (eds.), *FC 2014.* LNCS, vol. 8437. Christ Church, Barbados: Springer, Heidelberg, Germany. (Cited on page 178.)

[96] Bos, J. W., Montgomery, P. L., Shumow, D., and Zaverucha, G. M. 2014b. Montgomery Multiplication Using Vector Instructions. Pages 471–489 of: Lange, T., Lauter, K., and Lisonek, P. (eds.), *SAC 2013.* LNCS, vol. 8282. Burnaby, BC, Canada: Springer, Heidelberg, Germany. (Cited on pages 230, 231, 232, and 234.)

[97] Bos, J. W., Costello, C., and Naehrig, M. 2017. Scalar Multiplication and Exponentiation in Pairing Groups. Chap. 6, pages **6**–1–**6**–23 of: El Mrabet, N., and Joye, M. (eds.), *Guide to Pairing-Based Cryptography.* CRC Press. (Cited on page 310.)

[98] Bosma, W., Cannon, J., and Playoust, C. 1997. The Magma Algebra System I: The User Language. *Journal of Symbolic Computation,* **24**(3-4), 235–265. (Cited on page 128.)

[99] Bosma, W., Hutton, J., and Verheul, E. R. 2002. Looking beyond XTR. Pages 46–63 of: Zheng, Y. (ed.), *ASIACRYPT 2002.* LNCS, vol. 2501. Queenstown, New Zealand: Springer, Heidelberg, Germany. (Cited on page 305.)

[100] Boudot, F., Gaudry, P., Guillevic, A., Heninger, N., Emmanuel, T., and Zimmermann, P. 02/12/2019. *795-bit factoring and discrete logarithms.* NMBRTHRY list, https://lists.gforge.inria.fr/pipermail/cado-nfs-discuss/2019-December/001139.html. (Cited on page 137.)

[101] Boudot, F., Gaudry, P., Guillevic, A., Heninger, N., Thomé, E., and Zimmermann, P. 2020. Comparing the Difficulty of Factorization and Discrete Logarithm: A 240-Digit Experiment. Pages 62–91 of: Micciancio, D., and Ristenpart, T. (eds.), *CRYPTO 2020, Part II.* LNCS, vol. 12171. Santa Barbara, CA, USA: Springer, Heidelberg, Germany. (Cited on page 323.)

[102] Boudot, F., Gaudry, P., Guillevic, A., Heninger, N., Thomé, E., and Zimmermann, P. 28/02/2020. *Factorization of RSA-250.* NMBRTHRY list. (Cited on page 138.)

[103] Boyen, X. 2008. The Uber-Assumption Family (Invited Talk). Pages 39–56 of: Galbraith, S. D., and Paterson, K. G. (eds.), *PAIRING 2008.* LNCS, vol. 5209. Egham, UK: Springer, Heidelberg, Germany. (Cited on page 298.)

[104] Brakerski, Z., and Vaikuntanathan, V. 2011. Efficient Fully Homomorphic Encryption from (Standard) LWE. Pages 97–106 of: Ostrovsky, R. (ed.), *52nd FOCS.* Palm Springs, CA, USA: IEEE Computer Society Press. (Cited on page 16.)

[105] Brakerski, Z., Langlois, A., Peikert, C., Regev, O., and Stehlé, D. 2013. Classical Hardness of Learning with Errors. Pages 575–584 of: Boneh, D., Roughgarden, T., and Feigenbaum, J. (eds.), *45th ACM STOC.* Palo Alto, CA, USA: ACM Press. (Cited on page 16.)

[106] Brands, S. 1993. *An Efficient Off-line Electronic Cash System Based on the Representation Problem*. CWI Technical report, CS-R9323. (Cited on page 295.)

[107] Breitner, J., and Heninger, N. 2019. Biased Nonce Sense: Lattice Attacks Against Weak ECDSA Signatures in Cryptocurrencies. Pages 3–20 of: Goldberg, I., and Moore, T. (eds.), *FC 2019*. LNCS, vol. 11598. Frigate Bay, St. Kitts and Nevis: Springer, Heidelberg, Germany. (Cited on pages 177, 178, and 179.)

[108] Brengel, M., and Rossow, C. 2018. Identifying Key Leakage of Bitcoin Users. Pages 623–643 of: Bailey, M., Holz, T., Stamatogiannakis, M., and Ioannidis, S. (eds.), *Research in Attacks, Intrusions, and Defenses*. Cham: Springer International Publishing. (Cited on page 178.)

[109] Brent, R. P., and Kung, H. T. 1978. Fast Algorithms for Manipulating Formal Power Series. *J. ACM*, **25**(4), 581–595. (Cited on pages 278 and 280.)

[110] Brent, R. P., Gaudry, P., Thomé, E., and Zimmermann, P. 2008. Faster Multiplication in GF(2)[x]. Pages 153–166 of: van der Poorten, A. J., and Stein, A. (eds.), *Algorithmic Number Theory – ANTS-VIII*. LNCS, vol. 5011. Springer, Heidelberg, Germany. (Cited on page 285.)

[111] Brillhart, J., Lehmer, D. H., Selfridge, J. L., Tuckerman, B., and Wagstaff, Jr, S. S. 2002. *Factorizations of $b^n \pm 1$, b = 2, 3, 5, 6, 7, 10, 11, 12 up to High Powers*. Third edn. Contemporary Mathematics, vol. 22. American Mathematical Society. (Cited on page 77.)

[112] Brouwer, A. E., Pellikaan, R., and Verheul, E. R. 1999. Doing More with Fewer Bits. Pages 321–332 of: Lam, K.-Y., Okamoto, E., and Xing, C. (eds.), *ASIACRYPT'99*. LNCS, vol. 1716. Singapore: Springer, Heidelberg, Germany. (Cited on pages 123, 298, and 305.)

[113] BSI. 2013. *Kryptographische Verfahren: Empfehlungen und Schlüssellängen. BSI TR- 02102*. Version 2013.2. (Cited on page 315.)

[114] Buhler, J., Lenstra, Jr, H. W., and Pomerance, C. 1993. Factoring Integers with the Number Field Sieve. Pages 50–94 of: Lenstra, A. K., and Lenstra, Jr, H. W. (eds.), *The Development of the Number Field Sieve*. Lecture Notes in Mathematics, vol. 1554. Springer-Verlag, Berlin, Germany. (Cited on page 61.)

[115] Castellucci, R. 2015. *Cracking Cryptocurrency Brainwallets*. https://rya. nc/cracking_cryptocurrency_brainwallets.pdf. (Cited on page 177.)

[116] Castellucci, R., and Valsorda, F. 2016. *Stealing Bitcoin with Math*. https: //news.webamooz.com/wp-content/uploads/bot/offsecmag/151.pdf. (Cited on page 178.)

[117] Cavallar, S., Dodson, B., Lenstra, A. K., Leyland, P. C., Lioen, W. M., Montgomery, P. L., Murphy, B., te Riele, H., and Zimmermann, P. 1999. Factorization of RSA-140 Using the Number Field Sieve. Pages 195–207 of: Lam, K.-Y., Okamoto, E., and Xing, C. (eds.), *ASIACRYPT'99*. LNCS, vol. 1716. Singapore: Springer, Heidelberg, Germany. (Cited on pages 3 and 62.)

[118] Cavallar, S., Dodson, B., Lenstra, A. K., Lioen, W. M., Montgomery, P. L., Murphy, B., te Riele, H., Aardal, K., Gilchrist, J., Guillerm, G., Leyland, P. C., Marchand, J., Morain, F., Muffett, A., Putnam, C., Putnam, C., and Zimmermann, P. 2000. Factorization of a 512-Bit RSA Modulus. Pages 1–18 of: Preneel, B. (ed.), *EUROCRYPT 2000*. LNCS, vol. 1807. Bruges, Belgium: Springer, Heidelberg, Germany. (Cited on pages 3, 143, and 315.)

[119] Checkoway, S., Niederhagen, R., Everspaugh, A., Green, M., Lange, T., Ristenpart, T., Bernstein, D. J., Maskiewicz, J., Shacham, H., and Fredrikson, M. 2014. On the Practical Exploitability of Dual EC in TLS Implementations. Pages 319–335 of: Fu, K., and Jung, J. (eds.), *USENIX Security 2014*. San Diego, CA, USA: USENIX Association. (Cited on page 147.)

[120] Checkoway, S., Maskiewicz, J., Garman, C., Fried, J., Cohney, S., Green, M., Heninger, N., Weinmann, R.-P., Rescorla, E., and Shacham, H. 2016. A Systematic Analysis of the Juniper Dual EC Incident. Pages 468–479 of: Weippl, E. R., Katzenbeisser, S., Kruegel, C., Myers, A. C., and Halevi, S. (eds.), *ACM CCS 2016*. Vienna, Austria: ACM Press. (Cited on page 147.)

[121] Chen, Y. 2013. *Réduction de réseau et sécurité concrète du chiffrement complètement homomorphe*. PhD thesis, Paris 7. (Cited on page 23.)

[122] Chen, Y., and Nguyen, P. Q. 2011. BKZ 2.0: Better Lattice Security Estimates. Pages 1–20 of: Lee, D. H., and Wang, X. (eds.), *ASIACRYPT 2011*. LNCS, vol. 7073. Seoul, South Korea: Springer, Heidelberg, Germany. (Cited on pages 20, 23, 25, 27, and 330.)

[123] Cheon, J. H., Jeong, J., and Lee, C. 2016. An Algorithm for NTRU Problems and Cryptanalysis of the GGH Multilinear Map without a Low-Level Encoding of Zero. *LMS Journal of Computation and Mathematics*, **19**(A), 255–266. (Cited on pages 35 and 36.)

[124] Chevallier-Mames, B., Joye, M., and Paillier, P. 2003. Faster Double-Size Modular Multiplication from Euclidean Multipliers. Pages 214–227 of: Walter, C. D., Koç, Ç. K., and Paar, C. (eds.), *CHES 2003*. LNCS, vol. 2779. Cologne, Germany: Springer, Heidelberg, Germany. (Cited on page 236.)

[125] Chrome Platform Status. 2017. *Chrome Platform Status: Remove DHE-Based Ciphers*. https://www.chromestatus.com/feature/5128908798164992. (Cited on pages 157, 158, and 170.)

[126] Cohen, H., and Lenstra, A. K. 1987. Supplement to Implementation of a New Primality Test. *Mathematics of Computation*, **48**(177), S1–S4. (Cited on page 306.)

[127] Cohen, H. 1993. *A Course in Computational Algebraic Number Theory*. Springer, Heidelberg, Germany. (Cited on page 290.)

[128] Cohn, H., and Heninger, N. 2011. Ideal Forms of Coppersmith's Theorem and Guruswami-Sudan List Decoding. Pages 298–308 of: Chazelle, B. (ed.), *ICS 2011*. Tsinghua University, Beijing, China: Tsinghua University Press. (Cited on page 78.)

[129] Cohney, S., Kwong, A., Paz, S., Genkin, D., Heninger, N., Ronen, E., and Yarom, Y. 2020. Pseudorandom Black Swans: Cache Attacks on CTR_DRBG. Pages 1241–1258 of: *2020 IEEE Symposium on Security and Privacy*. San Francisco, CA, USA: IEEE Computer Society Press. (Cited on page 147.)

[130] Cohney, S. N., Green, M. D., and Heninger, N. 2018. Practical State Recovery Attacks against Legacy RNG Implementations. Pages 265–280 of: Lie, D., Mannan, M., Backes, M., and Wang, X. (eds.), *ACM CCS 2018*. Toronto, ON, Canada: ACM Press. (Cited on page 147.)

[131] Commeine, A., and Semaev, I. 2006. An Algorithm to Solve the Discrete Logarithm Problem with the Number Field Sieve. Pages 174–190 of: Yung, M.,

Dodis, Y., Kiayias, A., and Malkin, T. (eds.), *PKC 2006*. LNCS, vol. 3958. New York, NY, USA: Springer, Heidelberg, Germany. (Cited on page 159.)

[132] Comodo Fraud Incident. 2011. *Comodo Fraud Incident*. https://www.comodo.com/Comodo-Fraud-Incident-2011-03-23.html. (Cited on page 154.)

[133] Cooley, J. W., and Tukey, J. W. 1965. An Algorithm for the Machine Calculation of Complex Fourier Series. *Mathematics of Computation*, **19**, 297–301. (Cited on page 259.)

[134] Coppersmith, D. 1984a. Evaluating Logarithms in GF(2^n). Pages 201–207 of: *16th ACM STOC*. Washington, DC, USA: ACM Press. (Cited on pages 131 and 138.)

[135] Coppersmith, D. 1984b. Fast Evaluation of Logarithms in Fields of Characteristic Two. *IEEE Transactions on Information Theory*, **30**(4), 587–594. (Cited on pages 131 and 328.)

[136] Coppersmith, D. 1996a. Finding a Small Root of a Bivariate Integer Equation; Factoring with High Bits Known. Pages 178–189 of: Maurer, U. M. (ed.), *EUROCRYPT'96*. LNCS, vol. 1070. Saragossa, Spain: Springer, Heidelberg, Germany. (Cited on pages 4, 15, 74, 79, and 84.)

[137] Coppersmith, D. 1996b. Finding a Small Root of a Univariate Modular Equation. Pages 155–165 of: Maurer, U. M. (ed.), *EUROCRYPT'96*. LNCS, vol. 1070. Saragossa, Spain: Springer, Heidelberg, Germany. (Cited on pages 4, 74, and 79.)

[138] Coppersmith, D. 1997. Small Solutions to Polynomial Equations, and Low Exponent RSA Vulnerabilities. *Journal of Cryptology*, **10**(4), 233–260. (Cited on pages 74, 79, 84, 145, and 148.)

[139] Coppersmith, D., and Shamir, A. 1997. Lattice Attacks on NTRU. Pages 52–61 of: Fumy, W. (ed.), *EUROCRYPT'97*. LNCS, vol. 1233. Konstanz, Germany: Springer, Heidelberg, Germany. (Cited on page 35.)

[140] Courtois, N. T., Emirdag, P., and Valsorda, F. 2014. *Private Key Recovery Combination Attacks: On Extreme Fragility of Popular Bitcoin Key Management, Wallet and Cold Storage Solutions in Presence of Poor RNG Events*. Cryptology ePrint Archive, Report 2014/848. http://eprint.iacr.org/2014/848. (Cited on page 178.)

[141] Couveignes, J.-M. 2006. *Hard Homogeneous Spaces*. Cryptology ePrint Archive, Report 2006/291. http://eprint.iacr.org/2006/291. (Cited on page 297.)

[142] Couveignes, J. M., and Lercier, R. 2009. Elliptic Periods for Finite Fields. *Finite Fields and Their Applications*, **15**(1), 1–22. (Cited on page 136.)

[143] Cowie, J., Dodson, B., Elkenbracht-Huizing, R. M., Lenstra, A. K., Montgomery, P. L., and Zayer, J. 1996. A World Wide Number Field Sieve Factoring Record: On to 512 Bits. Pages 382–394 of: Kim, K., and Matsumoto, T. (eds.), *ASIACRYPT'96*. LNCS, vol. 1163. Kyongju, Korea: Springer, Heidelberg, Germany. (Cited on page 3.)

[144] Crandall, R., and Pomerance, C. 2005. *Prime Numbers: A Computational Perspective*. Second edn. New York: Springer-Verlag. (Cited on pages 58, 62, and 67.)

[145] Crandall, R. E. 1992 (October). *Method and Apparatus for Public Key Exchange in a Cryptographic System.* US Patent 5,159,632. (Cited on pages 239 and 242.)

[146] CrySyS Lab. 2012 (May 31,). *sKyWIper (a.k.a. Flame a.k.a. Flamer): A Complex Malware for Targeted Attacks.* Laboratory of Cryptography and System Security, Budapest University of Technology and Economics: http://www.crysys.hu/skywiper/skywiper.pdf. (Cited on page 198.)

[147] CVE. 2015. *CVE-2015-3240.* http://cve.mitre.org/cgi-bin/cvename. cgi?name=2015-3240. (Cited on page 165.)

[148] Dachman-Soled, D., Ducas, L., Gong, H., and Rossi, M. 2020. *LWE with Side Information: Attacks and Concrete Security Estimation.* Cryptology ePrint Archive, Report 2020/292. https://eprint.iacr.org/2020/292. (Cited on pages 33, 34, 35, and 40.)

[149] Damgård, I. 1990. A Design Principle for Hash Functions. Pages 416–427 of: Brassard, G. (ed.), *CRYPTO'89.* LNCS, vol. 435. Santa Barbara, CA, USA: Springer, Heidelberg, Germany. (Cited on page 183.)

[150] Dash, A., Sarmah, D., Behera, B. K., and Panigrahi, P. K. 2018. *Exact Search Algorithm to Factorize Large Biprimes and a Triprime on IBM Quantum Computer.* https://arxiv.org/abs/1805.10478. (Cited on page 330.)

[151] Daum, M., and Lucks, S. 2005 (June). *Attacking Hash Functions by Poisoned Messages, "The Story of Alice and her Boss".* https://web.archive.org/ web/20160713130211/http://th.informatik.uni-mannheim.de:80/ people/lucks/HashCollisions/. (Cited on page 205.)

[152] De Cannière, C., and Rechberger, C. 2006. Finding SHA-1 Characteristics: General Results and Applications. Pages 1–20 of: Lai, X., and Chen, K. (eds.), *ASIACRYPT 2006.* LNCS, vol. 4284. Shanghai, China: Springer, Heidelberg, Germany. (Cited on page 189.)

[153] Delignat-Lavaud, A. 2014. *Mozilla Foundation Security Advisory 2014-73:RSA Signature Forgery in NSS.* https://www.mozilla.org/en-US/security/ advisories/mfsa2014-73/. (Cited on page 153.)

[154] den Boer, B., and Bosselaers, A. 1994. Collisions for the Compressin Function of MD5. Pages 293–304 of: Helleseth, T. (ed.), *EUROCRYPT'93.* LNCS, vol. 765. Lofthus, Norway: Springer, Heidelberg, Germany. (Cited on page 183.)

[155] Dennis, J. E., and Schnabel, R. B. 1983. *Numerical Methods for Unconstrained Optimization and Nonlinear Equations.* Computational Mathematics. Prentice-Hall, Hoboken, NJ, USA. (Cited on page 77.)

[156] Denny, T. F., Dodson, B., Lenstra, A. K., and Manasse, M. S. 1994. On the Factorization of RSA-120. Pages 166–174 of: Stinson, D. R. (ed.), *CRYPTO'93.* LNCS, vol. 773. Santa Barbara, CA, USA: Springer, Heidelberg, Germany. (Cited on page 3.)

[157] Dickman, K. 1930. On the Frequency of Numbers Containing Prime Factors of a Certain Relative Magnitude. *Arkiv för Matatematik, Astronomi och Fysik,* **22A, 10**, 1–14. (Cited on pages 66 and 76.)

[158] Diem, C. 2011. On the Discrete Logarithm Problem in Elliptic Curves. *Compositio Mathematica,* **147**, 75–104. (Cited on pages 118 and 326.)

[159] Diem, C. 2013. On the discrete logarithm problem in elliptic curves II. *Algebra Number Theory,* **7**(6), 1281–1323. (Cited on page 118.)

[160] Dierks, T., and Rescorla, E. 2006 (Apr.). *The Transport Layer Security (TLS) Protocol Version 1.1*. RFC 4346. RFC Editor. https://www.rfc-editor.org/info/rfc4346. (Cited on page 141.)

[161] Dierks, T., and Rescorla, E. 2008 (Aug.). *The Transport Layer Security (TLS) Protocol Version 1.2*. RFC 5246. RFC Editor. https://www.rfc-editor.org/info/rfc5246. (Cited on pages 141, 150, 155, and 156.)

[162] Dierks, T., and Allen, C. 1999 (Jan.). *RFC 2246 - The TLS Protocol Version 1.0*. Internet Activities Board. (Cited on pages 149 and 154.)

[163] Diffie, W., and Hellman, M. E. 1976. New Directions in Cryptography. *IEEE Transactions on Information Theory*, **22**(6), 644–654. (Cited on pages 47, 48, 106, 122, 155, and 293.)

[164] Diffie, W., and Hellman, M. E. 1977. Special Feature Exhaustive Cryptanalysis of the NBS Data Encryption Standard. *IEEE Computer*, **10**(6), 74–84. (Cited on page 316.)

[165] Dixon, B., and Lenstra, A. K. 1994. Factoring Integers Using SIMD Sieves. Pages 28–39 of: Helleseth, T. (ed.), *EUROCRYPT'93*. LNCS, vol. 765. Lofthus, Norway: Springer, Heidelberg, Germany. (Cited on page 3.)

[166] Dixon, J. D. 1981. Asymptotically Fast Factorization of Integers. *Mathematics of Computation*, **36**, 255–260. (Cited on pages 68 and 69.)

[167] Dodson, B., and Lenstra, A. K. 1995. NFS with Four Large Primes: An Explosive Experiment. Pages 372–385 of: Coppersmith, D. (ed.), *CRYPTO'95*. LNCS, vol. 963. Santa Barbara, CA, USA: Springer, Heidelberg, Germany. (Cited on page 3.)

[168] Ducas, L., Durmus, A., Lepoint, T., and Lyubashevsky, V. 2013. Lattice Signatures and Bimodal Gaussians. Pages 40–56 of: Canetti, R., and Garay, J. A. (eds.), *CRYPTO 2013, Part I*. LNCS, vol. 8042. Santa Barbara, CA, USA: Springer, Heidelberg, Germany. (Cited on page 35.)

[169] Ducas, L., Kiltz, E., Lepoint, T., Lyubashevsky, V., Schwabe, P., Seiler, G., and Stehlé, D. 2018. CRYSTALS-Dilithium: A Lattice-Based Digital Signature Scheme. *IACR TCHES*, **2018**(1), 238–268. https://tches.iacr.org/index.php/TCHES/article/view/839. (Cited on pages 26 and 27.)

[170] Düll, M., Haase, B., Hinterwälder, G., Hutter, M., Paar, C., Sánchez, A. H., and Schwabe, P. 2015. High-speed Curve25519 on 8-bit, 16-bit and 32-bit Microcontrollers. *Design, Codes and Cryptography*, **77**(2), 493–514. (Cited on page 243.)

[171] Dumas, J.-G., Giorgi, P., and Pernet, C. 2008. Dense Linear Algebra over Word-Size Prime Fields: the FFLAS and FFPACK Packages. *ACM Transactions on Mathematical Software*, **35**(3), 1–42. (Cited on page 288.)

[172] Dummit, D. S., and Foote, R. M. 2004. *Abstract Algebra*. Third edn. New York: John Wiley & Sons. (Cited on page 58.)

[173] Duquesne, S., Mrabet, N. E., Haloui, S., Robert, D., and Rondepierre, F. 2017. Choosing Parameters. Chap. 10, pages **10**–1–**10**–22 of: El Mrabet, N., and Joye, M. (eds.), *Guide to Pairing-Based Cryptography*. CRC Press. (Cited on page 311.)

[174] Durumeric, Z., Wustrow, E., and Halderman, J. A. 2013. ZMap: Fast Internet-wide Scanning and Its Security Applications. Pages 605–620 of: King, S. T.

(ed.), *USENIX Security 2013*. Washington, DC, USA: USENIX Association. (Cited on page 146.)

[175] Durumeric, Z., Adrian, D., Mirian, A., Kasten, J., Bursztein, E., Lidzborski, N., Thomas, K., Eranti, V., Bailey, M., and Halderman, J. A. 2015. Neither Snow Nor Rain Nor MITM...: An Empirical Analysis of Email Delivery Security. Pages 27–39 of: *Proceedings of the 2015 Internet Measurement Conference*. IMC. New York, NY, USA: Association for Computing Machinery. (Cited on page 144.)

[176] Duursma, I. M., and Lee, H.-S. 2003. Tate Pairing Implementation for Hyperelliptic Curves $y^2 = x^p - x + d$. Pages 111–123 of: Laih, C.-S. (ed.), *ASIACRYPT 2003*. LNCS, vol. 2894. Taipei, Taiwan: Springer, Heidelberg, Germany. (Cited on page 310.)

[177] Duvillard, L. 2019. *Arjen Lenstra, la craie, le tableau noir et le tournevis*. https://actu.epfl.ch/news/arjen-lenstra-la-craie-le-tableau-noir-et-le-tourn. (Cited on page 8.)

[178] ECRYPT, and Smart, N. P. 2018. *Algorithms, Key Size and Protocols Report*. https://www.ecrypt.eu.org/csa/documents/D5.4-FinalAlgKeySizeProt.pdf. (Cited on pages 313 and 333.)

[179] Edge, J. 2014 (July). *A system call for random numbers: getrandom()*. https://lwn.net/Articles/606141/. (Cited on page 147.)

[180] Edlinger, B. 2019. *Change DH parameters to generate the order q subgroup instead of 2q*. https://github.com/openssl/openssl/pull/9363. (Cited on page 162.)

[181] ElGamal, T. 1984. A Public Key Cryptosystem and a Signature Scheme Based on Discrete Logarithms. Pages 10–18 of: Blakley, G. R., and Chaum, D. (eds.), *CRYPTO'84*. LNCS, vol. 196. Santa Barbara, CA, USA: Springer, Heidelberg, Germany. (Cited on page 156.)

[182] ElGamal, T. 1985. A Public Key Cryptosystem and a Signature Scheme Based on Discrete Logarithms. *IEEE Transactions on Information Theory*, **31**, 469–472. (Cited on pages 122 and 295.)

[183] Ernst, M., Jochemsz, E., May, A., and de Weger, B. 2005. Partial Key Exposure Attacks on RSA up to Full Size Exponents. Pages 371–386 of: Cramer, R. (ed.), *EUROCRYPT 2005*. LNCS, vol. 3494. Aarhus, Denmark: Springer, Heidelberg, Germany. (Cited on page 98.)

[184] Estibals, N. 2010. Compact Hardware for Computing the Tate Pairing over 128-Bit-Security Supersingular Curves. Pages 397–416 of: Joye, M., Miyaji, A., and Otsuka, A. (eds.), *PAIRING 2010*. LNCS, vol. 6487. Yamanaka Hot Spring, Japan: Springer, Heidelberg, Germany. (Cited on page 67.)

[185] Faugère, J.-C. 1999. A New Efficient Algorithm for Computing Gröbner Bases (F_4). *Journal of Pure and Applied Algebra*, **139 (1-3)**, 61–88. (Cited on page 129.)

[186] Ferguson, N., Schneier, B., and Kohno, T. 2010. *Cryptography Engineering: Design Principles and Practical Applications*. Wiley, New Jersey, United States. (Cited on page 236.)

[187] Fillinger, M., and Stevens, M. 2015. Reverse-Engineering of the Cryptanalytic Attack Used in the Flame Super-Malware. Pages 586–611 of: Iwata, T., and Cheon, J. H. (eds.), *ASIACRYPT 2015, Part II*. LNCS, vol. 9453. Auckland,

New Zealand: Springer, Heidelberg, Germany. (Cited on pages 188, 189, 212, and 219.)

[188] Finney, H. 2006. *Bleichenbacher's RSA Signature Forgery Based on Implementation Error.* https://mailarchive.ietf.org/arch/msg/openpgp/5rnE9ZRN1AokBVj3VqblGlP63QE. (Cited on pages 145 and 152.)

[189] Fischer, W., and Seifert, J.-P. 2003. Increasing the Bitlength of a Crypto-Coprocessor. Pages 71–81 of: Kaliski Jr, B. S., Koç, Ç. K., and Paar, C. (eds.), *CHES 2002.* LNCS, vol. 2523. Redwood Shores, CA, USA: Springer, Heidelberg, Germany. (Cited on pages 236 and 237.)

[190] Flynn, M. J. 1972. Some Computer Organizations and Their Effectiveness. *IEEE Transactions on Computers,* **C-21**(9), 948–960. (Cited on page 230.)

[191] Fousse, L., Hanrot, G., Lefèvre, V., Pélissier, P., and Zimmermann, P. 2007. MPFR: A Multiple-Precision Binary Floating-Point Library with Correct Rounding. *ACM Transactions on Mathematical Software,* **33**(2), 13. (Cited on page 291.)

[192] Franke, J., Kleinjung, T., Paar, C., Pelzl, J., Priplata, C., and Stahlke, C. 2005. SHARK: A Realizable Special Hardware Sieving Device for Factoring 1024-Bit Integers. Pages 119–130 of: Rao, J. R., and Sunar, B. (eds.), *CHES 2005.* LNCS, vol. 3659. Edinburgh, UK: Springer, Heidelberg, Germany. (Cited on page 324.)

[193] Frey, G., and Ruck, H. 1994. A Remark Considering *m*-Divisibility in the Divisor Class Group of Curves. *Mathematics of Computation,* **62**(206), 865–874. (Cited on page 116.)

[194] Fried, J. 2020. *Personal communication.* (Cited on page 160.)

[195] Fried, J., Gaudry, P., Heninger, N., and Thomé, E. 2017. A Kilobit Hidden SNFS Discrete Logarithm Computation. Pages 202–231 of: Coron, J.-S., and Nielsen, J. B. (eds.), *EUROCRYPT 2017, Part I.* LNCS, vol. 10210. Paris, France: Springer, Heidelberg, Germany. (Cited on pages 169, 170, and 323.)

[196] Friedl, M., Provos, N., and Simpson, W. 2006 (Mar.). *Diffie-Hellman Group Exchange for the Secure Shell (SSH) Transport Layer Protocol.* RFC 4419. RFC Editor. https://www.rfc-editor.org/info/rfc4419. (Cited on pages 156, 160, and 168.)

[197] Galbraith, S. D., Paterson, K. G., and Smart, N. P. 2006. *Pairings for Cryptographers.* Cryptology ePrint Archive, Report 2006/165. http://eprint.iacr.org/2006/165. (Cited on page 309.)

[198] Galbraith, S. D. 2001. Supersingular Curves in Cryptography. Pages 495–513 of: Boyd, C. (ed.), *ASIACRYPT 2001.* LNCS, vol. 2248. Gold Coast, Australia: Springer, Heidelberg, Germany. (Cited on page 310.)

[199] Galbraith, S. D. 2012. *Mathematics of Public Key Cryptography.* Cambridge University Press, Cambridge, UK. (Cited on page 304.)

[200] Galbraith, S. D., and Gebregiyorgis, S. W. 2014. Summation Polynomial Algorithms for Elliptic Curves in Characteristic Two. Pages 409–427 of: Meier, W., and Mukhopadhyay, D. (eds.), *INDOCRYPT 2014.* LNCS, vol. 8885. New Delhi, India: Springer, Heidelberg, Germany. (Cited on page 117.)

[201] Galbraith, S. D., and Scott, M. 2008. Exponentiation in Pairing-Friendly

Groups Using Homomorphisms. Pages 211–224 of: Galbraith, S. D., and Paterson, K. G. (eds.), *PAIRING 2008*. LNCS, vol. 5209. Egham, UK: Springer, Heidelberg, Germany. (Cited on page 311.)

[202] Galbraith, S. D., Hess, F., and Smart, N. P. 2002. Extending the GHS Weil Descent Attack. Pages 29–44 of: Knudsen, L. R. (ed.), *EUROCRYPT 2002*. LNCS, vol. 2332. Amsterdam, The Netherlands: Springer, Heidelberg, Germany. (Cited on page 117.)

[203] Galbraith, S. D., Hess, F., and Vercauteren, F. 2008. Aspects of Pairing Inversion. *IEEE Transactions of Information Theory*, **54**(12), 5719–5728. (Cited on page 309.)

[204] Gallant, R. P., Lambert, R. J., and Vanstone, S. A. 2001. Faster Point Multiplication on Elliptic Curves with Efficient Endomorphisms. Pages 190–200 of: Kilian, J. (ed.), *CRYPTO 2001*. LNCS, vol. 2139. Santa Barbara, CA, USA: Springer, Heidelberg, Germany. (Cited on page 301.)

[205] Gama, N., and Nguyen, P. Q. 2008a. Finding Short Lattice Vectors within Mordell's Inequality. Pages 207–216 of: Ladner, R. E., and Dwork, C. (eds.), *40th ACM STOC*. Victoria, BC, Canada: ACM Press. (Cited on pages 15 and 19.)

[206] Gama, N., and Nguyen, P. Q. 2008b. Predicting Lattice Reduction. Pages 31–51 of: Smart, N. P. (ed.), *EUROCRYPT 2008*. LNCS, vol. 4965. Istanbul, Turkey: Springer, Heidelberg, Germany. (Cited on pages 19 and 30.)

[207] Gardner, M. August, 1977. A New Kind of Cipher that would take Millions of Years to Break. *Scientific American*, 120–124. (Cited on pages 49 and 320.)

[208] Garg, S., Gentry, C., and Halevi, S. 2013. Candidate Multilinear Maps from Ideal Lattices. Pages 1–17 of: Johansson, T., and Nguyen, P. Q. (eds.), *EUROCRYPT 2013*. LNCS, vol. 7881. Athens, Greece: Springer, Heidelberg, Germany. (Cited on page 37.)

[209] Garman, C., Green, M., Kaptchuk, G., Miers, I., and Rushanan, M. 2016. Dancing on the Lip of the Volcano: Chosen Ciphertext Attacks on Apple iMessage. Pages 655–672 of: Holz, T., and Savage, S. (eds.), *USENIX Security 2016*. Austin, TX, USA: USENIX Association. (Cited on page 150.)

[210] Gaudry, P. 2009. Index Calculus for Abelian Varieties of small Dimension and the Elliptic Curve Discrete Logarithm Problem. *Journal of Symbolic Computation*, **44**, 1690–1702. (Cited on pages 117, 125, and 326.)

[211] Gaudry, P., and Schost, É. 2012. Genus 2 Point Counting over Prime Fields. *Journal of Symbolic Computation*, **47**(4), 368–400. (Cited on page 238.)

[212] Gaudry, P., Hess, F., and Smart, N. P. 2002. Constructive and Destructive Facets of Weil Descent on Elliptic Curves. *Journal of Cryptology*, **15**(1), 19–46. (Cited on pages 117 and 326.)

[213] Gauss, C. F. 1965. *Disquisitiones Arithmeticae*. Translated by A. A. Clarke. Yale University Press, New Haven, CT, USA. (Cited on page 106.)

[214] Gebhardt, M., Illies, G., and Schindler, W. 2006. A Note on the Practical Value of Single Hash Collisions for Special File Formats. Pages 333–344 of: *Sicherheit*. LNI, vol. P-77. GI. (Cited on pages 184 and 205.)

[215] Geiselmann, W., and Steinwandt, R. 2003. A Dedicated Sieving Hardware. Pages 254–266 of: Desmedt, Y. (ed.), *PKC 2003*. LNCS, vol. 2567. Miami, FL, USA: Springer, Heidelberg, Germany. (Cited on page 324.)

[216] Geiselmann, W., and Steinwandt, R. 2004. Yet Another Sieving Device. Pages 278–291 of: Okamoto, T. (ed.), *CT-RSA 2004*. LNCS, vol. 2964. San Francisco, CA, USA: Springer, Heidelberg, Germany. (Cited on page 324.)

[217] Gentry, C., and Szydlo, M. 2002. Cryptanalysis of the Revised NTRU Signature Scheme. Pages 299–320 of: Knudsen, L. R. (ed.), *EUROCRYPT 2002*. LNCS, vol. 2332. Amsterdam, The Netherlands: Springer, Heidelberg, Germany. (Cited on pages 35 and 36.)

[218] Gentry, C., Peikert, C., and Vaikuntanathan, V. 2008. Trapdoors for Hard Lattices and New Cryptographic Constructions. Pages 197–206 of: Ladner, R. E., and Dwork, C. (eds.), *40th ACM STOC*. Victoria, BC, Canada: ACM Press. (Cited on page 16.)

[219] Giacomelli, I., Madsen, J., and Orlandi, C. 2016. ZKBoo: Faster Zero-Knowledge for Boolean Circuits. Pages 1069–1083 of: Holz, T., and Savage, S. (eds.), *USENIX Security 2016*. Austin, TX, USA: USENIX Association. (Cited on page 332.)

[220] Gillmor, D. 2016 (Aug.). *Negotiated Finite Field Diffie-Hellman Ephemeral Parameters for Transport Layer Security (TLS)*. RFC 7919. RFC Editor. https://www.rfc-editor.org/info/rfc7919. (Cited on pages 154, 156, 158, 159, 160, and 170.)

[221] Gilmore (Ed.), J. 1998. *Cracking DES: Secrets of Encryption Research, Wiretap Politics and Chip Design*. Electronic Frontier Foundation, O'Reilly & Associates, Sebastopol, CA, USA. (Cited on page 316.)

[222] Girault, M., Toffin, P., and Vallée, B. 1990. Computation of Approximate L-th Roots Modulo n and Application to Cryptography. Pages 100–117 of: Goldwasser, S. (ed.), *CRYPTO '88*. LNCS, vol. 403. Santa Barbara, CA, USA: Springer, Heidelberg, Germany. (Cited on page 79.)

[223] Goldberg, D. 1991. What Every Computer Scientist Should Know About Floating Point Arithmetic. *ACM Computing Surveys*, **23**(1), 5–48. (Cited on page 252.)

[224] Golliver, R. A., Lenstra, A. K., and McCurley, K. S. 1994. Lattice Sieving and Trial Division. Pages 18–27 of: Adleman, L. M., and Huang, M. A. (eds.), *Algorithmic Number Theory, First International Symposium – ANTS-I*. LNCS, vol. 877. Springer, Heidelberg, Germany. (Cited on page 3.)

[225] Göloglu, F., and Joux, A. 2019. A Simplified Approach to Rigorous Degree 2 Elimination in Discrete Logarithm Algorithms. *Mathematics of Computation*, **88**(319), 2485–2496. (Cited on page 136.)

[226] Göloglu, F., Granger, R., McGuire, G., and Zumbrägel, J. 2013. On the Function Field Sieve and the Impact of Higher Splitting Probabilities — Application to Discrete Logarithms in $\mathbb{F}_{2^{1971}}$ and $\mathbb{F}_{2^{3164}}$. Pages 109–128 of: Canetti, R., and Garay, J. A. (eds.), *CRYPTO 2013, Part II*. LNCS, vol. 8043. Santa Barbara, CA, USA: Springer, Heidelberg, Germany. (Cited on pages 130, 132, 135, 137, and 328.)

[227] Göloglu, F., Granger, R., McGuire, G., and Zumbrägel, J. 2014. Solving a 6120-bit DLP on a Desktop Computer. Pages 136–152 of: Lange, T., Lauter, K., and Lisonek, P. (eds.), *SAC 2013*. LNCS, vol. 8282. Burnaby, BC, Canada: Springer, Heidelberg, Germany. (Cited on pages 130, 135, and 137.)

[228] Gong, G., and Harn, L. 1999. Public-Key Cryptosystems based on Cubic Finite Field Extensions. *IEEE Transactions on Information Theory*, **45**(7), 2601–2605. (Cited on page 298.)

[229] Gong, G., Harn, L., and Wu, H. 2001. The GH Public-Key Cryptosystem. Pages 284–300 of: Vaudenay, S., and Youssef, A. M. (eds.), *SAC 2001*. LNCS, vol. 2259. Toronto, Ontario, Canada: Springer, Heidelberg, Germany. (Cited on page 298.)

[230] Göpfert, F., and Yakkundimath, A. 2015. *Darmstadt LWE Challenges*. https://www.latticechallenge.org/lwe_challenge/challenge.php. Accessed: 15-08-2018. (Cited on page 33.)

[231] Gordon, D. M. 1998. A Survey of Fast Exponentiation Methods. *Journal of Algorithms*, **27**, 129–146. (Cited on page 226.)

[232] Gordon, D. M. 1993. Designing and Detecting Trapdoors for Discrete Log Cryptosystems. Pages 66–75 of: Brickell, E. F. (ed.), *CRYPTO'92*. LNCS, vol. 740. Santa Barbara, CA, USA: Springer, Heidelberg, Germany. (Cited on page 169.)

[233] Gordon, D. M., and McCurley, K. S. 1993. Massively Parallel Computation of Discrete Logarithms. Pages 312–323 of: Brickell, E. F. (ed.), *CRYPTO'92*. LNCS, vol. 740. Santa Barbara, CA, USA: Springer, Heidelberg, Germany. (Cited on page 138.)

[234] Granger, R., Page, D., and Stam, M. 2006a. On Small Characteristic Algebraic Tori in Pairing-Based Cryptography. *LMS Journal of Computation and Mathematics*, **9**, 64–85. (Cited on page 310.)

[235] Granger, R., and Lenstra, A. K. 2013. Personal communication between Robert Granger and Arjen K. Lenstra. (Cited on page 123.)

[236] Granger, R., and Scott, M. 2010. Faster Squaring in the Cyclotomic Subgroup of Sixth Degree Extensions. Pages 209–223 of: Nguyen, P. Q., and Pointcheval, D. (eds.), *PKC 2010*. LNCS, vol. 6056. Paris, France: Springer, Heidelberg, Germany. (Cited on page 307.)

[237] Granger, R., and Stam, M. 2005. Personal communication between Robert Granger and Martijn Stam. (Cited on page 123.)

[238] Granger, R., and Vercauteren, F. 2005. On the Discrete Logarithm Problem on Algebraic Tori. Pages 66–85 of: Shoup, V. (ed.), *CRYPTO 2005*. LNCS, vol. 3621. Santa Barbara, CA, USA: Springer, Heidelberg, Germany. (Cited on pages 124, 125, 126, 128, 130, 132, 313, and 328.)

[239] Granger, R., Kleinjung, T., Lenstra, A. K., Wesolowski, B., and Zumbrägel, J. 10/07/2019. *Discrete Logarithms in $GF(2^{30750})$*. NMBRTHRY list. (Cited on page 137.)

[240] Granger, R., Page, D., and Stam, M. 2004. A Comparison of CEILIDH and XTR. Pages 235–249 of: Buell, D. A. (ed.), *Algorithmic Number Theory – ANTS*. LNCS, vol. 3076. Springer, Heidelberg, Germany. (Cited on page 307.)

[241] Granger, R., Page, D., and Stam, M. 2005. Hardware and Software Normal Basis Arithmetic for Pairing-Based Cryptography in Characteristic Three. *IEEE Transactions on Computers*, **54**(7), 852–860. (Cited on page 310.)

[242] Granger, R., Page, D., and Smart, N. P. 2006b. High Security Pairing-Based Cryptography Revisited. Pages 480–494 of: Hess, F., Pauli, S., and Pohst,

M. E. (eds.), *Algorithmic Number Theory – ANTS*. LNCS, vol. 4076. Springer, Heidelberg, Germany. (Cited on page 309.)

[243] Granger, R., Kleinjung, T., and Zumbrägel, J. 2014a. Breaking '128-bit Secure' Supersingular Binary Curves - (Or How to Solve Discrete Logarithms in $\mathbb{F}_{2^{4 \cdot 1223}}$ and $\mathbb{F}_{2^{12 \cdot 367}}$). Pages 126–145 of: Garay, J. A., and Gennaro, R. (eds.), *CRYPTO 2014, Part II*. LNCS, vol. 8617. Santa Barbara, CA, USA: Springer, Heidelberg, Germany. (Cited on pages 130, 132, and 137.)

[244] Granger, R., Kleinjung, T., and Zumbrägel, J. 2014b. *On the Powers of 2*. Cryptology ePrint Archive, Report 2014/300. http://eprint.iacr.org/2014/300. (Cited on pages 130 and 137.)

[245] Granger, R., Kleinjung, T., and Zumbrägel, J. 2018a. Indiscreet Logarithms in Finite Fields of Small Characteristic. *Advances in Mathematics of Communications*, **12**(2), 263–286. (Cited on pages 122, 131, and 135.)

[246] Granger, R., Kleinjung, T., and Zumbrägel, J. 2018b. On the discrete logarithm problem in finite fields of fixed characteristic. *Transactions of the American Mathematical Society*, **370**, 3129–3145. (Cited on pages 130, 132, 136, 137, and 328.)

[247] Granlund, T., and the GMP development team. 2002. *GNU MP: The GNU Multiple Precision Arithmetic Library*. http://gmplib.org/. (Cited on page 252.)

[248] Granville, A. 2005. It is Easy to Determine whether a Given Integer is Prime. *Bulletin of the American Mathematical Society*, **42**, 3–38. (Cited on page 47.)

[249] Gruber, P. M., and Lekkerkerker, C. G. 1987. *Geometry of numbers*. Second edn. North-Holland Mathematical Library, vol. 37. North-Holland Publishing Co., Amsterdam. (Cited on page 82.)

[250] Gueron, S. 2003. Enhanced Montgomery Multiplication. Pages 46–56 of: Kaliski Jr, B. S., Koç, Ç. K., and Paar, C. (eds.), *CHES 2002*. LNCS, vol. 2523. Redwood Shores, CA, USA: Springer, Heidelberg, Germany. (Cited on page 234.)

[251] Guillevic, A. 2020a. *Pairing-Friendly Curves*. https://members.loria.fr/AGuillevic/pairing-friendly-curves. (Cited on page 311.)

[252] Guillevic, A. 2020b. A Short-List of Pairing-Friendly Curves Resistant to Special TNFS at the 128-Bit Security Level. Pages 535–564 of: Kiayias, A., Kohlweiss, M., Wallden, P., and Zikas, V. (eds.), *PKC 2020, Part II*. LNCS, vol. 12111. Edinburgh, UK: Springer, Heidelberg, Germany. (Cited on page 329.)

[253] Hachez, G., and Quisquater, J.-J. 2000. Montgomery Exponentiation with no Final Subtractions: Improved Results. Pages 293–301 of: Koç, Ç. K., and Paar, C. (eds.), *CHES 2000*. LNCS, vol. 1965. Worcester, Massachusetts, USA: Springer, Heidelberg, Germany. (Cited on page 234.)

[254] Halevi, S., and Shoup, V. 2014. Algorithms in HElib. Pages 554–571 of: Garay, J. A., and Gennaro, R. (eds.), *CRYPTO 2014, Part I*. LNCS, vol. 8616. Santa Barbara, CA, USA: Springer, Heidelberg, Germany. (Cited on page 281.)

[255] Hamburg, M. 2012. *Fast and Compact Elliptic-Curve Cryptography*. Cryptology ePrint Archive, Report 2012/309. http://eprint.iacr.org/2012/309. (Cited on page 241.)

[256] Han, D.-G., Lim, J., and Sakurai, K. 2004. On Security of XTR Public Key Cryptosystems Against Side Channel Attacks. Pages 454–465 of: Wang, H., Pieprzyk, J., and Varadharajan, V. (eds.), *ACISP 04*. LNCS, vol. 3108. Sydney, NSW, Australia: Springer, Heidelberg, Germany. (Cited on page 302.)

[257] Hanrot, G., Pujol, X., and Stehlé, D. 2011a. Analyzing Blockwise Lattice Algorithms Using Dynamical Systems. Pages 447–464 of: Rogaway, P. (ed.), *CRYPTO 2011*. LNCS, vol. 6841. Santa Barbara, CA, USA: Springer, Heidelberg, Germany. (Cited on pages 18 and 19.)

[258] Hanrot, G., Pujol, X., and Stehlé, D. 2011b. *Terminating BKZ*. Cryptology ePrint Archive, Report 2011/198. http://eprint.iacr.org/2011/198. (Cited on page 19.)

[259] Harkins, D., and Carrel, D. 1998. *The Internet Key Exchange (IKE)*. IETF RFC 2409 (Proposed Standard). (Cited on pages 152, 154, 155, 156, and 158.)

[260] Hars, L. 2004. Long Modular Multiplication for Cryptographic Applications. Pages 45–61 of: Joye, M., and Quisquater, J.-J. (eds.), *CHES 2004*. LNCS, vol. 3156. Cambridge, Massachusetts, USA: Springer, Heidelberg, Germany. (Cited on page 229.)

[261] Harvey, D. 2014. Faster Arithmetic for Number-Theoretic Transforms. *Journal of Symbolic Computation*, **60**, 113–119. (Cited on page 267.)

[262] Håstad, J. 1986. On Using RSA with Low Exponent in a Public Key Network. Pages 403–408 of: Williams, H. C. (ed.), *CRYPTO'85*. LNCS, vol. 218. Santa Barbara, CA, USA: Springer, Heidelberg, Germany. (Cited on pages 79, 81, 86, and 227.)

[263] Hastings, M., Fried, J., and Heninger, N. 2016. Weak Keys Remain Widespread in Network Devices. Pages 49–63 of: *Proceedings of the 2016 Internet Measurement Conference*. IMC. New York, NY, USA: Association for Computing Machinery. (Cited on page 148.)

[264] Hawkes, P., Paddon, M., and Rose, G. G. 2004. *Musings on the Wang et al. MD5 Collision*. Cryptology ePrint Archive, Report 2004/264. http://eprint.iacr.org/2004/264. (Cited on page 317.)

[265] Hayashi, T., Shinohara, N., Wang, L., Matsuo, S., Shirase, M., and Takagi, T. 2010. Solving a 676-Bit Discrete Logarithm Problem in GF(3^{6n}). Pages 351–367 of: Nguyen, P. Q., and Pointcheval, D. (eds.), *PKC 2010*. LNCS, vol. 6056. Paris, France: Springer, Heidelberg, Germany. (Cited on page 138.)

[266] Hayashi, T., Shimoyama, T., Shinohara, N., and Takagi, T. 2012. Breaking Pairing-Based Cryptosystems Using η_T Pairing over GF(3^{97}). Pages 43–60 of: Wang, X., and Sako, K. (eds.), *ASIACRYPT 2012*. LNCS, vol. 7658. Beijing, China: Springer, Heidelberg, Germany. (Cited on page 138.)

[267] Heffner, C. 2010. *LittleBlackBox: Database of Private SSL/SSH Keys for Embedded Devices*. http://code.google.com/p/littleblackbox. (Cited on page 146.)

[268] Hellman, M. E., and Reyneri, J. M. 1982. Fast Computation of Discrete Logarithms in $GF(q)$. Pages 3–13 of: Chaum, D., Rivest, R. L., and Sherman, A. T. (eds.), *CRYPTO'82*. Santa Barbara, CA, USA: Plenum Press, New York, USA. (Cited on page 328.)

[269] Heninger, N., Durumeric, Z., Wustrow, E., and Halderman, J. A. 2012. Mining Your Ps and Qs: Detection of Widespread Weak Keys in Network Devices.

Pages 205–220 of: Kohno, T. (ed.), *USENIX Security 2012*. Bellevue, WA, USA: USENIX Association. (Cited on pages 143, 145, 146, 147, 148, 149, 150, 175, and 178.)

[270] Herrmann, M., and May, A. 2009. Attacking Power Generators Using Unravelled Linearization: When Do We Output Too Much? Pages 487–504 of: Matsui, M. (ed.), *ASIACRYPT 2009*. LNCS, vol. 5912. Tokyo, Japan: Springer, Heidelberg, Germany. (Cited on pages 97 and 104.)

[271] Herrmann, M., and May, A. 2010. Maximizing Small Root Bounds by Linearization and Applications to Small Secret Exponent RSA. Pages 53–69 of: Nguyen, P. Q., and Pointcheval, D. (eds.), *PKC 2010*. LNCS, vol. 6056. Paris, France: Springer, Heidelberg, Germany. (Cited on pages 97, 98, 100, and 103.)

[272] Herstein, I. N. 1999. *Abstract Algebra*. Third edn. New York: John Wiley & Sons. (Cited on page 58.)

[273] Hess, F. 2003. The GHS Attack Revisited. Pages 374–387 of: Biham, E. (ed.), *EUROCRYPT 2003*. LNCS, vol. 2656. Warsaw, Poland: Springer, Heidelberg, Germany. (Cited on page 117.)

[274] Hoffstein, J., Pipher, J., and Silverman, J. H. 1996. *NTRU: A New High Speed Public Key Cryptosystem*. Draft Distributed at Crypto'96, available at http://web.securityinnovation.com/hubfs/files/ntru-orig.pdf. (Cited on pages 16, 17, and 35.)

[275] Hoffstein, J., Pipher, J., and Silverman, J. H. 1998. NTRU: A Ring-Based Public Key Cryptosystem. Pages 267–288 of: Buhler, J. (ed.), *Algorithmic Number Theory – ANTS*. LNCS, vol. 1423. Springer, Heidelberg, Germany. (Cited on page 35.)

[276] Horwitz, J., and Venkatesan, R. 2002. Random Cayley Digraphs and the Discrete Logarithm. Pages 416–430 of: Fieker, C., and Kohel, D. R. (eds.), *Algorithmic Number Theory*. Springer, Heidelberg, Germany. (Cited on page 108.)

[277] Howgrave-Graham, N. A., and Smart, N. P. 2001. Lattice Attacks on Digital Signature Schemes. *Designs, Codes and Cryptography*, **23**(3), 283–290. (Cited on pages 162 and 179.)

[278] Howgrave-Graham, N. 1997. Finding Small Roots of Univariate Modular Equations Revisited. Pages 131–142 of: Darnell, M. (ed.), *6th IMA International Conference on Cryptography and Coding*. LNCS, vol. 1355. Cirencester, UK: Springer, Heidelberg, Germany. (Cited on pages 81 and 84.)

[279] Howgrave-Graham, N. 2001. Approximate Integer Common Divisors. Pages 51–66 of: Silverman, J. H. (ed.), *Cryptography and Lattices*. Springer, Heidelberg, Germany. (Cited on pages 148 and 149.)

[280] Howgrave-Graham, N. 2007. A Hybrid Lattice-Reduction and Meet-in-the-Middle Attack Against NTRU. Pages 150–169 of: Menezes, A. (ed.), *CRYPTO 2007*. LNCS, vol. 4622. Santa Barbara, CA, USA: Springer, Heidelberg, Germany. (Cited on page 26.)

[281] ICSI. 2020. *The ICSI Certificate Notary*. https://web.archive.org/web/20200624025519/https://notary.icsi.berkeley.edu/. (Cited on pages 141, 171, and 174.)

[282] Information Technology Laboratory National Institute of Standards and Technology. 2000. *Digital Signature Standard (DSS)*. https://csrc.nist.

gov/CSRC/media/Publications/fips/186/2/archive/2001-10-05/documents/fips186-2-change1.pdf. (Cited on page 170.)

[283] Information Technology Laboratory National Institute of Standards and Technology. 2013. *Digital Signature Standard (DSS)*. https://nvlpubs.nist.gov/nistpubs/FIPS/NIST.FIPS.186-4.pdf. (Cited on pages 142, 162, 169, 174, 175, 176, and 238.)

[284] Ireland, K., and Rosen, M. 1998. *A Classical Introduction to Modern Number Theory*. Berlin, New York: Springer-Verlag. (Cited on page 58.)

[285] Ishai, Y., Kushilevitz, E., Ostrovsky, R., and Sahai, A. 2007. Zero-Knowledge from Secure Multiparty Computation. Pages 21–30 of: Johnson, D. S., and Feige, U. (eds.), *39th ACM STOC*. San Diego, CA, USA: ACM Press. (Cited on page 332.)

[286] Jager, T., Schwenk, J., and Somorovsky, J. 2015. Practical Invalid Curve Attacks on TLS-ECDH. Pages 407–425 of: Pernul, G., Ryan, P. Y. A., and Weippl, E. R. (eds.), *ESORICS 2015, Part I*. LNCS, vol. 9326. Vienna, Austria: Springer, Heidelberg, Germany. (Cited on page 173.)

[287] Jager, T., Kakvi, S. A., and May, A. 2018. On the Security of the PKCS#1 v1.5 Signature Scheme. Pages 1195–1208 of: Lie, D., Mannan, M., Backes, M., and Wang, X. (eds.), *ACM CCS 2018*. Toronto, ON, Canada: ACM Press. (Cited on page 152.)

[288] Janusz, G. 1998. *Algebraic Number Fields*. Second edn. American Mathematical Society. (Cited on page 58.)

[289] Jochemsz, E., and May, A. 2006. A Strategy for Finding Roots of Multivariate Polynomials with New Applications in Attacking RSA Variants. Pages 267–282 of: Lai, X., and Chen, K. (eds.), *ASIACRYPT 2006*. LNCS, vol. 4284. Shanghai, China: Springer, Heidelberg, Germany. (Cited on page 101.)

[290] Jochemsz, E., and May, A. 2007. A Polynomial Time Attack on RSA with Private CRT-Exponents Smaller Than $N^{0.073}$. Pages 395–411 of: Menezes, A. (ed.), *CRYPTO 2007*. LNCS, vol. 4622. Santa Barbara, CA, USA: Springer, Heidelberg, Germany. (Cited on pages 80 and 100.)

[291] Jonsson, J., and Kaliski, B. 2003 (Feb.). *Public-Key Cryptography Standards (PKCS) #1: RSA Cryptography Specification, Version 2.1*. RFC 3447. RFC Editor. https://www.rfc-editor.org/info/rfc3447. (Cited on page 225.)

[292] Joux, A. 2000. A One Round Protocol for Tripartite Diffie-Hellman. Pages 385–394 of: Bosma, W. (ed.), *Algorithmic Number Theory – ANTS*. LNCS, vol. 1838. Springer, Heidelberg, Germany. (Cited on page 327.)

[293] Joux, A. 2013. Faster Index Calculus for the Medium Prime Case Application to 1175-bit and 1425-bit Finite Fields. Pages 177–193 of: Johansson, T., and Nguyen, P. Q. (eds.), *EUROCRYPT 2013*. LNCS, vol. 7881. Athens, Greece: Springer, Heidelberg, Germany. (Cited on pages 131 and 132.)

[294] Joux, A. 2014. A New Index Calculus Algorithm with Complexity $L(1/4+o(1))$ in Small Characteristic. Pages 355–379 of: Lange, T., Lauter, K., and Lisonek, P. (eds.), *SAC 2013*. LNCS, vol. 8282. Burnaby, BC, Canada: Springer, Heidelberg, Germany. (Cited on pages 130, 132, 134, 135, 137, and 328.)

[295] Joux, A., and Lercier, R. 2002. The Function Field Sieve Is Quite Special. Pages 431–445 of: Fieker, C., and Kohel, D. R. (eds.), *Algorithmic Number*

Theory – ANTS. LNCS, vol. 2369. Springer, Heidelberg, Germany. (Cited on page 328.)

[296] Joux, A., and Lercier, R. 2006. The Function Field Sieve in the Medium Prime Case. Pages 254–270 of: Vaudenay, S. (ed.), *EUROCRYPT 2006.* LNCS, vol. 4004. St. Petersburg, Russia: Springer, Heidelberg, Germany. (Cited on pages 131, 132, 135, 138, and 328.)

[297] Joux, A., and Peyrin, T. 2007. Hash Functions and the (Amplified) Boomerang Attack. Pages 244–263 of: Menezes, A. (ed.), *CRYPTO 2007.* LNCS, vol. 4622. Santa Barbara, CA, USA: Springer, Heidelberg, Germany. (Cited on page 189.)

[298] Joux, A., and Pierrot, C. 2014. Improving the Polynomial time Precomputation of Frobenius Representation Discrete Logarithm Algorithms - Simplified Setting for Small Characteristic Finite Fields. Pages 378–397 of: Sarkar, P., and Iwata, T. (eds.), *ASIACRYPT 2014, Part I.* LNCS, vol. 8873. Kaoshiung, Taiwan, R.O.C.: Springer, Heidelberg, Germany. (Cited on page 137.)

[299] Joux, A., and Pierrot, C. 2016. Technical History of Discrete Logarithms in Small Characteristic Finite Fields – The Road from Subexponential to Quasi-Polynomial Complexity. *Designs, Codes and Cryptography,* **78**(1), 73–85. (Cited on page 131.)

[300] Joux, A., and Pierrot, C. 2019. *Algorithmic Aspects of Elliptic Bases in Finite Field Discrete Logarithm Algorithms.* Cryptology ePrint Archive, Report 2019/782. https://eprint.iacr.org/2019/782. (Cited on page 136.)

[301] Joux, A., and Vitse, V. 2012. Cover and Decomposition Index Calculus on Elliptic Curves Made Practical - Application to a Previously Unreachable Curve over $\mathbb{F}_{p^6}$. Pages 9–26 of: Pointcheval, D., and Johansson, T. (eds.), *EUROCRYPT 2012.* LNCS, vol. 7237. Cambridge, UK: Springer, Heidelberg, Germany. (Cited on page 117.)

[302] Joux, A., Lercier, R., Smart, N., and Vercauteren, F. 2006. The Number Field Sieve in the Medium Prime Case. Pages 326–344 of: Dwork, C. (ed.), *CRYPTO 2006.* LNCS, vol. 4117. Santa Barbara, CA, USA: Springer, Heidelberg, Germany. (Cited on pages 328 and 329.)

[303] Joux, A., Odlyzko, A., and Pierrot, C. 2014. The Past, Evolving Present, and Future of the Discrete Logarithm. Pages 5–36 of: *Open Problems in Mathematics and Computational Science.* Springer, Heidelberg, Germany. (Cited on page 131.)

[304] Joye, M. 2012. On Quisquater's Multiplication Algorithm. Pages 3–7 of: Naccache, D. (ed.), *Cryptography and Security: From Theory to Applications.* LNCS, vol. 6805. Springer, Heidelberg, Germany. (Cited on page 229.)

[305] Joye, M., and Yen, S.-M. 2003. The Montgomery Powering Ladder. Pages 291–302 of: Kaliski Jr, B. S., Koç, Ç. K., and Paar, C. (eds.), *CHES 2002.* LNCS, vol. 2523. Redwood Shores, CA, USA: Springer, Heidelberg, Germany. (Cited on page 302.)

[306] Joye, M., Lenstra, A. K., and Quisquater, J.-J. 1999. Chinese Remaindering Based Cryptosystems in the Presence of Faults. *Journal of Cryptology,* **12**(4), 241–245. (Cited on page 71.)

[307] Kakvi, S. A., Kiltz, E., and May, A. 2012. Certifying RSA. Pages 404–414 of:

Wang, X., and Sako, K. (eds.), *ASIACRYPT 2012*. LNCS, vol. 7658. Beijing, China: Springer, Heidelberg, Germany. (Cited on pages 80, 92, and 93.)

[308] Kaliski, B. 1998 (Mar.). *PKCS #1: RSA Encryption Version 1.5*. RFC 2313. RFC Editor. https://www.rfc-editor.org/info/rfc2313. (Cited on pages 149, 150, and 152.)

[309] Kaltofen, E., and Shoup, V. 1998. Subquadratic-Time Factoring of Polynomials over Finite Fields. *Mathematics of Computation*, **67**(223), 1179–1197. (Cited on page 281.)

[310] Kaminsky, D. 2005. MD5 to Be Considered Harmful Someday. Pages 323 – 337 of: Archibald, N., dedhed, Fogie, S., Hurley, C., Kaminsky, D., Long, J., McOmie (aka Pyr0), L., Meer, H., Potter, B., Temmingh, R., Wyler (aka Grifter), N. R., and Mullen (THOR), T. M. (eds.), *Aggressive Network Self-Defense*. Burlington: Syngress. Available at https://eprint.iacr.org/2004/357.pdf. (Cited on pages 184 and 207.)

[311] Kannan, R. 1987. Minkowski's Convex Body Theorem and Integer Programming. *Mathematics Of Operations Research*, **12**(3), 415–440. (Cited on page 30.)

[312] Kannan, R., Lenstra, A. K., and Lovász, L. 1984. Polynomial Factorization and Nonrandomness of Bits of Algebraic and Some Transcendental Numbers. Pages 191–200 of: *16th ACM STOC*. Washington, DC, USA: ACM Press. (Cited on page 2.)

[313] Karabina, K. 2013. Squaring in Cyclotomic Subgroups. *Mathematics of Computation*, **82**(281), 555–579. (Cited on pages 307 and 310.)

[314] Karabina, K., Knapp, E., and Menezes, A. 2013. Generalizations of Verheul's Theorem to Asymmetric Pairings. *Advances in Mathematics of Communications*, **7**(1), 103–111. (Cited on page 309.)

[315] Karatsuba, A., and Ofman, Y. 1962. Multiplication of Many-Digital Numbers by Automatic Computers. *Doklady Akademii Nauk SSSR*, **145**, 293–294. Translation in Physics-Doklady 7, 595-596, 1963. (Cited on pages 235 and 271.)

[316] Kaspersky Lab. 2012 (May 28,). *The Flame: Questions and Answers*. Securelist blog: https://www.securelist.com/en/blog/208193522/The_Flame_Questions_and_Answers. (Cited on page 198.)

[317] Kaufman, C., Hoffman, P., Nir, Y., Eronen, P., and Kivinen, T. 2014 (Oct.). *Internet Key Exchange Protocol Version 2 (IKEv2)*. RFC 7296. RFC Editor. https://www.rfc-editor.org/info/rfc7296. (Cited on pages 154, 155, 156, and 158.)

[318] Kedlaya, K. S., and Umans, C. 2011. Fast Polynomial Factorization and Modular Composition. *SIAM Journal on Computing*, **40**(6), 1767–1802. (Cited on page 278.)

[319] Kelsey, J., and Kohno, T. 2006. Herding Hash Functions and the Nostradamus Attack. Pages 183–200 of: Vaudenay, S. (ed.), *EUROCRYPT 2006*. LNCS, vol. 4004. St. Petersburg, Russia: Springer, Heidelberg, Germany. (Cited on pages 203 and 205.)

[320] Kiltz, E., Pietrzak, K., Stam, M., and Yung, M. 2009. A New Randomness Extraction Paradigm for Hybrid Encryption. Pages 590–609 of: Joux, A. (ed.), *EUROCRYPT 2009*. LNCS, vol. 5479. Cologne, Germany: Springer, Heidelberg, Germany. (Cited on page 303.)

[321] Kim, T., and Barbulescu, R. 2016. Extended Tower Number Field Sieve: A New Complexity for the Medium Prime Case. Pages 543–571 of: Robshaw, M., and Katz, J. (eds.), *CRYPTO 2016, Part I*. LNCS, vol. 9814. Santa Barbara, CA, USA: Springer, Heidelberg, Germany. (Cited on pages 123 and 329.)

[322] Kim, T., and Jeong, J. 2017. Extended Tower Number Field Sieve with Application to Finite Fields of Arbitrary Composite Extension Degree. Pages 388–408 of: Fehr, S. (ed.), *PKC 2017, Part I*. LNCS, vol. 10174. Amsterdam, The Netherlands: Springer, Heidelberg, Germany. (Cited on page 329.)

[323] Kipnis, A., Patarin, J., and Goubin, L. 1999. Unbalanced Oil and Vinegar Signature Schemes. Pages 206–222 of: Stern, J. (ed.), *EUROCRYPT'99*. LNCS, vol. 1592. Prague, Czech Republic: Springer, Heidelberg, Germany. (Cited on page 331.)

[324] Kirchner, P., and Fouque, P.-A. 2017. Revisiting Lattice Attacks on Overstretched NTRU Parameters. Pages 3–26 of: Coron, J.-S., and Nielsen, J. B. (eds.), *EUROCRYPT 2017, Part I*. LNCS, vol. 10210. Paris, France: Springer, Heidelberg, Germany. (Cited on pages 35, 36, 38, and 39.)

[325] Kleinjung, T. 2007. *Discrete Logarithms in GF(p) — 160 Digits.* https://listserv.nodak.edu/cgi-bin/wa.exe?A2=NMBRTHRY;1c737cf8.0702. (Cited on page 157.)

[326] Kleinjung, T., and Wesolowski, B. 2019. *Discrete Logarithms in Quasi-Polynomial Time in Finite Fields of Fixed Characteristic.* Cryptology ePrint Archive, Report 2019/751. https://eprint.iacr.org/2019/751. (Cited on pages 136 and 137.)

[327] Kleinjung, T., Aoki, K., Franke, J., Lenstra, A. K., Thomé, E., Bos, J. W., Gaudry, P., Kruppa, A., Montgomery, P. L., Osvik, D. A., te Riele, H. J. J., Timofeev, A., and Zimmermann, P. 2010. Factorization of a 768-Bit RSA Modulus. Pages 333–350 of: Rabin, T. (ed.), *CRYPTO 2010*. LNCS, vol. 6223. Santa Barbara, CA, USA: Springer, Heidelberg, Germany. (Cited on pages 3, 8, 137, and 157.)

[328] Kleinjung, T., Lenstra, A. K., Page, D., and Smart, N. P. 2012. Using the Cloud to Determine Key Strengths. Pages 17–39 of: Galbraith, S. D., and Nandi, M. (eds.), *INDOCRYPT 2012*. LNCS, vol. 7668. Kolkata, India: Springer, Heidelberg, Germany. (Cited on pages 4, 316, 317, and 325.)

[329] Kleinjung, T., Bos, J. W., and Lenstra, A. K. 2014. Mersenne Factorization Factory. Pages 358–377 of: Sarkar, P., and Iwata, T. (eds.), *ASIACRYPT 2014, Part I*. LNCS, vol. 8873. Kaoshiung, Taiwan, R.O.C.: Springer, Heidelberg, Germany. (Cited on pages 3, 61, and 138.)

[330] Kleinjung, T., Diem, C., Lenstra, A. K., Priplata, C., and Stahlke, C. 2017. Computation of a 768-Bit Prime Field Discrete Logarithm. Pages 185–201 of: Coron, J.-S., and Nielsen, J. B. (eds.), *EUROCRYPT 2017, Part I*. LNCS, vol. 10210. Paris, France: Springer, Heidelberg, Germany. (Cited on pages 6, 137, and 157.)

[331] Klima, V. 2005. *Finding MD5 Collisions on a Notebook PC Using Multi-message Modifications.* Cryptology ePrint Archive, Report 2005/102. http://eprint.iacr.org/2005/102. (Cited on page 184.)

[332] Klima, V. 2006. *Tunnels in Hash Functions: MD5 Collisions Within a Minute.*

Cryptology ePrint Archive, Report 2006/105. http://eprint.iacr.org/ 2006/105. (Cited on pages 184 and 189.)

[333] Klitzke, E. 2017. *Bitcoin Transaction Malleability*. https://eklitzke.org/ bitcoin-transaction-malleability. (Cited on page 176.)

[334] Klyachko, A. A. 1988. On the Rationality of Tori with Cyclic Splitting Field (Russian). *Arithmetic and Geometry of Varieties*, 73–78. (Cited on page 124.)

[335] Klyubin, A. 2013 (August). *Some SecureRandom Thoughts*. https:// android-developers.googleblog.com/2013/08/some-securerandom-thoughts.html. (Cited on page 178.)

[336] Knežević, M., Vercauteren, F., and Verbauwhede, I. 2010. Speeding Up Bipartite Modular Multiplication. Pages 166–179 of: Hasan, M. A., and Helleseth, T. (eds.), *Arithmetic of Finite Fields – WAIFI*. LNCS, vol. 6087. Springer, Heidelberg, Germany. (Cited on page 241.)

[337] Knuth, D. E., and Papadimitriou, C. H. 1981. Duality in addition chains. *Bulletin of the European Association for Theoretical Computer Science*, **13**, 2–4. (Cited on page 302.)

[338] Knuth, D. E. 1997. *Seminumerical Algorithms*. 3rd edn. The Art of Computer Programming, vol. 2. Reading, Massachusetts, USA: Addison-Wesley. (Cited on page 226.)

[339] Koblitz, A. H., Koblitz, N., and Menezes, A. 2011. Elliptic Curve Cryptography: The Serpentine Course of a Paradigm Shift. *Journal of Number Theory*, **131**(5), 781–814. (Cited on page 171.)

[340] Koblitz, N. 1987. Elliptic Curve Cryptosystems. *Mathematics of Computation*, **48**(177), 203–209. (Cited on pages 62, 171, 237, and 295.)

[341] Koblitz, N., and Menezes, A. 2016. A Riddle Wrapped in an Enigma. *IEEE Security & Privacy*, **14**(6), 34–42. (Cited on pages 171 and 172.)

[342] Kocher, P. 2020. *Personal communication*. (Cited on pages 143 and 148.)

[343] Kraitchik, M. 1922. *Théorie des nombres*. Vol. 1. Paris: Gauthier-Villars. (Cited on pages 54 and 109.)

[344] Kraitchik, M. 1924. *Recherches sur la théorie des nombres*. Vol. 1. Paris: Gauthier-Villars. (Cited on page 109.)

[345] Krawczyk, H. 2003. SIGMA: The "SIGn-and-MAc" Approach to Authenticated Diffie-Hellman and Its Use in the IKE Protocols. Pages 400–425 of: Boneh, D. (ed.), *CRYPTO 2003*. LNCS, vol. 2729. Santa Barbara, CA, USA: Springer, Heidelberg, Germany. (Cited on page 202.)

[346] Kronecker, L. 1882. Grundzüge einer arithmetischen theorie der algebraischen grössen. *Journal für die reine und angewandte Mathematik*, 1–122. (Cited on page 289.)

[347] Kuhn, T. S. 1996. *The Structure of Scientific Revolutions*. Chicago, IL: University of Chicago Press. (Cited on page 130.)

[348] Kumar, S., Paar, C., Pelzl, J., Pfeiffer, G., and Schimmler, M. 2006. Breaking Ciphers with COPACOBANA - A Cost-Optimized Parallel Code Breaker. Pages 101–118 of: Goubin, L., and Matsui, M. (eds.), *CHES 2006*. LNCS, vol. 4249. Yokohama, Japan: Springer, Heidelberg, Germany. (Cited on page 316.)

[349] Kunihiro, N., Shinohara, N., and Izu, T. 2012. A Unified Framework for Small Secret Exponent Attack on RSA. Pages 260–277 of: Miri, A., and Vaudenay, S.

(eds.), *SAC 2011*. LNCS, vol. 7118. Toronto, ON, Canada: Springer, Heidelberg, Germany. (Cited on pages 98 and 103.)

[350] Lagarias, J. C., and Odlyzko, A. M. 1983. Solving Low-Density Subset Sum Problems. Pages 1–10 of: *24th FOCS*. Tucson, AZ: IEEE Computer Society Press. (Cited on page 30.)

[351] Lamport, L. 1979 (Oct.). *Constructing Digital Signatures from a One-way Function*. Technical Report SRI-CSL-98. SRI International Computer Science Laboratory. (Cited on page 332.)

[352] Laphroaig, M. (ed.). 2017. *Pastor Laphroaig Screams High Five to the Heavens as the Whole World Goes Under*. PoC‖GTFO, vol. 0x14. Tract Association of POC‖GTFO and Friends. (Cited on page 211.)

[353] Laurie, B., Langley, A., and Kasper, E. 2013 (June). *Certificate Transparency*. RFC 6962. RFC Editor. https://www.rfc-editor.org/info/rfc6962. (Cited on page 154.)

[354] Lehmer, D. H., and Powers, R. E. 1931. On Factoring Large Numbers. *Bulletin of the American Mathematical Society*, **37**, 770–776. (Cited on page 54.)

[355] Lehmer, D. N. 1909. *Factor Table for the First Ten Millions Containing the Smallest Factor of Every Number not Divisible by 2, 3, 5, or 7 Between the Limits 0 and 10017000*. Carnegie Institute of Washington. (Cited on page 41.)

[356] Lehmer, D. H. 1928. The Mechanical Combination of Linear Forms. *American Mathematical Monthly*, **35**, 114–121. (Cited on page 323.)

[357] Lehmer, D. H. 1933. A Photo-Electric Number Sieve. *American Mathematical Monthly*, **40**, 401–406. (Cited on pages 323 and 324.)

[358] Lehmer, D. N. 1932. Hunting Big Game in the Theory of Numbers. *Scripta Mathematica I*, 229–235. (Cited on page 323.)

[359] Lenstra, A. K., and Lenstra Jr, H. W. 1987. *Algorithms in Number Theory*. Technical Report 87-008, University of Chicago. (Cited on page 109.)

[360] Lenstra, A. K., and Lenstra, Jr, H. W. 1993. *The Development of the Number Field Sieve*. Lecture Notes in Mathematics, vol. 1554. New York: Springer-Verlag. (Cited on pages 58 and 321.)

[361] Lenstra, A. K., and Verheul, E. R. 2001a. An Overview of the XTR Public Key System. Pages 151–180 of: *Public Key Cryptography and Computational Number Theory*. Verlages Walter de Gruyter. (Cited on page 123.)

[362] Lenstra, A. K. 1981. Lattices and Factorization of Polynomials. *SIGSAM Bulletin*, **15**(3), 15–16. (Cited on page 2.)

[363] Lenstra, A. K. 1982. Lattices and Factorization of Polynomials over Algebraic Number Fields. Pages 32–39 of: Calmet, J. (ed.), *European Computer Algebra Conference – EUROCAM*. LNCS, vol. 144. Springer. (Cited on page 2.)

[364] Lenstra, A. K. 1983. Factoring Multivariate Polynomials over Finite Fields (Extended Abstract). Pages 189–192 of: *15th ACM STOC*. Boston, MA, USA: ACM Press. (Cited on page 2.)

[365] Lenstra, A. K. 1984. *Polynomial-Time Algorithms for the Factorization of Polynomials*. PhD thesis, University of Amsterdam. (Cited on page 2.)

[366] Lenstra, A. K. 1997. Using Cyclotomic Polynomials to Construct Efficient Discrete Logarithm Cryptosystems over Finite Fields. Pages 127–138 of: Varadharajan, V., Pieprzyk, J., and Mu, Y. (eds.), *ACISP 97*. LNCS, vol. 1270. Syd-

ney, NSW, Australia: Springer, Heidelberg, Germany. (Cited on pages 6, 123, 130, 296, and 311.)

[367] Lenstra, A. K. 1998. Generating RSA Moduli with a Predetermined Portion. Pages 1–10 of: Ohta, K., and Pei, D. (eds.), *ASIACRYPT'98*. LNCS, vol. 1514. Beijing, China: Springer, Heidelberg, Germany. (Cited on pages 79, 228, and 241.)

[368] Lenstra, A. K. 2001. Unbelievable Security. Matching AES Security Using Public Key Systems (Invited Talk). Pages 67–86 of: Boyd, C. (ed.), *ASIACRYPT 2001*. LNCS, vol. 2248. Gold Coast, Australia: Springer, Heidelberg, Germany. (Cited on pages 4, 162, 238, 312, and 313.)

[369] Lenstra, A. K., and de Weger, B. 2005. On the Possibility of Constructing Meaningful Hash Collisions for Public Keys. Pages 267–279 of: Boyd, C., and Nieto, J. M. G. (eds.), *ACISP 05*. LNCS, vol. 3574. Brisbane, Queensland, Australia: Springer, Heidelberg, Germany. (Cited on pages 6, 184, and 191.)

[370] Lenstra, A. K., and Haber, S. 1991. *Comment on Proposed Digital Signature Standard*. Letter to NIST regarding DSS, 1991. (Cited on page 169.)

[371] Lenstra, A. K., and Manasse, M. S. 1990. Factoring by Electronic Mail. Pages 355–371 of: Quisquater, J.-J., and Vandewalle, J. (eds.), *EUROCRYPT'89*. LNCS, vol. 434. Houthalen, Belgium: Springer, Heidelberg, Germany. (Cited on pages 2, 3, 57, and 319.)

[372] Lenstra, A. K., and Manasse, M. S. 1991. Factoring with Two Large Primes. Pages 72–82 of: Damgård, I. (ed.), *EUROCRYPT'90*. LNCS, vol. 473. Aarhus, Denmark: Springer, Heidelberg, Germany. (Cited on pages 3 and 57.)

[373] Lenstra, A. K., and Shamir, A. 2000. Analysis and Optimization of the TWIN-KLE Factoring Device. Pages 35–52 of: Preneel, B. (ed.), *EUROCRYPT 2000*. LNCS, vol. 1807. Bruges, Belgium: Springer, Heidelberg, Germany. (Cited on pages 62 and 324.)

[374] Lenstra, A. K., and Shparlinski, I. 2002. Selective Forgery of RSA Signatures with Fixed-Pattern Padding. Pages 228–236 of: Naccache, D., and Paillier, P. (eds.), *PKC 2002*. LNCS, vol. 2274. Paris, France: Springer, Heidelberg, Germany. (Cited on page 8.)

[375] Lenstra, A. K., and Verheul, E. R. 2000a. Key Improvements to XTR. Pages 220–233 of: Okamoto, T. (ed.), *ASIACRYPT 2000*. LNCS, vol. 1976. Kyoto, Japan: Springer, Heidelberg, Germany. (Cited on pages 4 and 299.)

[376] Lenstra, A. K., and Verheul, E. R. 2000b. Selecting Cryptographic Key Sizes. Pages 446–465 of: Imai, H., and Zheng, Y. (eds.), *PKC 2000*. LNCS, vol. 1751. Melbourne, Victoria, Australia: Springer, Heidelberg, Germany. (Cited on pages 4, 314, and 315.)

[377] Lenstra, A. K., and Verheul, E. R. 2000c. The XTR Public Key System. Pages 1–19 of: Bellare, M. (ed.), *CRYPTO 2000*. LNCS, vol. 1880. Santa Barbara, CA, USA: Springer, Heidelberg, Germany. (Cited on pages 4, 123, 297, 300, 303, 306, 311, and 327.)

[378] Lenstra, A. K., and Verheul, E. R. 2001b. Fast Irreducibility and Subgroup Membership Testing in XTR. Pages 73–86 of: Kim, K. (ed.), *PKC 2001*. LNCS, vol. 1992. Cheju Island, South Korea: Springer, Heidelberg, Germany. (Cited on pages 4, 299, and 301.)

[379] Lenstra, A. K., and Verheul, E. R. 2001c. Selecting Cryptographic Key Sizes. *Journal of Cryptology*, **14**(4), 255–293. (Cited on pages 4, 157, 235, 238, and 314.)

[380] Lenstra, A. K., Lenstra, Jr., H. W., and Lovász, L. 1982. Factoring Polynomials with Rational Coefficients. *Mathematische Annalen*, **261**(4), 515–534. (Cited on pages 2, 15, 71, 72, 73, 78, 79, 179, 290, and 330.)

[381] Lenstra, A. K., Lenstra Jr., H. W., Manasse, M. S., and Pollard, J. M. 1990. The Number Field Sieve. Pages 564–572 of: *22nd ACM STOC*. Baltimore, MD, USA: ACM Press. (Cited on pages 3 and 58.)

[382] Lenstra, A. K., Lenstra, Jr, H. W., Manasse, M. S., and Pollard, J. M. 1993. The Factorization of the Ninth Fermat Number. *Mathematics of Computation*, **61**(203), 319–349. (Cited on page 3.)

[383] Lenstra, A. K., Shamir, A., Tomlinson, J., and Tromer, E. 2002. Analysis of Bernstein's Factorization Circuit. Pages 1–26 of: Zheng, Y. (ed.), *ASIACRYPT 2002*. LNCS, vol. 2501. Queenstown, New Zealand: Springer, Heidelberg, Germany. (Cited on page 324.)

[384] Lenstra, A. K., Tromer, E., Shamir, A., Kortsmit, W., Dodson, B., Hughes, J. P., and Leyland, P. C. 2003. Factoring Estimates for a 1024-Bit RSA Modulus. Pages 55–74 of: Laih, C.-S. (ed.), *ASIACRYPT 2003*. LNCS, vol. 2894. Taipei, Taiwan: Springer, Heidelberg, Germany. (Cited on pages 3 and 324.)

[385] Lenstra, A. K., Wang, X., and de Weger, B. 2005. *Colliding X.509 Certificates*. Cryptology ePrint Archive, Report 2005/067. http://eprint.iacr.org/2005/067. (Cited on page 6.)

[386] Lenstra, A. K., Hughes, J. P., Augier, M., Bos, J. W., Kleinjung, T., and Wachter, C. 2012. Public Keys. Pages 626–642 of: Safavi-Naini, R., and Canetti, R. (eds.), *CRYPTO 2012*. LNCS, vol. 7417. Santa Barbara, CA, USA: Springer, Heidelberg, Germany. (Cited on pages 8, 70, 80, 143, 145, 146, 147, 148, 155, 164, 174, and 175.)

[387] Lenstra, A. K., Kleinjung, T., and Thomé, E. 2013. Universal Security - From Bits and Mips to Pools, Lakes - and Beyond. Pages 121–124 of: Fischlin, M., and Katzenbeisser, S. (eds.), *Number Theory and Cryptography - Papers in Honor of Johannes Buchmann on the Occasion of His 60th Birthday*. LNCS, vol. 8260. Springer, Heidelberg, Germany. (Cited on page 333.)

[388] Lenstra, Jr, H. W. 1987. Factoring Integers with Elliptic Curves. *Annals of Mathematics*, **126**(3), 649–673. (Cited on pages 3, 62, 64, 65, 163, and 233.)

[389] Lepinski, M., and Kent, S. 2008. *Additional Diffie-Hellman Groups for Use with IETF Standards*. IETF RFC 5114. (Cited on pages 162, 166, and 170.)

[390] Leurent, G., and Peyrin, T. 2019. From Collisions to Chosen-Prefix Collisions Application to Full SHA-1. Pages 527–555 of: Ishai, Y., and Rijmen, V. (eds.), *EUROCRYPT 2019, Part III*. LNCS, vol. 11478. Darmstadt, Germany: Springer, Heidelberg, Germany. (Cited on pages 185, 189, 201, and 318.)

[391] Leurent, G., and Peyrin, T. 2020. SHA-1 is a Shambles: First Chosen-Prefix Collision on SHA-1 and Application to the PGP Web of Trust. Pages 1839–1856 of: Capkun, S., and Roesner, F. (eds.), *USENIX Security 2020*. USENIX Association. (Cited on pages 185 and 201.)

[392] Leyland, P. C., Lenstra, A. K., Dodson, B., Muffett, A., and Wagstaff, Jr, S. S. 2002. MPQS with Three Large Primes. Pages 446–460 of: Fieker, C., and

Kohel, D. R. (eds.), *Algorithmic Number Theory, 5th International Symposium – ANTS-V*. LNCS, vol. 2369. Springer, Heidelberg, Germany. (Cited on page 3.)

[393] Lido, G. 2016. *Discrete Logarithm over Finite Fields of Small Characteristic*. Master's thesis, Universita di Pisa. (Cited on page 136.)

[394] Lim, C. H., and Lee, P. J. 1996. *Generating Efficient Primes for Discrete Log Cryptosystems*. http://citeseerx.ist.psu.edu/viewdoc/summary? doi=10.1.1.43.8261. (Cited on page 169.)

[395] Lim, C. H., and Lee, P. J. 1997. A Key Recovery Attack on Discrete Log-based Schemes Using a Prime Order Subgroup. Pages 249–263 of: Kaliski Jr., B. S. (ed.), *CRYPTO '97*. LNCS, vol. 1294. Santa Barbara, CA, USA: Springer, Heidelberg, Germany. (Cited on pages 166 and 167.)

[396] Liu, M., and Nguyen, P. Q. 2013. Solving BDD by Enumeration: An Update. Pages 293–309 of: Dawson, E. (ed.), *CT-RSA 2013*. LNCS, vol. 7779. San Francisco, CA, USA: Springer, Heidelberg, Germany. (Cited on page 33.)

[397] Livitt, C. D. 2016. *Preliminary Expert Report of Carl D. Livitt*. https:// medsec.com/stj_expert_witness_report.pdf. (Cited on page 144.)

[398] Lochter, M., and Merkle, J. 2010 (Mar.). *Elliptic Curve Cryptography (ECC) Brainpool Standard Curves and Curve Generation*. RFC 5639. RFC Editor. https://www.rfc-editor.org/info/rfc5639. (Cited on page 172.)

[399] López-Alt, A., Tromer, E., and Vaikuntanathan, V. 2012. On-the-Fly Multiparty Computation on the Cloud via Multikey Fully Homomorphic Encryption. Pages 1219–1234 of: Karloff, H. J., and Pitassi, T. (eds.), *44th ACM STOC*. New York, NY, USA: ACM Press. (Cited on page 37.)

[400] Lovorn, R. 1992. *Rigorous Subexponential Algorithms for Discrete Logarithms over Finite Fields*. PhD thesis, University of Georgia. (Cited on page 131.)

[401] Lyubashevsky, V., Peikert, C., and Regev, O. 2010. On Ideal Lattices and Learning with Errors over Rings. Pages 1–23 of: Gilbert, H. (ed.), *EUROCRYPT 2010*. LNCS, vol. 6110. French Riviera: Springer, Heidelberg, Germany. (Cited on page 16.)

[402] Manger, J. 2001. A Chosen Ciphertext Attack on RSA Optimal Asymmetric Encryption Padding (OAEP) as Standardized in PKCS #1 v2.0. Pages 230–238 of: Kilian, J. (ed.), *CRYPTO 2001*. LNCS, vol. 2139. Santa Barbara, CA, USA: Springer, Heidelberg, Germany. (Cited on page 152.)

[403] Matsumoto, T., and Imai, H. 1988. Public Quadratic Polynominal-Tuples for Efficient Signature-Verification and Message-Encryption. Pages 419–453 of: Günther, C. G. (ed.), *EUROCRYPT '88*. LNCS, vol. 330. Davos, Switzerland: Springer, Heidelberg, Germany. (Cited on page 331.)

[404] Maurer, U. M. 1994. Towards the Equivalence of Breaking the Diffie-Hellman Protocol and Computing Discrete Algorithms. Pages 271–281 of: Desmedt, Y. (ed.), *CRYPTO '94*. LNCS, vol. 839. Santa Barbara, CA, USA: Springer, Heidelberg, Germany. (Cited on page 155.)

[405] Mavrogiannopoulos, N., Vercauteren, F., Velichkov, V., and Preneel, B. 2012. A Cross-Protocol Attack on the TLS Protocol. Pages 62–72 of: Yu, T., Danezis, G., and Gligor, V. D. (eds.), *ACM CCS 2012*. Raleigh, NC, USA: ACM Press. (Cited on page 167.)

[406] May, A. 2002. Cryptanalysis of Unbalanced RSA with Small CRT-Exponent. Pages 242–256 of: Yung, M. (ed.), *CRYPTO 2002*. LNCS, vol. 2442. Santa Barbara, CA, USA: Springer, Heidelberg, Germany. (Cited on pages 80 and 99.)

[407] May, A. 2004. Computing the RSA Secret Key Is Deterministic Polynomial Time Equivalent to Factoring. Pages 213–219 of: Franklin, M. (ed.), *CRYPTO 2004*. LNCS, vol. 3152. Santa Barbara, CA, USA: Springer, Heidelberg, Germany. (Cited on pages 71 and 73.)

[408] May, A. 2010. Using LLL-Reduction for Solving RSA and Factorization Problems. *In:* [457]. (Cited on pages 83 and 84.)

[409] May, A., and Ritzenhofen, M. 2008. Solving Systems of Modular Equations in One Variable: How Many RSA-Encrypted Messages Does Eve Need to Know? Pages 37–46 of: Cramer, R. (ed.), *PKC 2008*. LNCS, vol. 4939. Barcelona, Spain: Springer, Heidelberg, Germany. (Cited on pages 88, 89, and 104.)

[410] May, A., and Silverman, J. H. 2001. Dimension Reduction Methods for Convolution Modular Lattices. Pages 110–125 of: Silverman, J. H. (ed.), *Cryptography and Lattices – CaLC*. LNCS, vol. 2146. Springer, Heidelberg, Germany. (Cited on page 35.)

[411] McCurley, K. S. 1990. The Discrete Logarithm Problem. Pages 49–74 of: *Cryptology and computational number theory, Proceedings of Symposia in Applied Mathematics*, vol. 42. American Mathematical Society, Providence, Rhode Island, USA. (Cited on page 109.)

[412] McDonald, C., Hawkes, P., and Pieprzyk, J. 2009. *Differential Path for SHA-1 with complexity $O(2^{52})$*. Cryptology ePrint Archive, Report 2009/259. http://eprint.iacr.org/2009/259. (Cited on page 185.)

[413] McEliece, R. J. 1978. A Public-Key Cryptosystem Based On Algebraic Coding Theory. *DSN Progress Report*, **44**, 114–116. (Cited on page 331.)

[414] McLean, T. 2015. *Critical Vulnerabilities in JSON Web Token Libraries*. https://auth0.com/blog/critical-vulnerabilities-in-json-web-token-libraries/. (Cited on page 153.)

[415] Mendel, F., Rechberger, C., and Rijmen, V. 2007. *Update on SHA-1*. Rump session of CRYPTO 2007. http://rump2007.cr.yp.to/09-rechberger.pdf. (Cited on page 185.)

[416] Menezes, A. 1993. *Elliptic Curve Public Key Cryptosystems*. Boston, MA: Kluwer. (Cited on page 308.)

[417] Menezes, A., and Vanstone, S. 2000. *ECSTR (XTR): Elliptic Curve Singular Trace Representation*. Presented at the Rump Session of Crypto 2000. (Cited on page 308.)

[418] Menezes, A., Okamoto, T., and Vanstone, S. 1993. Reducing Elliptic Curve Logarithms to Logarithms in a Finite Field. *IEEE Transactions on Information Theory*, **39**(5), 1639–1646. (Cited on page 116.)

[419] Menezes, A., and Qu, M. 2001. Analysis of the Weil Descent Attack of Gaudry, Hess and Smart. Pages 308–318 of: Naccache, D. (ed.), *CT-RSA 2001*. LNCS, vol. 2020. San Francisco, CA, USA: Springer, Heidelberg, Germany. (Cited on page 117.)

[420] Menezes, A., and Wu, Y. 1997. The Discrete Logarithm Problem in $GL(n, q)$. *Ars Combinatoria*, **47**. (Cited on page 118.)

[421] Menezes, A., Vanstone, S. A., and Okamoto, T. 1991. Reducing Elliptic Curve Logarithms to Logarithms in a Finite Field. Pages 80–89 of: *23rd ACM STOC*. New Orleans, LA, USA: ACM Press. (Cited on pages 67, 308, and 327.)

[422] Merkle, R. C. 1990a. A Certified Digital Signature. Pages 218–238 of: Brassard, G. (ed.), *CRYPTO'89*. LNCS, vol. 435. Santa Barbara, CA, USA: Springer, Heidelberg, Germany. (Cited on page 332.)

[423] Merkle, R. C. 1990b. One Way Hash Functions and DES. Pages 428–446 of: Brassard, G. (ed.), *CRYPTO'89*. LNCS, vol. 435. Santa Barbara, CA, USA: Springer, Heidelberg, Germany. (Cited on page 183.)

[424] Meyer, C., Somorovsky, J., Weiss, E., Schwenk, J., Schinzel, S., and Tews, E. 2014. Revisiting SSL/TLS Implementations: New Bleichenbacher Side Channels and Attacks. Pages 733–748 of: Fu, K., and Jung, J. (eds.), *USENIX Security 2014*. San Diego, CA, USA: USENIX Association. (Cited on page 151.)

[425] Micciancio, D., and Walter, M. 2016. Practical, Predictable Lattice Basis Reduction. Pages 820–849 of: Fischlin, M., and Coron, J.-S. (eds.), *EUROCRYPT 2016, Part I*. LNCS, vol. 9665. Vienna, Austria: Springer, Heidelberg, Germany. (Cited on page 15.)

[426] Michaelis, K., Meyer, C., and Schwenk, J. 2013. Randomly Failed! The State of Randomness in Current Java Implementations. Pages 129–144 of: Dawson, E. (ed.), *CT-RSA 2013*. LNCS, vol. 7779. San Francisco, CA, USA: Springer, Heidelberg, Germany. (Cited on page 178.)

[427] Microsoft. 2012 (June 6,). *Flame Malware Collision Attack Explained*. Security Research & Defense, Microsoft TechNet Blog: `https://msrc-blog.microsoft.com/2012/06/06/flame-malware-collision-attack-explained`. (Cited on pages 198 and 199.)

[428] Miele, A., Bos, J. W., Kleinjung, T., and Lenstra, A. K. 2014. Cofactorization on Graphics Processing Units. Pages 335–352 of: Batina, L., and Robshaw, M. (eds.), *CHES 2014*. LNCS, vol. 8731. Busan, South Korea: Springer, Heidelberg, Germany. (Cited on page 3.)

[429] Mikle, O. 2004. *Practical Attacks on Digital Signatures Using MD5 Message Digest*. Cryptology ePrint Archive, Report 2004/356. `http://eprint.iacr.org/2004/356`. (Cited on pages 184 and 207.)

[430] Miller, D. 2020 (May). *OpenSSH 8.3 released (and ssh-rsa deprecation notice)*. `https://lwn.net/Articles/821544/`. (Cited on page 141.)

[431] Miller, G. L. 1976. Riemann's Hypothesis and Tests for Primality. *Journal of Computer and System Sciences*, **13**(3), 300–317. (Cited on pages 91, 93, and 141.)

[432] Miller, S. D., Narayanan, B., and Venkatesan, R. 2017. *Coppersmith's Lattices and "Focus Groups": an Attack on Small-Exponent RSA*. Cryptology ePrint Archive, Report 2017/835. `http://eprint.iacr.org/2017/835`. (Cited on page 102.)

[433] Miller, V. S. 1986. Use of Elliptic Curves in Cryptography. Pages 417–426 of: Williams, H. C. (ed.), *CRYPTO'85*. LNCS, vol. 218. Santa Barbara, CA, USA: Springer, Heidelberg, Germany. (Cited on pages 62, 171, 237, and 295.)

[434] Miller, V. S. 2004. The Weil Pairing, and Its Efficient Calculation. *Journal of Cryptology*, **17**(4), 235–261. (Cited on page 116.)

[435] Mironov, I. 2012. *Factoring RSA Moduli. Part II.* https:// windowsontheory.org/2012/05/17/factoring-rsa-moduli-part-ii/. (Cited on page 142.)

[436] Möller, N., and Granlund, T. 2011. Improved Division by Invariant Integers. *IEEE Transactions on Computers*, **60**(2), 165–175. (Cited on page 270.)

[437] Montgomery, P. L. 1983. *Evaluating Recurrences of Form* $X_{m+n} = f(X_m, X_n, X_{m-n})$ *via Lucas Chains*. Revised (1992) version from ftp.cw.nl: /pub/pmontgom/Lucas.ps.gs. (Cited on pages 300 and 302.)

[438] Montgomery, P. L. 1985. Modular Multiplication without Trial Division. *Mathematics of Computation*, **44**(170), 519–521. (Cited on pages 224, 237, and 270.)

[439] Montgomery, P. L. 1994. *Vectorization of the Elliptic Curve Method*. Technical Report. Centrum Wiskunde & Informatica (CWI). (Cited on page 233.)

[440] Montgomery, P. L., and Murphy, B. 1999. Improved polynomial selection for the Number Field Sieve. In: *The Mathematics of Public Key Cryptography Conference*. Toronto: Fields Institute. (Cited on page 61.)

[441] Moore, G. E. 1965. Cramming more Components onto Integrated Circuits. *Electronics Magazine*, **38**(8), 114–117. (Cited on page 230.)

[442] Morain, F. 2004. La primalité en temps polynomial (d'après Adleman, Huang; Agrawal, Kayal, Saxena). *Astérisque*, **294**, 205–230. (Cited on page 47.)

[443] Moriarty, K., Kaliski, B., Jonsson, J., and Rusch, A. 2016 (Nov.). *PKCS #1: RSA Cryptography Specifications Version 2.2*. https://www.rfc-editor.org/info/rfc8017. (Cited on page 149.)

[444] Morrison, M. A., and Brillhart, J. 1975. A Method of Factoring and the Factorization of F_7. *Mathematics of Computation*, **29**, 183–205. (Cited on pages 53, 54, and 319.)

[445] Mozilla. 2017. *HTTP Public Key Pinning (HPKP)*. https://developer.mozilla.org/en-US/docs/Web/HTTP/Public_Key_Pinning. (Cited on page 154.)

[446] Müller, W., and Nöbauer, R. 1981. Some Remarks on Public-Key Cryptosystems. *Studia Scientiarum Mathematicarum Hungarica*, **16**, 71–76. (Cited on page 298.)

[447] National Institute of Standards and Technology. 2012. *Special Publication 800-57. Recommendation for Key Management Part 1: General (Revision 3)*. (Cited on page 315.)

[448] National Security Agency. 2020. *Patch Critical Cryptographic Vulnerability in Microsoft Windows Clients and Servers*. https://media.defense.gov/2020/Jan/14/2002234275/-1/-1/0/CSA-WINDOWS-10-CRYPT-LIB-20190114.PDF. (Cited on page 176.)

[449] National Security Agency Central Security Service. 2015. *Cryptography Today*. https://web.archive.org/web/20151123081120/https://www.nsa.gov/ia/programs/suiteb_cryptography. (Cited on page 171.)

[450] Nechaev, V. I. 1994. On the Complexity of a Deterministic Algorithm for a Discrete Logarithm. *Matematicheskie Zametki*, **55**, 91–101. (Cited on page 107.)

[451] Nemec, M., Sýs, M., Svenda, P., Klinec, D., and Matyas, V. 2017. The Return of Coppersmith's Attack: Practical Factorization of Widely Used RSA Moduli. Pages 1631–1648 of: Thuraisingham, B. M., Evans, D., Malkin, T., and Xu, D. (eds.), *ACM CCS 2017*. Dallas, TX, USA: ACM Press. (Cited on pages 15, 80, and 143.)

[452] Neumaier, A., and Stehlé, D. 2016. Faster LLL-type reduction of Lattice Bases. Pages 373–380 of: Abramov, S. A., Zima, E. V., and Gao, X. (eds.), *International Symposium on Symbolic and Algebraic Computation – ISSAC*. Association for Computing Machinery, New York, USA. (Cited on page 82.)

[453] Nguyen, P. Q. 1999. Cryptanalysis of the Goldreich-Goldwasser-Halevi Cryptosystem from Crypto'97. Pages 288–304 of: Wiener, M. J. (ed.), *CRYPTO'99*. LNCS, vol. 1666. Santa Barbara, CA, USA: Springer, Heidelberg, Germany. (Cited on page 290.)

[454] Nguyen, P. Q. 2010. Hermite's Constant and Lattice Algorithms. *In:* [457]. (Cited on page 18.)

[455] Nguyen, P. Q., and Stehlé, D. 2005. Floating-Point LLL Revisited. Pages 215–233 of: Cramer, R. (ed.), *EUROCRYPT 2005*. LNCS, vol. 3494. Aarhus, Denmark: Springer, Heidelberg, Germany. (Cited on page 291.)

[456] Nguyen, P. Q., and Stern, J. 1998. Cryptanalysis of the Ajtai-Dwork Cryptosystem. Pages 223–242 of: Krawczyk, H. (ed.), *CRYPTO'98*. LNCS, vol. 1462. Santa Barbara, CA, USA: Springer, Heidelberg, Germany. (Cited on page 290.)

[457] Nguyen, P. Q., and Vallée, B. (eds.). 2010. *The LLL Algorithm - Survey and Applications.* ISC. Springer, Heidelberg, Germany. (Cited on pages 2, 291, 366, 369, and 375.)

[458] NMBRTHRY. 2020. NumberTheoryList: `https://listserv.nodak.edu/cgi-bin/wa.exe?A0=NMBRTHRY`. (Cited on page 138.)

[459] Nyberg, K., and Rueppel, R. A. 1993. A New Signature Scheme Based on the DSA Giving Message Recovery. Pages 58–61 of: Denning, D. E., Pyle, R., Ganesan, R., Sandhu, R. S., and Ashby, V. (eds.), *ACM CCS 93*. Fairfax, Virginia, USA: ACM Press. (Cited on page 303.)

[460] Odlyzko, A. M. 1985. Discrete Logarithms in Finite Fields and Their Cryptographic Significance. Pages 224–314 of: Beth, T., Cot, N., and Ingemarsson, I. (eds.), *EUROCRYPT'84*. LNCS, vol. 209. Paris, France: Springer, Heidelberg, Germany. (Cited on pages 109 and 131.)

[461] Odoni, R., Varadharajan, V., and Sanders, P. 1984. Public Key Distribution in Matrix Rings. *Electronics Letters*, **20**(9), 386–387. (Cited on page 118.)

[462] OpenSSH. 2020. *OpenSSH Legacy Options.* `https://www.openssh.com/legacy.html`. (Cited on page 158.)

[463] Oracle. 2014. *JDK 8 Security Enhancements.* `https://docs.oracle.com/javase/8/docs/technotes/guides/security/enhancements-8.html`. (Cited on page 157.)

[464] Orman, H. 1998 (Nov.). *RFC 2412 - The Oakley Key Determination Protocol.* Internet Engineering Task Force. `http://www.ietf.org/rfc/rfc2412.txt`. (Cited on pages 158 and 169.)

[465] Page, D., and Smart, N. P. 2004. Parallel Cryptographic Arithmetic using a Redundant Montgomery Representation. *IEEE Transactions on Computers*, **53**(11), 1474–1482. (Cited on pages 233 and 234.)

[466] Page, D., and Stam, M. 2004. On XTR and Side-Channel Analysis. Pages 54–68 of: Handschuh, H., and Hasan, A. (eds.), *SAC 2004*. LNCS, vol. 3357. Waterloo, Ontario, Canada: Springer, Heidelberg, Germany. (Cited on page 302.)

[467] Paillier, P. 1999. Low-Cost Double-Size Modular Exponentiation or How to Stretch Your Cryptoprocessor. Pages 223–234 of: Imai, H., and Zheng, Y. (eds.), *PKC'99*. LNCS, vol. 1560. Kamakura, Japan: Springer, Heidelberg, Germany. (Cited on page 236.)

[468] Parhami, B. 2000. *Computer Arithmetic: Algorithms and Hardware Designs*. 1 edn. Oxford University Press, Oxford, UK. (Cited on page 227.)

[469] Pataki, G., and Tural, M. 2008. *On sublattice determinants in reduced bases*. https://arxiv.org/abs/0804.4014. (Cited on page 38.)

[470] Patarin, J. 1995. Cryptanalysis of the Matsumoto and Imai Public Key Scheme of Eurocrypt'88. Pages 248–261 of: Coppersmith, D. (ed.), *CRYPTO'95*. LNCS, vol. 963. Santa Barbara, CA, USA: Springer, Heidelberg, Germany. (Cited on page 331.)

[471] Patarin, J. 1997. *The Oil and Vinegar Algorithm for Signatures*. Presented at the Dagstuhl Workshop on Cryptography. (Cited on page 331.)

[472] Pohlig, S. C., and Hellman, M. E. 1978. An Improved Algorithm for Computing Logarithms over GF(p) and its Cryptographic Significance (Corresp.). *IEEE Transactions on Information Theory*, **24**(1), 106–110. (Cited on pages 107, 161, 163, and 294.)

[473] Pollard, J. M. 1975. A Monte Carlo Method for Factorization. *Nordisk Tidskrift for Informationsbehandling (BIT)*, **15**, 331–335. (Cited on page 43.)

[474] Pollard, J. M. 1993. *Factoring with Cubic Integers*. Lecture Notes in Mathematics, vol. 1554. New York: Springer-Verlag. Pages 4–10. (Cited on pages 3 and 58.)

[475] Pollard, J. M. 1974. Theorems on Factorization and Primality Testing. Pages 521–528 of: *Mathematical Proceedings of the Cambridge Philosophical Society*, vol. 76. Cambridge University Press, Cambridge, UK. (Cited on pages 43, 67, 142, 161, and 319.)

[476] Pollard, J. M. 1978. Monte Carlo Methods for Index Computation (mod p). *Mathematics of Computation*, **32**, 918–924. (Cited on pages 108, 160, 163, 294, and 325.)

[477] Pollard, J. M. 2000. Kangaroos, Monopoly and Discrete Logarithms. *Journal of Cryptology*, **13**, 437–447. (Cited on page 325.)

[478] Pomerance, C. 1982. Analysis and Comparison of Some Integer Factoring Algorithms. Pages 89–139 of: Lenstra, Jr, H. W., and Tijdeman, R. (eds.), *Computational Methods in Number Theory, Part 1*. Mathematical Centre Tracts, vol. 154. (Cited on pages 3, 53, and 57.)

[479] Pomerance, C. 1994. The Number Field Sieve. Pages 465–480 of: *Mathematics of Computation 1943–1993: a Half-Century of Computational Mathematics*. Proceedings of Symposia in Applied Mathematics, vol. 48. American Mathematics Society, Providence, USA. (Cited on page 58.)

[480] Pomerance, C. 2010. Primality Testing: Variations on a Theme of Lucas. *Congressus Numerantium*, **201**, 301–312. (Cited on page 46.)

[481] Pomerance, C., Selfridge, J. L., and Wagstaff, Jr, S. S. 1980. The Pseudoprimes to $25 \cdot 10^9$. *Mathematics of Computation*, **35**, 1003–1026. (Cited on page 46.)

[482] Pomerance, C., Smith, J. W., and Wagstaff, S. S. 1983. New Ideas for Factoring Large Integers. Pages 81–85 of: Chaum, D. (ed.), *CRYPTO '83*. Santa Barbara, CA, USA: Plenum Press, New York, USA. (Cited on pages 54 and 323.)

[483] Pornin, T. 2013 (Aug.). *Deterministic Usage of the Digital Signature Algorithm (DSA) and Elliptic Curve Digital Signature Algorithm (ECDSA)*. RFC 6979. RFC Editor. https://www.rfc-editor.org/info/rfc6979. (Cited on pages 174, 176, and 179.)

[484] Priest, D. M. 1992 (11). *On Properties of Floating Point Arithmetics: Numerical Stability and the Cost of Accurate Computations*. PhD thesis, University of California, Berkeley. ftp://ftp.icsi.berkeley.edu/pub/theory/priest-thesis.ps.Z. (Cited on page 291.)

[485] Primmer, R., and D'Halluin, C. 2005. *Collision and Preimage Resistance of the Centera Content Address*. Technical Report: http://citeseerx.ist.psu.edu/viewdoc/download?doi=10.1.1.140.2189&rep=rep1&type=pdf. EMC Corporation. (Cited on page 210.)

[486] Qualys. 2020. *SSL Pulse*. https://www.ssllabs.com/ssl-pulse/. (Cited on page 144.)

[487] Quisquater, J.-J. 1992. *Encoding System According to the So-Called RSA Method, by Means of a Microcontroller and Arrangement Implementing this System*. U.S. Patent 5,166,979. (Cited on page 229.)

[488] Quisquater, J.-J., and Standaert, F. 2005. Exhaustive key search of the DES: Updates and refinements. *Special-purpose Hardware for Attacking Cryptographic Systems – SHARCS*. (Cited on page 316.)

[489] Rabin, M. 1979. *Digitized Signatures and Public-Key Functions as Intractable as Factoring*. Technical Report LCS/TR-212. M.I.T. Lab for Computer Science. (Cited on page 49.)

[490] Rabin, M. O. 1980. Probabilistic Algorithm for Testing Primality. *Journal of Number Theory*, **12**(1), 128–138. (Cited on page 141.)

[491] Rankin, R. A. 1953. On Positive Definite Quadratic Forms. *Journal of the London Mathematical Society*, **s1-28**(3), 309–314. (Cited on page 18.)

[492] Razavi, K., Gras, B., Bosman, E., Preneel, B., Giuffrida, C., and Bos, H. 2016. Flip Feng Shui: Hammering a Needle in the Software Stack. Pages 1–18 of: Holz, T., and Savage, S. (eds.), *USENIX Security 2016*. Austin, TX, USA: USENIX Association. (Cited on page 163.)

[493] Regev, O. 2005. On Lattices, Learning with Errors, Random Linear Codes, and Cryptography. Pages 84–93 of: Gabow, H. N., and Fagin, R. (eds.), *37th ACM STOC*. Baltimore, MA, USA: ACM Press. (Cited on page 16.)

[494] Regev, O. 2009. On Lattices, Learning with Errors, Random Linear Codes, and Cryptography. *Journal of the ACM*, **56**(6), 1–40. (Cited on page 16.)

[495] Rescorla, E. 2018 (Aug.). *The Transport Layer Security (TLS) Protocol Version 1.3*. RFC 8446. RFC Editor. https://www.rfc-editor.org/info/rfc8446. (Cited on pages 155 and 156.)

[496] Rescorla, E. 2015. *NSS Accepts Export-Length DHE Keys with Regular DHE Cipher Suites ("Logjam")*. https://bugzilla.mozilla.org/show_bug.cgi?id=1138554. (Cited on page 158.)

[497] rico666. 2016. *Large Bitcoin Collider*. https://lbc.cryptoguru.org/. (Cited on page 177.)

[498] Rieger, G. 2016. *Socat security advisory 7 - Created new 2048bit DH modulus.* https://www.openwall.com/lists/oss-security/2016/02/01/4. (Cited on page 164.)

[499] Rivest, R. L. 1980. A Description of a Single-Chip Implementation of the RSA Cipher. *LAMBDA Mazazine*, **1**(3), 14–18. (Cited on page 226.)

[500] Rivest, R. L., Shamir, A., and Adleman, L. 1983. *Cryptographic Communications System and Method.* U.S.A. Patent 4,405,829. (Cited on page 227.)

[501] Rivest, R. L., Shamir, A., and Adleman, L. M. 1978. A Method for Obtaining Digital Signatures and Public-Key Cryptosystems. *Communications of the Association for Computing Machinery*, **21**(2), 120–126. (Cited on pages 3, 48, 70, 93, 141, 225, 226, 229, 234, and 319.)

[502] Rogaway, P. 2006. Formalizing Human Ignorance. Pages 211–228 of: Nguyen, P. Q. (ed.), *Progress in Cryptology - VIETCRYPT 06.* LNCS, vol. 4341. Hanoi, Vietnam: Springer, Heidelberg, Germany. (Cited on page 183.)

[503] Rubin, K., and Silverberg, A. 2003. Torus-Based Cryptography. Pages 349–365 of: Boneh, D. (ed.), *CRYPTO 2003.* LNCS, vol. 2729. Santa Barbara, CA, USA: Springer, Heidelberg, Germany. (Cited on pages 124, 125, 126, 128, and 304.)

[504] Rubin, K., and Silverberg, A. 2008. Compression in Finite Fields and Torus-Based Cryptography. *SIAM Journal on Computing*, **37**(5), 1401–1428. (Cited on page 304.)

[505] Rudd, W. G., Buell, D. A., and Chiarulli, D. M. 1984. A High Performance Factoring Machine. Pages 297–300 of: Agrawal, D. P. (ed.), *Symposium on Computer Architecture.* ACM, New York, USA. (Cited on page 323.)

[506] Sakai, R., Ohgishi, K., and Kasahara, M. 2000 (Jan.). Cryptosystems Based on Pairing. Pages 26–28 of: *Symposium on Cryptography and Information Security – SCIS.* (Cited on page 327.)

[507] Sakemi, Y., Kobayashi, T., Saito, T., and Wahby, R. S. 2020. *Pairing-Friendly Curves.* Internet-Draft draft-irtf-cfrg-pairing-friendly-curves-08. Internet Engineering Task Force. Work in Progress, https://datatracker.ietf.org/doc/html/draft-irtf-cfrg-pairing-friendly-curves-08. (Cited on page 311.)

[508] Sarkar, P., and Singh, S. 2016. A General Polynomial Selection Method and New Asymptotic Complexities for the Tower Number Field Sieve Algorithm. Pages 37–62 of: Cheon, J. H., and Takagi, T. (eds.), *ASIACRYPT 2016, Part I.* LNCS, vol. 10031. Hanoi, Vietnam: Springer, Heidelberg, Germany. (Cited on page 329.)

[509] Satoh, T., and Araki, K. 1998. Fermat Quotients and the Polynomial Time Discrete Log Algorithm for Anomalous Elliptic Curves. *Commentarii mathematici Universitatis Sancti Pauli*, **47**, 81–92. (Cited on page 326.)

[510] Schirokauer, O. 2000. Using Number Fields to Compute Logarithms in Finite Fields. *Mathematics of Computation*, **69**(231), 1267–1283. (Cited on pages 322 and 329.)

[511] Schneier, B. 2005 (august). *The MD5 Defense.* Schneier on Security blog. https://www.schneier.com/blog/archives/2005/08/the_md5_defense.html. (Cited on page 210.)

[512] Schnorr, C.-P. 1987. A Hierarchy of Polynomial Time Lattice Basis Reduction Algorithms. *Theoretical Computer Science*, **53**, 201–224. (Cited on pages 15, 18, 179, and 330.)

[513] Schnorr, C.-P. 1990a. Efficient Identification and Signatures for Smart Cards. Pages 239–252 of: Brassard, G. (ed.), *CRYPTO'89*. LNCS, vol. 435. Santa Barbara, CA, USA: Springer, Heidelberg, Germany. (Cited on page 294.)

[514] Schnorr, C.-P. 1990b. Factoring Integers and Computing Discrete Logarithms via Diophantine Approximation. Pages 171–181 of: *Advances in Computational Complexity Theory*. American Mathematical Society, New York, USA. (Cited on page 76.)

[515] Schnorr, C.-P. 1991a. Efficient Signature Generation by Smart Cards. *Journal of Cryptology*, **4**(3), 161–174. (Cited on page 122.)

[516] Schnorr, C.-P. 1991b. Factoring Integers and Computing Discrete Logarithms via Diophantine Approximations. Pages 281–293 of: Davies, D. W. (ed.), *EUROCRYPT'91*. LNCS, vol. 547. Brighton, UK: Springer, Heidelberg, Germany. (Cited on page 76.)

[517] Schnorr, C. 1994. Block Reduced Lattice Bases and Successive Minima. *Combinatorics, Probability & Computing*, **3**, 507–522. (Cited on page 19.)

[518] Schnorr, C. 2003. Lattice Reduction by Random Sampling and Birthday Methods. Pages 145–156 of: Alt, H., and Habib, M. (eds.), *Symposium on Theoretical Aspects of Computer Science – STACS*. LNCS, vol. 2607. Springer, Heidelberg, Germany. (Cited on page 21.)

[519] Schnorr, C., and Euchner, M. 1991. Lattice Basis Reduction: Improved Practical Algorithms and Solving Subset Sum Problems. Pages 68–85 of: Budach, L. (ed.), *Fundamentals of Computation Theory – FCT*. LNCS, vol. 529. Springer, Heidelberg, Germany. (Cited on pages 290 and 291.)

[520] Schnorr, C., and Euchner, M. 1994. Lattice Basis Reduction: Improved Practical Algorithms and Solving Subset Sum Problems. *Mathematical Programming*, **66**, 181–199. (Cited on pages 15, 18, and 179.)

[521] Schönhage, A., and Strassen, V. 1971. Schnelle Multiplikation großer Zahlen. *Computing*, **7**(3-4), 281–292. (Cited on page 271.)

[522] Schoof, R. 1995. Counting Points on Elliptic Curves over Finite Fields. *Journal de théorie des nombres de Bordeaux*, **7**(1), 219–254. (Cited on pages 112 and 295.)

[523] Scientific Working Group on Digital Evidence. 2019 (September). *SWGDE Position on the Use of MD5 and SHA1 Hash Algorithms in Digital and Multimedia Forensics*. (Cited on page 209.)

[524] Scott, M., and Barreto, P. S. L. M. 2004. Compressed Pairings. Pages 140–156 of: Franklin, M. (ed.), *CRYPTO 2004*. LNCS, vol. 3152. Santa Barbara, CA, USA: Springer, Heidelberg, Germany. (Cited on page 309.)

[525] SECG. 2000. SEC 2: Recommended Elliptic Curve Domain Parameters. *Standards for Efficient Cryptography Group, Certicom Corp.* (Cited on pages 172 and 175.)

[526] Secure Sockets Layer. 2015. *Secure Sockets Layer (SSL)/Transport Layer Security (TLS)*. http://www.spiegel.de/media/media-35511.pdf. (Cited on page 154.)

[527] Selinger, P. 2006. http://www.mathstat.dal.ca/~selinger/md5collision/. (Cited on page 208.)

[528] Semaev, I. A. 1998. Evaluation of Discrete Logarithms in a Group of p-torsion Points of an Elliptic Curve in Characteristic p. *Mathematics of Computation*, **67**(221), 353–356. (Cited on pages 121 and 326.)

[529] Semaev, I. 2004. *Summation Polynomials and the Discrete Logarithm Problem on Elliptic Curves*. Cryptology ePrint Archive, Report 2004/031. http://eprint.iacr.org/2004/031. (Cited on pages 117 and 326.)

[530] Shallit, J., Williams, H. C., and Morain, F. 1995. Discovery of a Lost Factoring Machine. *The Mathematical Intelligencer*, 41–47. (Cited on page 323.)

[531] Shamir, A. 1979. Factoring Numbers in O(log n) Arithmetic Steps. *Information Processing Letters*, **8**, 28–31. (Cited on page 69.)

[532] Shamir, A. 1999. Factoring Large Numbers with the Twinkle Device (Extended Abstract). Pages 2–12 of: Koç, Çetin Kaya., and Paar, C. (eds.), *CHES'99*. LNCS, vol. 1717. Worcester, Massachusetts, USA: Springer, Heidelberg, Germany. (Cited on pages 62 and 324.)

[533] Shamir, A., and Tromer, E. 2003. Factoring Large Number with the TWIRL Device. Pages 1–26 of: Boneh, D. (ed.), *CRYPTO 2003*. LNCS, vol. 2729. Santa Barbara, CA, USA: Springer, Heidelberg, Germany. (Cited on page 324.)

[534] Shanks, D. 1971. Class Number, a Theory of Factorization, and Genera. Pages 415–440 of: *1969 Number Theory Institute (Proceedings of Symposia in Pure Mathematics, Vol. XX, State University New York, Stony Brook, NY, 1969)*. American Mathematical Society, New York, USA. (Cited on pages 68, 294, and 319.)

[535] Sherman, A. T. 1989. *VLSI Placement and Routing: The PI Project*. Springer-Verlag New York. (Cited on page 227.)

[536] Shirase, M., Han, D.-G., Hibino, Y., Kim, H. W., and Takagi, T. 2007. Compressed XTR. Pages 420–431 of: Katz, J., and Yung, M. (eds.), *ACNS 07*. LNCS, vol. 4521. Zhuhai, China: Springer, Heidelberg, Germany. (Cited on page 310.)

[537] Shirase, M., Han, D., Hibino, Y., Kim, H., and Takagi, T. 2008. A More Compact Representation of XTR Cryptosystem. *IEICE Fundamentals of Electronics, Communications and Computer Sciences*, **91-A**(10), 2843–2850. (Cited on page 310.)

[538] Shor, P. W. 1994. Algorithms for Quantum Computation: Discrete Logarithms and Factoring. Pages 124–134 of: *35th FOCS*. Santa Fe, NM, USA: IEEE Computer Society Press. (Cited on pages 77 and 329.)

[539] Shor, P. W. 1997. Polynomial-Time Algorithms for Prime Factorization and Discrete Logarithms on a Quantum Computer. *SIAM Journal on Computing*, **26**(5), 1484–1509. (Cited on page 139.)

[540] Shoup, V. 1993. Factoring Polynomials over Finite Fields: Asymptotic Complexity vs. Reality. Pages 124–129 of: *Proceedings of the IMACS Symposium*. (Cited on page 253.)

[541] Shoup, V. 1995. A New Polynomial Factorization Algorithm and its Implementation. *Journal of Symbolic Computation*, **20**, 363–397. (Cited on page 277.)

[542] Shoup, V. 1997. Lower Bounds for Discrete Logarithms and Related Problems. Pages 256–266 of: Fumy, W. (ed.), *EUROCRYPT'97*. LNCS, vol. 1233. Konstanz, Germany: Springer, Heidelberg, Germany. (Cited on page 107.)

[543] Shoup, V. 2001a. OAEP Reconsidered. Pages 239–259 of: Kilian, J. (ed.), *CRYPTO 2001*. LNCS, vol. 2139. Santa Barbara, CA, USA: Springer, Heidelberg, Germany. (Cited on pages 79 and 86.)

[544] Shoup, V. 2001b. *A Proposal for an ISO Standard for Public Key Encryption (version 2.1),.* https://www.shoup.net/papers/iso-2_1.pdf. (Cited on page 154.)

[545] Sica, F., Ciet, M., and Quisquater, J.-J. 2003. Analysis of the Gallant-Lambert-Vanstone Method Based on Efficient Endomorphisms: Elliptic and Hyperelliptic Curves. Pages 21–36 of: Nyberg, K., and Heys, H. M. (eds.), *SAC 2002*. LNCS, vol. 2595. St. John's, Newfoundland, Canada: Springer, Heidelberg, Germany. (Cited on page 301.)

[546] Silverman, R. D., and Wagstaff, Jr, S. S. 1993. A Practical Analysis of the Elliptic Curve Factoring Algorithm. *Mathematics of Computation*, **61**, 445–462. (Cited on page 66.)

[547] Silverman, R. D. 1987. The Multiple Polynomial Quadratic Sieve. *Mathematics of Computation*, **48**, 329–339. (Cited on pages 57 and 319.)

[548] Singh, S. 1999. *The Code Book: The Secret History of Codes and Code-Breaking*. Fourth Estate, New York, USA. (Cited on page 226.)

[549] Smart, N. P. 2014. *Algorithms, Key Size and Parameters Report*. Technical Report. European Union Agency for Network and Information Security (ENISA). http://www.enisa.europa.eu. (Cited on pages 235 and 236.)

[550] Smart, N. P. 1999. The Discrete Logarithm Problem on Elliptic Curves of Trace One. *Journal of Cryptology*, **12**(3), 193–196. (Cited on pages 121 and 326.)

[551] Smeets, I., Lenstra, A. K., Lenstra, H., Lovász, L., and van Emde Boas, P. 2010. The History of the LLL-Algorithm. *In:* [457]. (Cited on page 79.)

[552] Smith, J. W., and Wagstaff, Jr, S. S. 1983. An Extended Precision Operand Computer. Pages 209–216 of: *Proceedings of the Twenty-First Southeast Region ACM Conference*. ACM, New York, USA. (Cited on page 54.)

[553] Smith, P., and Skinner, C. 1995. A Public-Key Cryptosystem and a Digital Signature System Based on the Lucas Function Analogue to Discrete Logarithms. Pages 357–364 of: Pieprzyk, J., and Safavi-Naini, R. (eds.), *ASIACRYPT'94*. LNCS, vol. 917. Wollongong, Australia: Springer, Heidelberg, Germany. (Cited on pages 122 and 298.)

[554] Smith, P. J., and Lennon, M. J. J. 1993. LUC: A New Public Key System. Pages 103–117 of: Dougall, E. G. (ed.), *Computer Security, Proceedings of the IFIP TC11, Conference on Information Security, IFIP/Sec*. IFIP Transactions, vol. A-37. North-Holland. (Cited on page 298.)

[555] Solinas, J. A. 1999. *Generalized Mersenne numbers*. Technical Report CORR 99-39. Centre for Applied Cryptographic Research. http://www.cacr.math.uwaterloo.ca/techreports/1999/corr99-39.pdf. (Cited on pages 238 and 239.)

[556] Sotirov, A. 2012 (June). *Analyzing the MD5 Collision in Flame*. SummerCon 2020, New York, USA, https://trailofbits.files.wordpress.com/2012/06/flame-md5.pdf. (Cited on pages 198 and 199.)

[557] Springall, D., Durumeric, Z., and Halderman, J. A. 2016. Measuring the Security Harm of TLS Crypto Shortcuts. Pages 33–47 of: *Proceedings of the 2016 Internet Measurement Conference*. IMC 2016. New York, NY, USA: Association for Computing Machinery. (Cited on page 161.)

[558] Stam, M. 2003. *Speeding Up Subgroup Cryptosystems*. PhD thesis, Technische Universiteit Eindhoven. (Cited on pages 298, 300, 301, and 302.)

[559] Stam, M., and Lenstra, A. K. 2001. Speeding Up XTR. Pages 125–143 of: Boyd, C. (ed.), *ASIACRYPT 2001*. LNCS, vol. 2248. Gold Coast, Australia: Springer, Heidelberg, Germany. (Cited on pages 4, 300, and 301.)

[560] Stam, M., and Lenstra, A. K. 2003. Efficient Subgroup Exponentiation in Quadratic and Sixth Degree Extensions. Pages 318–332 of: Kaliski Jr, B. S., Koç, Ç. K., and Paar, C. (eds.), *CHES 2002*. LNCS, vol. 2523. Redwood Shores, CA, USA: Springer, Heidelberg, Germany. (Cited on pages 4, 306, and 307.)

[561] Stehlé, D., Steinfeld, R., Tanaka, K., and Xagawa, K. 2009. Efficient Public Key Encryption Based on Ideal Lattices. Pages 617–635 of: Matsui, M. (ed.), *ASIACRYPT 2009*. LNCS, vol. 5912. Tokyo, Japan: Springer, Heidelberg, Germany. (Cited on page 16.)

[562] Stein, W. 2019. *Sage Mathematics Software Version 9.0*. The Sage Development Team. http://www.sagemath.org. (Cited on page 20.)

[563] Stevens, D. 2009a. http://blog.didierstevens.com/2009/01/17/. (Cited on page 208.)

[564] Stevens, M. 2006. *Fast Collision Attack on MD5*. Cryptology ePrint Archive, Report 2006/104. http://eprint.iacr.org/2006/104. (Cited on page 184.)

[565] Stevens, M. 2007 (June). *On Collisions for MD5*. MPhil thesis, Eindhoven University of Technology. (Cited on page 184.)

[566] Stevens, M. 2009b (June). *GitHub: Project HashClash - MD5 & SHA-1 cryptanalysis*. https://github.com/cr-marcstevens/hashclash. (Cited on pages 189, 200, 201, 203, and 217.)

[567] Stevens, M. 2012 (June). *Attacks on Hash Functions and Applications*. PhD thesis, Leiden University. (Cited on page 189.)

[568] Stevens, M. 2013a. Counter-Cryptanalysis. Pages 129–146 of: Canetti, R., and Garay, J. A. (eds.), *CRYPTO 2013, Part I*. LNCS, vol. 8042. Santa Barbara, CA, USA: Springer, Heidelberg, Germany. (Cited on pages 200, 212, and 219.)

[569] Stevens, M. 2013b. New Collision Attacks on SHA-1 Based on Optimal Joint Local-Collision Analysis. Pages 245–261 of: Johansson, T., and Nguyen, P. Q. (eds.), *EUROCRYPT 2013*. LNCS, vol. 7881. Athens, Greece: Springer, Heidelberg, Germany. (Cited on page 185.)

[570] Stevens, M., Lenstra, A. K., and de Weger, B. 2007a. Chosen-Prefix Collisions for MD5 and Colliding X.509 Certificates for Different Identities. Pages 1–22 of: Naor, M. (ed.), *EUROCRYPT 2007*. LNCS, vol. 4515. Barcelona, Spain: Springer, Heidelberg, Germany. (Cited on pages 6, 184, 186, 189, 191, 192, 217, 219, and 317.)

[571] Stevens, M., Lenstra, A., and de Weger, B. 2007b. *Predicting the Winner of the 2008 US Presidential Elections Using a Sony PlayStation 3*. http://www.win.tue.nl/hashclash/Nostradamus/. (Cited on page 206.)

[572] Stevens, M., Lenstra, A., and de Weger, B. 2007c. *Vulnerability of Software Integrity and Code Signing.* http://www.win.tue.nl/hashclash/ SoftIntCodeSign/. (Cited on page 208.)

[573] Stevens, M., Sotirov, A., Appelbaum, J., Lenstra, A. K., Molnar, D., Osvik, D. A., and de Weger, B. 2009. Short Chosen-Prefix Collisions for MD5 and the Creation of a Rogue CA Certificate. Pages 55–69 of: Halevi, S. (ed.), *CRYPTO 2009.* LNCS, vol. 5677. Santa Barbara, CA, USA: Springer, Heidelberg, Germany. (Cited on pages 6, 184, 188, 189, 191, 201, 212, 217, and 220.)

[574] Stevens, M., Lenstra, A. K., and Weger, B. D. 2012. Chosen-prefix Collisions for MD5 and Applications. *International Journal of Applied Cryptography,* **2**(4), 322–359. (Cited on pages 6, 205, and 206.)

[575] Stevens, M., Bursztein, E., Karpman, P., Albertini, A., and Markov, Y. 2017. The First Collision for Full SHA-1. Pages 570–596 of: Katz, J., and Shacham, H. (eds.), *CRYPTO 2017, Part I.* LNCS, vol. 10401. Santa Barbara, CA, USA: Springer, Heidelberg, Germany. (Cited on pages 185, 189, and 201.)

[576] Strassen, V. 1976/77. Einige Resultate über Berechnungskomplexität. *Jahresbericht der Deutschen Mathematiker-Vereinigung,* **78**, 1–8. (Cited on page 67.)

[577] Strassen, V. 1969. Gaussian Elimination is not Optimal. *Numerische Mathematik,* **13**, 354–356. (Cited on page 286.)

[578] Svenda, P., Nemec, M., Sekan, P., Kvasnovský, R., Formánek, D., Komárek, D., and Matyás, V. 2016. The Million-Key Question - Investigating the Origins of RSA Public Keys. Pages 893–910 of: Holz, T., and Savage, S. (eds.), *USENIX Security 2016.* Austin, TX, USA: USENIX Association. (Cited on pages 142 and 149.)

[579] Takayasu, A., and Kunihiro, N. 2014. Partial Key Exposure Attacks on RSA: Achieving the Boneh-Durfee Bound. Pages 345–362 of: Joux, A., and Youssef, A. M. (eds.), *SAC 2014.* LNCS, vol. 8781. Montreal, QC, Canada: Springer, Heidelberg, Germany. (Cited on page 98.)

[580] Takayasu, A., and Kunihiro, N. 2015. Partial Key Exposure Attacks on CRT-RSA: Better Cryptanalysis to Full Size Encryption Exponents. Pages 518–537 of: Malkin, T., Kolesnikov, V., Lewko, A. B., and Polychronakis, M. (eds.), *ACNS 15.* LNCS, vol. 9092. New York, NY, USA: Springer, Heidelberg, Germany. (Cited on page 98.)

[581] Takayasu, A., and Kunihiro, N. 2016. How to Generalize RSA Cryptanalyses. Pages 67–97 of: Cheng, C.-M., Chung, K.-M., Persiano, G., and Yang, B.-Y. (eds.), *PKC 2016, Part II.* LNCS, vol. 9615. Taipei, Taiwan: Springer, Heidelberg, Germany. (Cited on pages 98 and 103.)

[582] Takayasu, A., and Kunihiro, N. 2017. A Tool Kit for Partial Key Exposure Attacks on RSA. Pages 58–73 of: Handschuh, H. (ed.), *CT-RSA 2017.* LNCS, vol. 10159. San Francisco, CA, USA: Springer, Heidelberg, Germany. (Cited on page 98.)

[583] Takayasu, A., Lu, Y., and Peng, L. 2017. Small CRT-Exponent RSA Revisited. Pages 130–159 of: Coron, J.-S., and Nielsen, J. B. (eds.), *EUROCRYPT 2017, Part II.* LNCS, vol. 10211. Paris, France: Springer, Heidelberg, Germany. (Cited on pages 80 and 100.)

[584] Takayasu, A., Lu, Y., and Peng, L. 2019. Small CRT-Exponent RSA Revisited. *Journal of Cryptology,* **32**(4), 1337–1382. (Cited on pages 100 and 103.)

[585] Team, B. 2015. *Android Wallet Security Update.* https://blog. blockchain.com/2015/05/28/android-wallet-security-update/. (Cited on page 179.)

[586] thatch45, and basepi. 2013. *Change key generation seq.* https://github.com/saltstack/salt/commit/ 5dd304276ba5745ec21fc1e6686a0b28da29e6fc. (Cited on page 145.)

[587] The FPLLL Development Team. 2019a. *FPLLL, a Lattice Reduction Library.* Available at https://github.com/fplll/fplll. (Cited on pages 20 and 291.)

[588] The FPLLL Development Team. 2019b. *FPyLLL, a Python Interface to FPLLL.* Available at https://github.com/fplll/fpylll. (Cited on page 24.)

[589] The Washington Post. 2012 (June). *U.S., Israel Developed Flame Computer Virus to Slow Iranian Nuclear Efforts, Officials Say.* Ellen Nakashima, Greg Miller and Julie Tate (http://articles.washingtonpost.com/2012-06-19/world/35460741_1_stuxnet-computer-virus-malware). (Cited on page 198.)

[590] tranogatha. 2015. *Establish deprecation date for DHE cipher suites in WebRTC.* https://bugzilla.mozilla.org/show_bug.cgi?id=1227519. (Cited on pages 158 and 170.)

[591] Trappe, W., and Washington, L. C. 2002. *Introduction to Cryptography with Coding Theory.* Prentice Hall, Hoboken, NJ, USA. (Cited on page 67.)

[592] Tsiounis, Y., and Yung, M. 1998. On the Security of ElGamal Based Encryption. Pages 117–134 of: Imai, H., and Zheng, Y. (eds.), *PKC'98.* LNCS, vol. 1431. Pacifico Yokohama, Japan: Springer, Heidelberg, Germany. (Cited on page 295.)

[593] Valenta, L., Cohney, S., Liao, A., Fried, J., Bodduluri, S., and Heninger, N. 2016. Factoring as a Service. Pages 321–338 of: Grossklags, J., and Preneel, B. (eds.), *FC 2016.* LNCS, vol. 9603. Christ Church, Barbados: Springer, Heidelberg, Germany. (Cited on pages 143 and 144.)

[594] Valenta, L., Adrian, D., Sanso, A., Cohney, S., Fried, J., Hastings, M., Halderman, J. A., and Heninger, N. 2017. Measuring Small Subgroup Attacks against Diffie-Hellman. In: *NDSS 2017.* San Diego, CA, USA: The Internet Society. (Cited on pages 160, 162, 164, 165, 166, 167, 172, and 173.)

[595] van der Hoeven, J. 2004. The Truncated Fourier Transform and Applications. Pages 290–296 of: Gutierrez, J. (ed.), *Proceedings of the International Symposium on Symbolic and Algebraic Computation, ISSAC'04.* New York, NY, USA: Association for Computing Machinery. (Cited on page 267.)

[596] van der Hoeven, J., and Lecerf, G. 2020. Fast multivariate multi-point evaluation revisited. *Journal of Complexity*, **56**, 101405. (Cited on page 278.)

[597] van Dijk, M., and Woodruff, D. P. 2004. Asymptotically Optimal Communication for Torus-Based Cryptography. Pages 157–178 of: Franklin, M. (ed.), *CRYPTO 2004.* LNCS, vol. 3152. Santa Barbara, CA, USA: Springer, Heidelberg, Germany. (Cited on pages 305, 311, and 313.)

[598] van Dijk, M., Granger, R., Page, D., Rubin, K., Silverberg, A., Stam, M., and Woodruff, D. P. 2005. Practical Cryptography in High Dimensional Tori. Pages

234–250 of: Cramer, R. (ed.), *EUROCRYPT 2005*. LNCS, vol. 3494. Aarhus, Denmark: Springer, Heidelberg, Germany. (Cited on pages 130 and 306.)

[599] van Hoeij, M. 2002. Factoring Polynomials and the Knapsack Problem. *Journal of Number Theory*, **95**(2), 167–189. (Cited on page 290.)

[600] van Oorschot, P. C., and Wiener, M. J. 1994. Parallel Collision Search with Application to Hash Functions and Discrete Logarithms. Pages 210–218 of: Denning, D. E., Pyle, R., Ganesan, R., and Sandhu, R. S. (eds.), *ACM CCS 94*. Fairfax, VA, USA: ACM Press. (Cited on page 295.)

[601] van Oorschot, P. C., and Wiener, M. J. 1996. On Diffie-Hellman Key Agreement with Short Exponents. Pages 332–343 of: Maurer, U. M. (ed.), *EURO-CRYPT'96*. LNCS, vol. 1070. Saragossa, Spain: Springer, Heidelberg, Germany. (Cited on page 163.)

[602] van Oorschot, P. C., and Wiener, M. J. 1999. Parallel Collision Search with Cryptanalytic Applications. *Journal of Cryptology*, **12**(1), 1–28. (Cited on pages 109, 183, 186, 216, and 325.)

[603] Vandersypen, L. M. K., Steffen, M., Breyta, G., Yannoni, C. S., Sherwood, M. H., and Chuang, I. L. 2001. Experimental realization of Shor's quantum factoring algorithm using nuclear magnetic resonance. *Nature*, **414**(6866), 883–887. (Cited on page 330.)

[604] Vanstone, S. A., and Zuccherato, R. J. 1995. Short RSA Keys and Their Generation. *Journal of Cryptology*, **8**(2), 101–114. (Cited on page 228.)

[605] Verheul, E. R. 2001. Evidence that XTR Is More Secure than Supersingular Elliptic Curve Cryptosystems. Pages 195–210 of: Pfitzmann, B. (ed.), *EURO-CRYPT 2001*. LNCS, vol. 2045. Innsbruck, Austria: Springer, Heidelberg, Germany. (Cited on page 309.)

[606] Verheul, E. R. 2004. Evidence that XTR Is More Secure than Supersingular Elliptic Curve Cryptosystems. *Journal of Cryptology*, **17**(4), 277–296. (Cited on page 309.)

[607] Villatte, B. 2011. *Interview Intégrale du Lauréat de la Polysphère d'Or*. (Cited on page 6.)

[608] Voskresenskiĭ, V. E. 1998. *Algebraic Groups and Their Birational Invariants*. Translations of Mathematical Monographs, **179**, American Mathematical Society. (Cited on page 124.)

[609] VPN Exploitation Process. 2010 (Sept.). *Intro to the VPN Exploitation Process*. Media leak. http://www.spiegel.de/media/media-35515.pdf. (Cited on page 154.)

[610] Wagner, D., and Schneier, B. 1996. Analysis of the SSL 3.0 protocol. In: Tygar, D. (ed.), *Proceedings of the Second USENIX Workshop on Electronic Commerce*, vol. 1. USENIX Association, Berkeley, CA. (Cited on page 153.)

[611] Wagstaff, Jr, S. S. 2013. *The Joy of Factoring*. Student Mathematical Library, vol. 68. Providence, RI: American Mathematical Society. (Cited on pages 9, 41, and 75.)

[612] Wagstaff, Jr, S. S., and Smith, J. W. 1987. Methods of Factoring Large Integers. Pages 281–303 of: Chudnovsky, D. V., Chudnovsky, G. V., Cohn, H., and Nathanson, M. B. (eds.), *Number Theory, New York, 1984–1985*. Lecture Notes in Mathematics, vol. 1240. Springer, Heidelberg, Germany. (Cited on page 54.)

[613] Walter, C. D. 1999a. Montgomery Exponentiation Needs No Final Subtractions. *Electronics Letters*, **35**(21), 1831–1832. (Cited on page 234.)

[614] Walter, C. D. 1992. Faster Modular Multiplication by Operand Scaling. Pages 313–323 of: Feigenbaum, J. (ed.), *CRYPTO'91*. LNCS, vol. 576. Santa Barbara, CA, USA: Springer, Heidelberg, Germany. (Cited on page 229.)

[615] Walter, C. D. 1999b. Montgomery's Multiplication Technique: How to Make It Smaller and Faster (Invited Talk). Pages 80–93 of: Koç, Çetin Kaya., and Paar, C. (eds.), *CHES'99*. LNCS, vol. 1717. Worcester, Massachusetts, USA: Springer, Heidelberg, Germany. (Cited on page 234.)

[616] Wang, X., and Yu, H. 2005. How to Break MD5 and Other Hash Functions. Pages 19–35 of: Cramer, R. (ed.), *EUROCRYPT 2005*. LNCS, vol. 3494. Aarhus, Denmark: Springer, Heidelberg, Germany. (Cited on pages 183, 184, 189, 200, and 317.)

[617] Wang, X., Yao, A. C., and Yao, F. 2005a. *Cryptanalysis on SHA-1*. NIST Cryptographic Hash Workshop. http://csrc.nist.gov/groups/ST/hash/documents/Wang_SHA1-New-Result.pdf. (Cited on page 185.)

[618] Wang, X., Yin, Y. L., and Yu, H. 2005b. Finding Collisions in the Full SHA-1. Pages 17–36 of: Shoup, V. (ed.), *CRYPTO 2005*. LNCS, vol. 3621. Santa Barbara, CA, USA: Springer, Heidelberg, Germany. (Cited on page 185.)

[619] Washington, L. C. 2003. *Elliptic Curves: Number Theory and Cryptography*. Boca Raton, FL: Chapman & Hall/CRC Press. (Cited on page 67.)

[620] Weimerskirch, A., Stebila, D., and Shantz, S. C. 2003. Generic GF(2^m) Arithmetic in Software and Its Application to ECC. Pages 79–92 of: Safavi-Naini, R., and Seberry, J. (eds.), *ACISP 03*. LNCS, vol. 2727. Wollongong, NSW, Australia: Springer, Heidelberg, Germany. (Cited on page 284.)

[621] Wiener, M. J. 1990a. Cryptanalysis of Short RSA Secret Exponents. *IEEE Transactions on Information Theory*, **36**(3), 553–558. (Cited on pages 71 and 227.)

[622] Wiener, M. J. 1990b. Cryptanalysis of Short RSA Secret Exponents (Abstract). Page 372 of: Quisquater, J.-J., and Vandewalle, J. (eds.), *EUROCRYPT'89*. LNCS, vol. 434. Houthalen, Belgium: Springer, Heidelberg, Germany. (Cited on pages 80 and 95.)

[623] Williams, H. C. 1980. A Modification of the RSA Public-Key Encryption Procedure. *IEEE Transactions on Information Theory*, **26**(6), 726–729. (Cited on page 49.)

[624] Wilson, K. 2014. *Phasing out Certificates with 1024-bit RSA Keys*. https://blog.mozilla.org/security/2014/09/08/phasing-out-certificates-with-1024-bit-rsa-keys/. (Cited on page 144.)

[625] Wunderer, T. 2019. A detailed analysis of the hybrid lattice-reduction and meet-in-the-middle attack. *Journal of Mathemathical Cryptology*, **13**(1), 1–26. (Cited on page 26.)

[626] Xie, T., Liu, F., and Feng, D. 2008. *Could The 1-MSB Input Difference Be The Fastest Collision Attack For MD5?* Cryptology ePrint Archive, Report 2008/391. http://eprint.iacr.org/2008/391. (Cited on page 184.)

[627] Xu, J., Sarkar, S., Hu, L., Wang, H., and Pan, Y. 2019. New Results on Modular Inversion Hidden Number Problem and Inversive Congruential Generator. Pages 297–321 of: Boldyreva, A., and Micciancio, D. (eds.), *CRYPTO 2019*,

Part I. LNCS, vol. 11692. Santa Barbara, CA, USA: Springer, Heidelberg, Germany. (Cited on pages 80, 104, and 105.)

[628] Yilek, S., Rescorla, E., Shacham, H., Enright, B., and Savage, S. 2009. When Private Keys Are Public: Results from the 2008 Debian OpenSSL Vulnerability. Pages 15–27 of: Feldmann, A., and Mathy, L. (eds.), *Proceedings of the 9th ACM SIGCOMM Conference on Internet Measurement.* IMC '09. New York, NY, USA: Association for Computing Machinery. (Cited on pages 146 and 147.)

[629] Ylonen, T., and Lonvick, C. 2006 (Jan.). *The Secure Shell (SSH) Transport Layer Protocol.* RFC 4253. RFC Editor. https://www.rfc-editor.org/info/rfc4253. (Cited on pages 141, 154, 155, 156, 157, 158, and 174.)

[630] Yoshino, M., Okeya, K., and Vuillaume, C. 2007. Double-Size Bipartite Modular Multiplication. Pages 230–244 of: Pieprzyk, J., Ghodosi, H., and Dawson, E. (eds.), *ACISP 07.* LNCS, vol. 4586. Townsville, Australia: Springer, Heidelberg, Germany. (Cited on page 236.)

[631] Yu, Y., and Ducas, L. 2017. Second Order Statistical Behavior of LLL and BKZ. Pages 3–22 of: Adams, C., and Camenisch, J. (eds.), *SAC 2017.* LNCS, vol. 10719. Ottawa, ON, Canada: Springer, Heidelberg, Germany. (Cited on page 24.)

[632] Zassenhaus, H. 1969. On Hensel Factorization, I. *Journal of Number Theory,* **1**(3), 291–311. (Cited on page 290.)

[633] Zhang, Z., Chen, C., Hoffstein, J., Whyte, W., Schanck, J. M., Hulsing, A., Rijneveld, J., Schwabe, P., and Danba, O. 2019. *NTRUEncrypt.* Technical Report. National Institute of Standards and Technology. Available at https://csrc.nist.gov/projects/post-quantum-cryptography/round-2-submissions. (Cited on page 35.)

[634] Zimmermann, P., and Dodson, B. 2006. 20 Years of ECM. Pages 525–542 of: *Algorithmic Number Theory, Proceedings ANTS 2006.* LNCS, vol. 4076. Berlin: Springer. (Cited on page 66.)

Index

Printed in the United States
by Baker & Taylor Publisher Services